VOLUME FIVE HUNDRED AND SIXTY ONE

METHODS IN ENZYMOLOGY

Metabolic Analysis Using Stable Isotopes

METHODS IN ENZYMOLOGY

VOLUME FIVE HUNDRED AND SIXTY ONE

METHODS IN ENZYMOLOGY

Metabolic Analysis Using Stable Isotopes

Edited by

CHRISTIAN M. METALLO
Department of Bioengineering
University of California, San Diego

ELSEVIER

AMSTERDAM • BOSTON • HEIDELBERG • LONDON
NEW YORK • OXFORD • PARIS • SAN DIEGO
SAN FRANCISCO • SINGAPORE • SYDNEY • TOKYO
Academic Press is an imprint of Elsevier

Academic Press is an imprint of Elsevier
225 Wyman Street, Waltham, MA 02451, USA
525 B Street, Suite 1800, San Diego, CA 92101–4495, USA
125 London Wall, London, EC2Y 5AS, UK
The Boulevard, Langford Lane, Kidlington, Oxford OX5 1GB, UK

First edition 2015

ISBN: 978-0-12-802293-1
ISSN: 0076-6879

For information on all Academic Press publications
visit our website at http://store.elsevier.com/

DEDICATION

For Barbara and her unending support

CONTENTS

CONTRIBUTORS

T.E. Angel
KineMed Inc., Emeryville, California, USA

Vinay Bulusu
Cancer Research UK Beatson Institute & Institute of Cancer Sciences, University of Glasgow, Glasgow, United Kingdom

Nicholas Carpenter
Advanced Imaging Research Center, University of Texas Southwestern Medical Center, Dallas, Texas, USA

Jose Castro-Perez
Merck Research Laboratories, Kenilworth, New Jersey, USA

Myriam M. Chaumeil
Department of Radiology and Biomedical Imaging, University of California, San Francisco, California, USA

Michele A. Cleary
Merck Research Laboratories, Kenilworth, New Jersey, USA

Ralph J. DeBerardinis
Children's Medical Center Research Institute, University of Texas Southwestern Medical Center, Dallas, Texas, USA

Jianhai Du
Departments of Biochemistry, and Department of Ophthalmology, University of Washington, Seattle, Washington, USA

Roselle Gélinas
Montreal Heart Institute, and Department of Medicine, Université de Montréal, Montreal, Quebec, Canada

Eyal Gottlieb
Cancer Research UK Beatson Institute, Glasgow, UK

M.K. Hellerstein
KineMed Inc., Emeryville, and Department of Nutritional Sciences and Toxicology, University of California, Berkeley, Berkeley, California, USA

Kithsiri Herath
Merck Research Laboratories, Kenilworth, New Jersey, USA

Karsten Hiller
Luxembourg Centre for Systems Biomedicine, University of Luxembourg, Esch-Belval, Luxembourg

W.E. Holmes
KineMed Inc., Emeryville, California, USA

Brian K. Hubbard
Merck Research Laboratories, Kenilworth, New Jersey, USA

James B. Hurley
Departments of Biochemistry, and Department of Ophthalmology, University of Washington, Seattle, Washington, USA

Douglas G. Johns
Merck Research Laboratories, Kenilworth, New Jersey, USA

Jurre J. Kamphorst
Cancer Research UK Beatson Institute & Institute of Cancer Sciences, University of Glasgow, Glasgow, United Kingdom

Joanne K. Kelleher
Department of Pharmacology and Physiology, George Washington University Medical School, Washington, and Department of Chemical Engineering, MIT Cambridge, Cambridge, Massachusetts, USA

Benjamin Lauzier
Institut du thorax, Université de Nantes, Nantes, France

K.W. Li
KineMed Inc., Emeryville, California, USA

Jonathan D. Linton
Departments of Biochemistry, and Department of Ophthalmology, University of Washington, Seattle, Washington, USA

Lloyd Lumata
Department of Physics, University of Texas at Dallas, Richardson, Texas, USA

Gillian M. Mackay
Cancer Research UK Beatson Institute, Glasgow, UK

Ablatt Mahsut
Merck Research Laboratories, Kenilworth, New Jersey, USA

David G. McLaren
Merck Research Laboratories, Kenilworth, New Jersey, USA

Matthew E. Merritt
Advanced Imaging Research Center, University of Texas Southwestern Medical Center, Dallas, Texas, USA

Chloé Najac
Department of Radiology and Biomedical Imaging, University of California, San Francisco, California, USA

Gary B. Nickol
Department of Pharmacology and Physiology, George Washington University Medical School, Washington, USA

Stephen F. Previs
Merck Research Laboratories, Kenilworth, New Jersey, USA

Mukundan Ragavan
Advanced Imaging Research Center, University of Texas Southwestern Medical Center, Dallas, Texas, USA

Thomas P. Roddy
Merck Research Laboratories, Kenilworth, New Jersey, USA

Rory J. Rohm
Merck Research Laboratories, Kenilworth, New Jersey, USA

Sabrina M. Ronen
Department of Radiology and Biomedical Imaging, University of California, San Francisco, California, USA

Christine Des Rosiers
Department of Nutrition; Montreal Heart Institute, and Department of Medicine, Université de Montréal, Montreal, Quebec, Canada

Matthieu Ruiz
Department of Nutrition, and Montreal Heart Institute, Université de Montréal, Montreal, Quebec, Canada

Vinit Shah
Merck Research Laboratories, Kenilworth, New Jersey, USA

Steven J. Stout
Merck Research Laboratories, Kenilworth, New Jersey, USA

Sergey Tumanov
Cancer Research UK Beatson Institute & Institute of Cancer Sciences, University of Glasgow, Glasgow, United Kingdom

Fanny Vaillant
IHU Institut de Rythmologie et Modélisation Cardiaque, Fondation Bordeaux, and Inserm U1045 Centre de Recherche Cardio-Thoracique de Bordeaux, Université de Bordeaux, Bordeaux, France

Niels J.F. van den Broek
Cancer Research UK Beatson Institute, Glasgow, UK

Sheng-Ping Wang
Merck Research Laboratories, Kenilworth, New Jersey, USA

André Wegner
Luxembourg Centre for Systems Biomedicine, University of Luxembourg, Esch-Belval, Luxembourg

Daniel Weindl
Luxembourg Centre for Systems Biomedicine, University of Luxembourg, Esch-Belval, Luxembourg

Chendong Yang
Children's Medical Center Research Institute, University of Texas Southwestern Medical Center, Dallas, Texas, USA

Liang Zheng
Cancer Research UK Beatson Institute, Glasgow, UK

Wendy Zhong
Merck Research Laboratories, Kenilworth, New Jersey, USA

Haihong Zhou
Merck Research Laboratories, Kenilworth, New Jersey, USA

PREFACE

Biochemical reactions are the driving force for virtually all living systems. In the last century, biochemists have painstakingly mapped the diverse set of chemical reactions catalyzed in cells (often one enzyme at a time). Though far from complete, we now have comprehensive maps and databases outlining the substrates, products, cofactors, and enzymes that comprise *metabolism*.

Concomitant with the description of metabolic pathways that occurred over the last ~100 years, we have also seen technological advances that now enable precise characterization and quantitation of chemical processes. Chromatographic separation and detection using high-mass resolution mass spectrometers or alternatively nuclear magnetic resonance are emerging as powerful tools for metabolic analyses, particularly when combined with the application of chemically synthesized stable isotope tracers. Finally, computational modeling and simulation allows more reliable interpretation of the data generated using such technologies. These advances offer amazing opportunities for metabolic biochemists but also create challenges in terms of learning the technical details, which often require diverse skills that range from animal surgery to analytical chemistry to programming. In this volume of *Methods in Enzymology*, I have brought together experts who commonly apply stable isotope tracers to study metabolism in mammalian systems. Although the details of each protocol are available in a number of formats, I have attempted to include descriptive methods in a more integrated manner that makes the protocols more accessible to newcomers and young scientists.

Chapters 1 and 2 of this volume provide comprehensive protocols for the use of hyperpolarized metabolic tracers in tissues, cells, and isolated enzymes. The enhanced sensitivity afforded by hyperpolarization enables real-time metabolic imaging via magnetic resonance spectroscopy. In Chapter 1, Chaumeil et al. outline the theory, methodologies, and important considerations for application of dynamic nuclear polarization (DNP) to *in vitro* and *in vivo* systems in great detail. In Chapter 2, Lumata et al. provide additional insights into DNP methods with a particular focus on its application to perfused organs (i.e., liver and heart). Chapters 3 and 4 focus on the analysis of intact tissues using mass spectrometry-based mass isotopomer (or isotopologue) analysis, outlining approaches to study metabolism in the heart and retina, respectively. Ruiz et al. detail methods for the surgical preparation, setup, perfusion, and analysis of *ex vivo* hearts in both the

Langendorff and working modes. Next, Du et al. describe methods for probing metabolism in the retina, another tissue that exhibits high metabolic activity as well as unique compartmentalization and architecture.

Chapters 5 and 6 address the analysis of metabolism in cultured cells using mass spectrometry-based approaches. MacKay et al. first provide detailed methods and considerations on the application of liquid chromatography coupled to mass spectrometry (LC-MS) and variations therein (i.e., tandem mass spectrometry) to cultured cells. Tumanov et al. next specify details on the application of LC-MS as well as gas chromatography-mass spectrometry to fatty acid metabolism.

In Chapter 7, Holmes et al. provide a comprehensive protocol and discussion on the application of ^2H_2O (heavy water) to study proteome dynamics *in vivo*. Here, proteins are metabolically labeled via amino acid metabolism, and LC-MS/MS-based detection is applied to quantify isotope incorporation in peptides. Mass isotopomer distribution analysis is subsequently applied to estimate protein turnover. In Chapter 8, Weindl et al. outline an experimental approach and data analysis pipeline that enable detection of tracer fates in a nontargeted manner. Theoretical considerations of this method are described in great detail. Kelleher and Nickol then describe the early application of nonlinear modeling in metabolic research as embodied by isotopomer spectral analysis in Chapter 9, illustrating specific findings that were ascertained using this model-based approach. Finally, in Chapter 10 Previs et al. outline a model and associated calculations that can be employed to estimate how error propagates during the *in vivo* application of ^2H_2O for quantitation of lipid and protein metabolism.

The methods outlined herein only scratch the surface of technologies that can be applied to study metabolic pathways. Chromatography, mass spectrometry, and NMR will continue to improve rapidly such that larger and more comprehensive datasets are becoming available. As these data become available, researchers are beginning to apply quantitative, genome-scale models of metabolism based on flux balancing or metabolic tracing data to both microbial and mammalian systems. Detailed methods for such modeling approaches are available elsewhere, though I sincerely apologize to the many excellent scientists whose contributions could not be included here. There are much too many individuals who have directly or indirectly influenced my work in metabolism in the last decade to list. I hope the protocols included here will facilitate the recruitment of more highly skilled researchers in the area of metabolic research.

Christian M. Metallo

CHAPTER ONE

Studies of Metabolism Using ^{13}C MRS of Hyperpolarized Probes

Myriam M. Chaumeil, Chloé Najac, Sabrina M. Ronen[1]

Department of Radiology and Biomedical Imaging, University of California, San Francisco, California, USA

[1]Corresponding author: e-mail address: sabrina.ronen@ucsf.edu

Contents

Abstract

First described in 2003, the dissolution dynamic nuclear polarization (DNP) technique, combined with ^{13}C magnetic resonance spectroscopy (MRS), has since been used in numerous metabolic studies and has become a valuable metabolic imaging method. DNP dramatically increases the level of polarization of ^{13}C-labeled compounds resulting in an increase in the signal-to-noise ratio (SNR) of over 50,000 fold for the MRS spectrum

Methods in Enzymology, Volume 561
ISSN 0076-6879
http://dx.doi.org/10.1016/bs.mie.2015.04.001

of hyperpolarized compounds. The high SNR enables rapid real-time detection of metabolism in cells, tissues, and *in vivo*.

This chapter will present a comprehensive review of the DNP approaches that have been used to monitor metabolism in living systems. First, the list of ^{13}C DNP probes developed to date will be presented, with a particular focus on the most commonly used probe, namely [1-^{13}C] pyruvate. In the next four sections, we will then describe the different factors that need to be considered when designing ^{13}C DNP probes for metabolic studies, conducting *in vitro* or *in vivo* hyperpolarized experiments, as well as acquiring, analyzing, and modeling hyperpolarized ^{13}C data.

1. INTRODUCTION

Magnetic resonance spectroscopy (MRS, also known as nuclear magnetic resonance or NMR spectroscopy) is an analytical method that generates a characteristic fingerprint for molecules containing atomic nuclei with nonzero nuclear spin. Among these nuclei, ^{1}H and ^{31}P are biologically relevant and have a high natural abundance. As a result, ^{1}H and ^{31}P MRS can be used to probe the steady-state levels of ^{1}H and ^{31}P—containing molecules within cells, tissues, and *in vivo* (Gillies & Morse, 2005). MRS is also a useful method for probing cellular metabolism dynamically and in real time. This is achieved by using ^{13}C MRS to longitudinally monitor the metabolic fate of ^{13}C-labeled metabolic precursors. Because the naturally abundant ^{12}C nucleus, which has a spin of zero, is MRS invisible, ^{13}C-labeling approaches provide specific metabolic information with little confounding background from metabolites that are not directly associated with the metabolic pathways under investigation. However, conventional ^{13}C MRS is limited by the relatively low sensitivity of the method. The small gyromagnetic ratio of the ^{13}C nucleus leads to a relatively small thermal polarization (difference in number of spins aligned parallel or antiparallel to the external magnetic field) and therefore requires long acquisition times and/or large concentrations (or voxels), in order to obtain ^{13}C spectra with adequate signal-to-noise ratios (SNRs). For example, monitoring metabolism in a bioreactor system required $\sim 5 \times 10^7$ cells and several hours (Brandes, Ward, & Ronen, 2010), whereas a clinical study required a voxel size of ~50 ml and an acquisition time of well over an hour, to obtain a temporal sequence of quantifiable spectra (Gruetter et al., 2003; Wijnen et al., 2010). However, this challenge has recently been overcome, at least for some metabolic reactions, by the technological development of hyperpolarization associated with liquid dissolution methods. In this review, we will focus on one of these methods, namely the dissolution dynamic nuclear polarization (DNP) technique, which has been used in the majority of metabolic studies to date (for a review of other hyperpolarization methods, see, for example, the recent review by Comment, 2013).

Dissolution DNP was first described in 2003 and has since been used to increase the levels of polarization of several ^{13}C-labeled compounds, referred to as DNP probes. It leads to dramatic improvements in the *SNRs* by up to 50,000 fold (Ardenkjaer-Larsen et al., 2003) and thus allows for rapid detection of cellular metabolism in cells, tissues, and *in vivo* (Brindle, Bohndiek, Gallagher, & Kettunen, 2011; Kurhanewicz et al., 2011; Nelson et al., 2013). We will first present a comprehensive review of DNP approaches that have been used to date to monitor metabolism. A list of DNP probes as well as their chemical characteristics and physiological applications will be presented. We will then describe the different factors that need to be considered when designing a DNP probe, conducting *in vitro* or *in vivo* hyperpolarized experiments, and acquiring, analyzing, and modeling ^{13}C hyperpolarized data. Some familiarity of MRS methodologies and associated terminology is assumed.

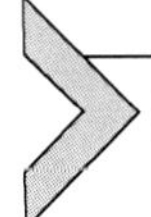

2. ^{13}C MRS OF DNP PROBES: CURRENT APPROACHES TO STUDYING METABOLISM

In this section, we briefly describe the dissolution DNP method. We then present a more detailed review of the use and applications of the most commonly used DNP probe, namely [1-^{13}C] pyruvate. Finally, we will describe other DNP probes that have been developed for *in vivo* assessment of enzymatic reactions and other physiological processes.

2.1 DNP Probes for *In Vivo* Applications: An Overview

2.1.1 The Dissolution DNP Method

DNP is based on the concept that, at low temperature and high magnetic field, electron spins can be used to enhance the polarization of nuclei with nonzero spin quantum numbers, such as ^{13}C nuclei (Carver & Slichter, 1953; Overhauser, 1953). Whereas the DNP process has been known for decades, the full potential of the technique for noninvasive investigations of metabolism in living systems was only realized by Ardenkjaer-Larsen et al. (2003), who first described dissolution DNP and the components of a "hyperpolarizer."

Briefly, the dissolution DNP process is as follows. A ^{13}C-labeled compound and a radical with an unpaired electron are homogenously mixed in solution. The mixture is placed in a sample cup and cooled to <2 K in a liquid helium bath located inside a magnet. The cooled radical/^{13}C-labeled compound mixture must form an amorphous glass. A microwave source is then used to saturate the electron spin resonance and induce a transfer of polarization from the radical electrons to the ^{13}C nuclei. After a build-up time τ (~1 h), the frozen ^{13}C-labeled sample becomes "hyperpolarized." Ardenkjaer-Larsen et al. showed that the hyperpolarized ^{13}C compound can then be quickly dissolved into an aqueous solution at room temperature while retaining its hyperpolarization. The MR signal of the resulting hyperpolarized compound can be increased by up to 50,000 fold compared to thermal polarization, and this technological advance has enabled the subsequent development and implementation of hyperpolarized studies *in vivo*.

2.1.2 Polarization Level P_{hp} and Relaxation Time T1

For a DNP probe to be successfully applied to the study of living systems, a large number of considerations need to be balanced, as described in detail in Section 3. However, the ultimate goal is to achieve the highest possible level of polarization P_{hp} and the longest possible spin-lattice relaxation time T1, which are defined as follows:

1. The *polarization level* P_{hp}, expressed in %, is a direct measure of the efficiency of the DNP process. Without hyperpolarization, at a magnetic field of 1.5 T, for example, the polarization of ^{13}C nuclei at body temperature is only approximately 1 part per million (ppm). Using DNP, and as mentioned above, one can increase this polarization to 10–50% (Ardenkjaer-Larsen et al., 2003).
2. The *spin-lattice relaxation time T1* characterizes how fast the hyperpolarized signal decays following dissolution. Specifically, upon dissolution, as the system returns to equilibrium, the hyperpolarized signal decays exponentially with a time constant T1. This means that after 1T1, only 37% of the hyperpolarized signal is left; after 2T1, 13.5%; after 3T1, 5%; etc.

Importantly, the maximum achievable polarization P_{hp} and T1 are dependent on the chemical structure of the DNP probe and its preparation, including solvent/glassing agent and radical (cf. Section 3). Table 1 summarizes the ^{13}C-labeled probes that have been hyperpolarized using DNP and applied to the study of enzymatic and/or living processes, *in vitro*, in perfused cells, organs, or *in vivo*. The values of the polarization level P_{hp} and the

Table 1 Technical Characteristics of ^{13}C DNP Probes

DNP Probe	Polarization Level P_{HP} (%)	T1 (s)	Solvent/Glassing Agent	Radical	References
[1-^{13}C] pyruvate	13.9–40	55 [1.5T]; 67 [3T]	None (neat)	OX063 (T)	Golman, Zandt, et al. (2006); Golman et al. (2008)
		48 [11.7T]; 44 [14.1T]	Water/glycerol		Merritt et al. (2008); Wilson et al. (2010)
					Koelsch et al. (2013a, 2013b)
Ethyl [1-^{13}C] pyruvate	28–35	45 [3T]	Ethanol	OX063 (T)	Hurd et al. (2010)
[1-^{13}C] lactate	6.9–31	39, 45, 50 [3T]	Water/DMSO	TEMPO (N)	Chen et al. (2008, 2009)
		34 [14.1T]	Water/glycerol	OX063 (T)	Mayer et al. (2012); Bastiaansen et al. (2014)
[1-^{13}C] alanine	13	41 [3T]	NaOH/DMSO	OX063 (T)	Hu, Zhu, et al. (2011)
[2-^{13}C] pyruvate	27	38.6 [3T]	None (neat)	OX063 (T)	Schroeder et al. (2009); Hu et al. (2012)
					Josan et al. (2014)
[1,2-$^{13}C_2$] pyruvate	14	44 [3T]	None (neat)	OX063 (T)	Chen et al., (2012)

Continued

Table 1 Technical Characteristics of ^{13}C DNP Probes—cont'd

DNP Probe	Polarization Level P_{HP} (%)	T1 (s)	Solvent/Glassing Agent	Radical	References
[U-^{13}C, U-^{2}H] glucose	13–21	9.5 [3T]; 11, 8.9 [7T]	Water	OX063 (T)	Meier, Jensen, et al. (2011); Meier, Karlsson, et al. (2011)
		13.5 for C1-5, 10 for C6 [11.7T]	D_20/DMSO-d_6		Harris et al. (2013); Allouche-Arnon et al. (2013)
		8.9 [11.8T]; 12 for C1-5, 10 for C6 [14.1T]			Rodrigues et al. (2014)
[2-^{13}C] fructose	13–22	14 [3T]; 16 [11.7T]; 27 [14.1T]	Water	OX063 (T)	Keshari et al. (2009); Meier, Karlsson, et al. (2011)
[2-^{13}C] dihydroxyacetone	–	32 [9.4T]	Dimethyl sulfoxide	OX063 (T)	Moreno et al. (2014)
[5-^{13}C] glutamine	5–28	8 [3T]; 15-16.1 [9.4T]	NaOH	OX063 (T)	Gallagher, Kettunen, Day, Hu, et al. (2008); Dafni et al. (2010)
		20 [14.1T]	NaOH/glycerol		Qu et al. (2011); Chiavazza et al. (2013)
			CsOH/DMSO		Cabella et al. (2013); Canape et al. (2014)
[1-^{13}C] glutamate	28	33.9 [9.4T]	Water/Trizma base	OX063 (T)	Gallagher, Kettunen, et al. (2011)

2-Keto[1-^{13}C] isocaproate	30–32	55 [9.4T]	None (neat)	OX063 (T)	Karlsson et al. (2010); Butt et al. (2012)
[1-^{13}C] α-ketoglutarate	16	52 [3T]; 19 [11.7T]	Water/glycerol	OX063 (T)	Chaumeil et al. (2013, 2014)
[1-^{13}C] acetate	9-17	44-58 [1.05T]; 40 [9.4T]	Water/glycerol	OX063 (T)	Comment et al. (2007)
			Water/ethanol	TEMPO (N)	Comment, Rentsch, et al. (2008)
			D_20/ethanol-d_6		Jensen, Peitersen, et al. (2009); Flori et al. (2014)
[1-^{13}C] propionate	–	–	Water	OX063 (T)	Jensen, Peitersen, et al. (2009)
[1-^{13}C] butyric acid	7	20 [11.7T]	DMSO	OX063 (T)	Ball et al. (2014)
[1,4-$^{13}C_2$] fumarate	12–35	24.1 [9.4T]; 29.3 [11.7T]	DMSO	AH111501 (T)	Gallagher, Kettunen, Hu, et al. (2009); Jensen, Peitersen, et al. (2009)
				OX063 (T)	Wilson et al. (2010)
[1-^{13}C] urea	11.6–26	40 [2.35T]; 47 [3T]	Glycerol	OX063 (T)	Ardenkjaer-Larsen et al. (2003); Wilson et al. (2010)
		44 [11.7T]			von Morze et al. (2011)

Continued

Table 1 Technical Characteristics of ^{13}C DNP Probes—cont'd

DNP Probe	Polarization Level P_{HP} (%)	T1 (s)	Solvent/Glassing Agent	Radical	References
[2-^{13}C, 1,2-$^{2}H_4$] choline	24	60 [7T]; 54 [9.4T]	Water	OX063 (T)	Allouche-Arnon, Gamliel, et al. (2011); Allouche-Arnon, Lerche, et al. (2011)
		53 [11.8T]; 47 [14.1T]			Friesen-Waldner et al. (2014)
[1-^{13}C] bicarbonate	12.7–19	34, 50 [3T]	Water/glycerol	OX063 (T)	Gallagher, Kettunen, Day, Hu, et al. (2008); Wilson et al. (2010)
		48.7 [11.7T]	D_2O		Chen et al. (2012); Scholz et al. (2015)
[1-^{13}C] dehydroascorbic acid	5.9–8.2	56.5 [3T]; 20.5 [9.4T]	DMSO-d_6	OX063 (T)	Bohndiek et al. (2011); Keshari et al. (2011)
		20.7 [11.7]	Dimethylacetamide		Keshari, Sai, et al. (2013); Keshari et al. (2015)

The list of ^{13}C DNP probes that have been polarized using the DNP technique and been applied to the study of enzymatic and/or living processes, either *in vitro*, perfused cells, or organs or *in vivo*, is presented in Column 1. The range of polarization levels P_{hp} achieved is reported in Column 2. The values of relaxation time T1 are listed in the third column for every field strength reported in the literature. Finally, for each DNP probe preparation, the solvent/glassing agent/s and radical/s that have been tested to date are also described (Columns 4 and 5). *Note: radical nomenclature: T = trityl; N = nitroxide.*

relaxation time T1 that have been achieved are reported, as well as the solvent/glassing agent and radical used.

2.2 The "Star Probe": [1-^{13}C] Pyruvate

2.2.1 Metabolism of [1-^{13}C] Pyruvate

[1-^{13}C] pyruvate is by far the most common DNP probe, thanks mostly to its advantageous chemical properties (long longitudinal relaxation time T1 and high polarization level P_{hp}, cf. Table 1). Pyruvate is the product of glucose metabolism through glycolysis and is at the key intersection of several metabolic pathways, making it a highly biologically relevant probe.

Upon injection in a living system, [1-^{13}C] pyruvate can be metabolized to [1-^{13}C] lactate and [1-^{13}C] alanine by the enzymes lactate dehydrogenase A (LDH-A) and alanine transaminase (ALT), respectively. Additionally, [1-^{13}C] pyruvate can be converted by the enzyme pyruvate dehydrogenase (PDH) into ^{13}C-labeled CO_2 and acetyl-CoA, serving as a readout of PDH activity and flux toward the tricarboxylic acid (TCA) cycle. It is important to note that $^{13}CO_2$ is in rapid equilibrium with [1-^{13}C] bicarbonate, and that this interconversion, catalyzed by the enzyme carbonic anhydrase, is pH-dependent (see below for more details). In addition, a few studies have probed pyruvate production, using [1-^{13}C] lactate (Bastiaansen, Yoshihara, Takado, Gruetter, & Comment, 2014; Chen et al., 2008; Mayer et al., 2012) and [1-^{13}C] alanine (Hu, Zhu, et al., 2011) as hyperpolarized agents to probe the backward reactions catalyzed by dehydrogenase B (LDH-B) and alanine transaminase (ALT), respectively.

An important biological advantage of pyruvate is the existence of a dedicated family of transmembrane transporters, namely monocarboxylate transporters or MCTs, that facilitate the cellular uptake of pyruvate (primarily MCT1) as well as the efflux of lactate (primarily MCT4) (Keshari, Sriram, Koelsch, et al., 2013; Keshari, Sriram, Van Criekinge, et al., 2013; Lodi, Woods, & Ronen, 2013). In the context of the brain in which the presence of the blood–brain barrier (BBB) can limit pyruvate delivery, [1-^{13}C] ethyl pyruvate, a lipophilic analog of [1-^{13}C] pyruvate that diffuses faster through the BBB (Hurd et al., 2010), has also been used as an alternative probe for studies of brain metabolism.

2.2.2 Applications to Pathology

The conversion of hyperpolarized [1-^{13}C] pyruvate to hyperpolarized [1-^{13}C] lactate is without question the "poster child" of DNP-MR. This metabolic reaction, catalyzed by LDH-A with NADH as its cofactor, has been investigated in many living systems. However, its most significant contribution has been in the studies of cancer (Albers et al., 2008; Chen, Albers, et al., 2007; Chen, Hurd, et al., 2007; Dafni & Ronen, 2010; Golman, Zandt, Lerche, Pehrson, & Ardenkjaer-Larsen, 2006; Harris, Eliyahu, Frydman, & Degani, 2009; Hu, Balakrishnan, et al., 2011; Hu et al., 2010; Hurd et al., 2010; Kettunen et al., 2010; Larson et al., 2010, 2013; Park et al., 2010; Schilling et al., 2013; Zierhut et al., 2010). In tumor cells, an increase in the conversion of pyruvate into lactate, as compared to normal tissue, can serve as a readout of the Warburg effect (Warburg, 1956b), considered a metabolic hallmark of cancer (Gillies, Robey, & Gatenby, 2008; Vander Heiden, Cantley, & Thompson, 2009; Warburg, 1956a, 1956b). High levels of hyperpolarized [1-^{13}C] lactate produced from injected hyperpolarized [1-^{13}C] pyruvate have been reported in the transgenic prostate adenocarcinoma of mouse prostate (TRAMP) model (Albers et al., 2008; Chen, Albers, et al., 2007; Chen, Hurd, et al., 2007; Hu et al., 2010; Larson et al., 2010, 2013; Zierhut et al., 2010), in Tet-o-MYC/LAP-tTA double transgenic mouse model of liver cancer (Hu et al., 2010), in P22 sarcoma rat cancer (Golman, Zandt, et al., 2006), EL4 lymphoma cancer (Day et al., 2007; Kettunen et al., 2010), and in U87 and U251 human glioblastoma (Park et al., 2010). In the TRAMP model, the level of lactate was also correlated to tumor grade, showing a significant increase in lactate for high-grade compared to normal and low-grade tissues.

Studies have also used [1-^{13}C] pyruvate to study normal heart and heart disease in excised heart or *in vivo* (Golman et al., 2008; Merritt, Harrison, Storey, Sherry, & Malloy, 2008; Schroeder et al., 2008), as well as other organs under normal or pathological conditions (Kohler et al., 2007; Laustsen et al., 2013; Lee et al., 2013; Leftin, Degani, & Frydman, 2013; Leftin, Roussel, & Frydman, 2014; Merritt, Harrison, Sherry, Malloy, & Burgess, 2011; Xu et al., 2011). Variations in hyperpolarized lactate, alanine, and bicarbonate have been linked to modulations in expression or activities of LDH-A, ALT, and PDH. Studies have also reported an increase in lactate-to-pyruvate ratio in inflamed plantar tissues of rat hind paw (MacKenzie et al., 2011) and radiation-induced lung inflammation (Thind et al., 2013, 2014).

2.2.3 Monitoring of Therapy Response

[1-^{13}C] pyruvate was also shown as a useful probe to monitor response to anticancer therapies. A decrease in the pyruvate to lactate flux, associated with depletion of the LDH-A cofactor NADH, was measured in EL4 lymphoma tumors in mice following treatment with the chemotherapeutic agent etoposide (Day et al., 2007). In breast cancer cells, treatment with a mitogen-activated protein kinase inhibitor induced a decrease in the pyruvate to lactate conversion that was linked to drop in MCT1 (Lodi et al., 2013). In PC-3MM2 prostate tumors treated with the tyrosine-kinase inhibitor imatinib, the drop in lactate-to-pyruvate ratio was associated with a decrease in the expression of LDH-A and its two transcription factors HIF-1α and c-Myc (Dafni & Ronen, 2010). In glioblastoma and breast cancer cells, treatment with LY294002 or everolimus, which targets the phosphoinositide 3-kinase/mammalian target of rapamycin (PI3K/mTOR) pathway, resulted in a decrease in hyperpolarized lactate which correlated with a drop in LDH-A activity, expression, and protein levels (Venkatesh et al., 2012; Ward et al., 2010), a result that was further confirmed in an *in vivo* study on orthotopic glioma tumors in rats (Chaumeil et al., 2012). The chemotherapeutic agent Temozolomide also led to a decrease in lactate production mediated by a drop in pyruvate kinase M2 activity in orthotopic models of human glioma (Park, Mukherjee, et al., 2014). Finally, the PDH activator dichloroacetate induced a decrease in the ratio of lactate to bicarbonate (Hu et al., 2012; Park et al., 2013).

2.2.4 Clinical Translation

Given the success of the preclinical studies described above, hyperpolarized [1-^{13}C] pyruvate underwent clinical translation and was tested in the first phase I clinical trial recently completed at the University of California, San Francisco (Nelson et al., 2013). In this study, both the safety and feasibility of using [1-^{13}C] pyruvate as a probe were assessed in prostate cancer patients under active surveillance. Consistent with preclinical findings, an increase in lactate signal and lactate-to-pyruvate ratio was observed in tumor compared to uninvolved tissue. Importantly, no dose-limiting toxicities were observed up to a dose of 230 m*M*. Further clinical trials using hyperpolarized [1-^{13}C] pyruvate are now underway.

2.3 Probing Enzymatic Pathways in Living Systems

2.3.1 TCA Cycle

As described above, pyruvate ^{13}C-labeled in the C1 position has been widely used using DNP-MR and has enabled noninvasive monitoring of lactate production via LDH-A as well as bicarbonate production via PDH. However, it is also important to note that pyruvate can be labeled in the C2 position, hyperpolarized, and serve to monitor the TCA cycle beyond

PDH. In particular, [2-^{13}C] pyruvate and its metabolism to [5-^{13}C] glutamate and/or [1-^{13}C] acetylcarnitine has been detected in the perfused heart (Schroeder et al., 2009, 2012), in a model of heart failure *in vivo* (Schroeder et al., 2013) as well as in the normal rat heart and abdomen *in vivo* (Hu et al., 2012b; Josan, Park, et al., 2013). Treatment with the PDH activator dichloroacetate resulted in detectable modulation of hyperpolarized [5-^{13}C] glutamate production (Hu et al., 2012; Josan, Park, et al., 2013; Park et al., 2013; Schroeder et al., 2012).

The lower *SNRs* of the products detected from [2-^{13}C] pyruvate (compared to [1-^{13}C] pyruvate-derived lactate), as well as their wide spectral dispersion, have limited the *in vivo* development of this probe. However, using complex dedicated acquisition schemes, two recent studies have demonstrated localized, dynamic data of [5-^{13}C] glutamate production in the rat heart and brain *in vivo* on a clinical scanner (Josan et al., 2014; Park et al., 2013). These studies could open the door to further uses of [2-^{13}C] pyruvate as an imaging probe.

Chen et al. have used a double-labeled pyruvate probe, namely [1,2-^{13}C] pyruvate, in order to obtain simultaneous information on glycolysis and oxidative fluxes (Chen et al., 2012). In this case though, the resonance splitting due to ^{13}C–^{13}C coupling induces decreased *SNRs* and spectral overlaps, thus limiting the interpretation of the data.

2.3.2 Carbohydrate Metabolism

In an effort to noninvasively monitor the glycolytic pathway in a living system using dissolution DNP-MR, [U-^{13}C, U-^{2}H] glucose was hyperpolarized for the first time in 2011 and its catabolism into multiple downstream metabolites detected in suspensions of *Escherichia coli* bacteria (Meier, Jensen, & Duus, 2011, 2012) as well as in *Saccharomyces cerevisiae* yeast (Meier, Karlsson, Jensen, Lerche, & Duus, 2011).

Imaging glucose metabolism is of particular interest in the context of cancer for monitoring the Warburg effect, in addition to pyruvate (Warburg, 1956b). In this context, hyperpolarized [U-^{13}C, U-^{2}H] glucose was characterized *in vivo* (Allouche-Arnon et al., 2013), and the conversion of [U-^{13}C, U-^{2}H] glucose to hyperpolarized ^{13}C-labeled lactate observed in perfused tumor cells (Harris, Degani, & Frydman, 2013) and in tumors *in vivo* (Rodrigues et al., 2014; Timm et al., 2014). Interestingly, intermediates of the pentose phosphate pathway were also detected (Harris et al., 2013; Timm et al., 2014). Furthermore, one study also showed that the glycolytic flux as detected by DNP-MR was decreased following chemotherapeutic treatment (Rodrigues et al., 2014). However, despite the use of deuteration to increase the T1 values of ^{13}C-labeled glucose (cf. Section 3), the *SNR*s of the detected metabolites, e.g., ^{13}C-labeled lactate, were still fairly low, requiring the use of dedicated nonlocalized acquisition strategies (e.g., surface coil) to observe glucose metabolism *in vivo*.

In addition to glucose, another monosaccharide of interest, [2-^{13}C] fructose, has been successfully polarized, its metabolism detected in yeast suspensions (Meier, Karlsson, et al., 2011), and its uptake detected in a prostate tumor model *in vivo* (Keshari et al., 2009). However, the *in vivo* limitation of this probe is the low chemical shift difference between [2-^{13}C] fructose and its phosphorylated counterpart, preventing spectral differentiation of substrate and its first metabolic product.

Finally, [2-^{13}C] dihydroxyacetone was recently polarized and applied to the detection of hepatic gluconeogenic and glycogenolytic states in the perfused liver (Moreno et al., 2014).

2.3.3 Glutaminolysis

Glutamine metabolism is particularly important in actively proliferating cells and most notably in tumors. Cancer cells have been shown to be glutamine avid (DeBerardinis et al., 2007), and glutamine addiction is being considered a promising therapeutic target (Wise & Thompson, 2010). These considerations explain the huge interest of the DNP-MR community in imaging glutamine metabolism. However, detection of [5-^{13}C] glutamate following injection of hyperpolarized [5-^{13}C] glutamine was only reported in three cell studies (in human hepatoma cells (Gallagher, Kettunen, Day, Lerche, & Brindle, 2008) and prostate cancer cells (Canape et al., 2014; Dafni et al., 2010)), and only one *in vivo* study on a preclinical model of liver cancer in rats (Cabella et al., 2013). This limited number of reports reflects the current challenges associated with hyperpolarization of glutamine and detection of its metabolism. First, glutamine is unstable in aqueous solutions and tends to rapidly degrade into two by-products, pyroglutamate and glutamate (Cabella et al., 2013). Glutamate is also the product of glutamine metabolism via glutaminase, and glutamate from the injected solution is indiscernible from the subsequent metabolic product of glutamine. This complicates the interpretation of data. Another limitation of hyperpolarized glutamine is that its cellular uptake is fairly slow (typically 2–3 nmol/min/mg of protein (Comment & Merritt, 2014)), limiting detection of its metabolism during the hyperpolarization lifetime. Finally, as discussed in detail in Section 3, scalar coupling relaxation processes drastically decrease the polarization level of [5-^{13}C] glutamine during transport from the polarizer to the MR system (Chiavazza et al., 2013). Nonetheless, efforts are ongoing to further optimize this probe. To limit by-product formation, alternative DNP preparations have been developed (see Section 3) (Cabella et al., 2013), and deuteration schemes have been proposed in order to increase T1 values (Qu et al., 2011).

2.3.4 Transamination and Oncometabolite Detection

The reversible transamination of glutamate to α-ketoglutarate is an important reaction, which represents a bridge between glutaminolysis and the TCA cycle. Importantly, this reaction can be catalyzed by several enzymes, including alanine transaminase (ALT) and branched chain amino acid transaminase 1 (BCAT1). In 2011, Gallagher et al. demonstrated that, following injection of hyperpolarized [1-^{13}C] glutamate in hepatoma cells and *in vivo* in a murine lymphoma model, formation of [1-^{13}C] α-ketoglutarate can be detected and that its level is increased when coinjecting hyperpolarized [1-^{13}C] glutamate with pyruvate (Gallagher, Kettunen, et al., 2011). The authors concluded that, in this model, the observed conversion of hyperpolarized [1-^{13}C] glutamate to α-ketoglutarate was mainly catalyzed by ALT, which uses pyruvate as a cofactor.

In the brain, BCAT1 catalyzes the transamination of glutamate to α-ketoglutarate through conversion of branched chained amino acids (BCAAs) to branched chained ketoacids (BCKAs) and plays a critical role in the glutamate/glutamine cycle and brain nitrogen homeostasis. 2-Keto [1-^{13}C] isocaproate (KIC), a BCKA, was hyperpolarized and its conversion to the BCAA leucine by BCAT1 was detected *in vivo* in the normal rat brain (Butt et al., 2012). In addition, because BCAT1 is modulated in several cancer types, hyperpolarized [1-^{13}C] KIC was also used to noninvasively detect changes in BCAT1 activity through detection of hyperpolarized leucine in murine lymphoma (EL4) and rat mammary adenocarcinoma (R3230AC) tumors *in vivo* (Karlsson et al., 2010).

The reverse reaction, namely α-ketoglutarate-to-glutamate conversion, has also been detected in a preclinical model of glioma at clinical field strength using DNP-MR. In this study, our group demonstrated that formation of hyperpolarized [1-^{13}C] glutamate from hyperpolarized [1-^{13}C] α-ketoglutarate was inhibited by the presence of the isocitrate dehydrogenase 1 (IDH1) mutation, a critical mutation in glioma pathogenesis, which had been shown to induce a drop in BCAT1 expression and activity (Chaumeil et al., 2014). The conversion of hyperpolarized [1-^{13}C] α-ketoglutarate to the oncometabolite 2-hydroxyglutarate, catalyzed by the mutant IDH1 enzyme, was also detected in a preclinical model of IDH1 mutant tumors at clinical field strength (Chaumeil et al., 2013).

2.3.5 Fatty Acid/Ketone Metabolism

Fatty acids, especially short-chain fatty acids, are the main energetic substrate of normal heart and, as such, DNP probes of choice for the study of cardiac metabolism. To date, two short-chain fatty acids have been successfully polarized and applied to heart studies. [1-^{13}C] acetate was first polarized by Jensen, Meier, et al. (2009), and its subsequent metabolism was investigated *in vivo* in murine (Jensen, Peitersen, et al., 2009) and swine hearts (Flori et al., 2014). Additionally, [1-^{13}C] butyrate metabolism was successfully probed in the perfused rat heart (Ball et al., 2014).

In the brain, where glial cells take up acetate, the cerebral conversion of hyperpolarized [1-^{13}C] acetate to [1-^{13}C] α-ketoglutarate was successfully detected using a dedicated polarization transfer scheme in rats (Mishkovsky, Comment, & Gruetter, 2012). In addition to hyperpolarized [1-^{13}C] acetate, a study reported the use of [1-^{13}C] propionate, another short-chain fatty acids, for the investigation of metabolism in mouse heart, liver, and skeletal muscle pre- and postischemia–reperfusion (Jensen, Peitersen, et al., 2009).

2.3.6 Necrosis

The potential of hyperpolarized [1,4-$^{13}C_2$] fumarate, an intermediate of the TCA cycle, for assessment of necrosis has been demonstrated in tumor cells (Gallagher, Kettunen, Hu, et al., 2009), in implanted tumors *in vivo* following chemotherapeutic (Witney et al., 2010) and antiangiogenic (Bohndiek, Kettunen, Hu, & Brindle, 2012) treatments, as well as in a model of early tubular necrosis in mice (Clatworthy et al., 2012). The use of hyperpolarized [1,4-$^{13}C_2$] fumarate for detection of necrosis is based on the fact that, in normal cells, fumarate uptake and metabolism are very limited within the lifetime of the hyperpolarization. However, when cells become necrotic, their plasma membrane permeability is compromised leading to detectable production of hyperpolarized [1,4-$^{13}C_2$] malate by the enzyme fumarase.

2.4 Applications to Other Physiological Processes

2.4.1 Perfusion

As compared to conventional perfusion imaging techniques, MR perfusion imaging of DNP probes presents several advantages, including higher diffusibility than many other commonly used contrast agents, as well as potentially reduced nephrotoxicity. In this context, multiple DNP probes have been developed for vascular imaging.

Among these, hyperpolarized [^{13}C] urea, an endogenous and metabolically inert metabolite, has been the most commonly used to date (Ardenkjaer-Larsen et al., 2003; Golman et al., 2003; Pages et al., 2013; Patrick et al., 2015; Reed et al., 2014; von Morze, Bok, et al., 2014a, 2014b; von Morze, Bok, Sands, Kurhanewicz, & Vigneron, 2012; von Morze et al., 2011; von Morze, Larson, et al., 2012; von Morze et al., 2013). It constitutes an excellent agent to monitor tumor perfusion (Bahrami, Swisher, Von Morze, Vigneron, & Larson, 2014; von Morze et al., 2011; Wilson et al., 2010), kidney physiology (von Morze, Bok, et al., 2012; von Morze et al., 2013), as well as other processes (Pages et al., 2013; Patrick et al., 2015). When copolarized and coinjected with other hyperpolarized probes, such as [1-^{13}C] pyruvate, hyperpolarized [^{13}C] urea can also be used to directly monitor delivery and uptake in the target tissue (Bahrami et al., 2014).

Recently, hyperpolarized [1, 1, 2, 2-D_4, 1-^{13}C] choline chloride was used *in vivo* to image perfusion and choline distribution in real time (Friesen-Waldner et al., 2014). Additional DNP probes enabling perfusion have been recently developed for numerous applications (von Morze et al., 2014), but are outside the scope of this metabolism-focused review.

2.4.2 pH

In vivo imaging of pH has always been of interest, especially in the context of cancer. As previously mentioned, the enzyme carbonic anhydrase (CA) catalyzes the rapid interconversion of bicarbonate and CO_2, and this exchange is pH-dependent. As a consequence, studies have demonstrated the potential of using hyperpolarized [1-^{13}C] bicarbonate as a probe to measure extracellular pH and its modulation in pathological conditions (Gallagher, Kettunen, Day, Hu, et al., 2008; Scholz et al., 2015; Wilson et al., 2010). Using the Henderson–Hasselbalch equation, pH can be determined from the bicarbonate-to-CO_2 signal ratio (De Graaf, 2007). In EL4 lymphoma and TRAMP mouse models, extracellular acidification was observed in tumor tissues (Gallagher, Kettunen, Day, Hu, et al., 2008; Wilson et al., 2010). Similarly, a decrease in pH was detected in acute inflammation induced by Concanavalin A (Scholz et al., 2015).

2.4.3 Redox Status

Reduction and oxidation (redox) reactions are involved not only in numerous normal physiological processes but are also often modulated in pathological conditions. In order to probe redox process *in vivo*, [1-^{13}C] dehydroascorbate (DHA) has been hyperpolarized and its reduction to [1-^{13}C] vitamin C detected in multiple organs in mice, including liver, kidneys, and brain, as well as in perfused lymphoma cells and in lymphoma and prostate tumors *in vivo* (Allouche-Arnon et al., 2013; Bohndiek et al., 2011; Keshari et al., 2011). In two studies using hyperpolarized [1-^{13}C] DHA to investigate redox status in preclinical models of prostate cancer (Keshari, Sai, et al., 2013) and diabetic renal injury (Keshari et al., 2015), Keshari et al. showed that the tumor contrast obtained using hyperpolarized [1-^{13}C] DHA correlates with the concentration of glutathione, an important antioxidant and scavenger of reactive oxygen species.

3. REQUIREMENTS OF ^{13}C DNP PROBES

With only about a decade of research, metabolic studies using DNP-hyperpolarized compounds are still in their infancy and further applications are likely to emerge in the years to come. This section describes in detail the major biological and chemical requirements that have to be taken into account when designing a ^{13}C DNP probe for studies of metabolism. We would like to point out that this section should be viewed as a guide based on experience in the field, and not as a dogmatic recipe. Figure 1 presents an overview of the entire section for the reader's convenience.

3.1 Biological Considerations

3.1.1 Reaction Speed

When considering a DNP probe, a major consideration is how its T1 relates to the speed of the physiological process that it is intended to probe. As described in Section 2, the window of opportunity to detect metabolic reactions is less than 3 T1s or, based on the T1 of probes developed to date, 30–180 s (after which time only 5% of the hyperpolarized signal remains, and little metabolism can be detected). As a consequence, physiological phenomena that are slower than 2–3 min are more challenging to probe. In that regard, the presence of dedicated transporters, allowing for fast transport of the hyperpolarized agent from the extracellular space to the cytoplasm, seems to confer an advantage for detection of metabolism (e.g., MCT1 in the case of [1-^{13}C] pyruvate) (Lodi et al., 2013). Also see Section 5 for emerging alternative strategies aimed at preserving the polarization for longer time frames.

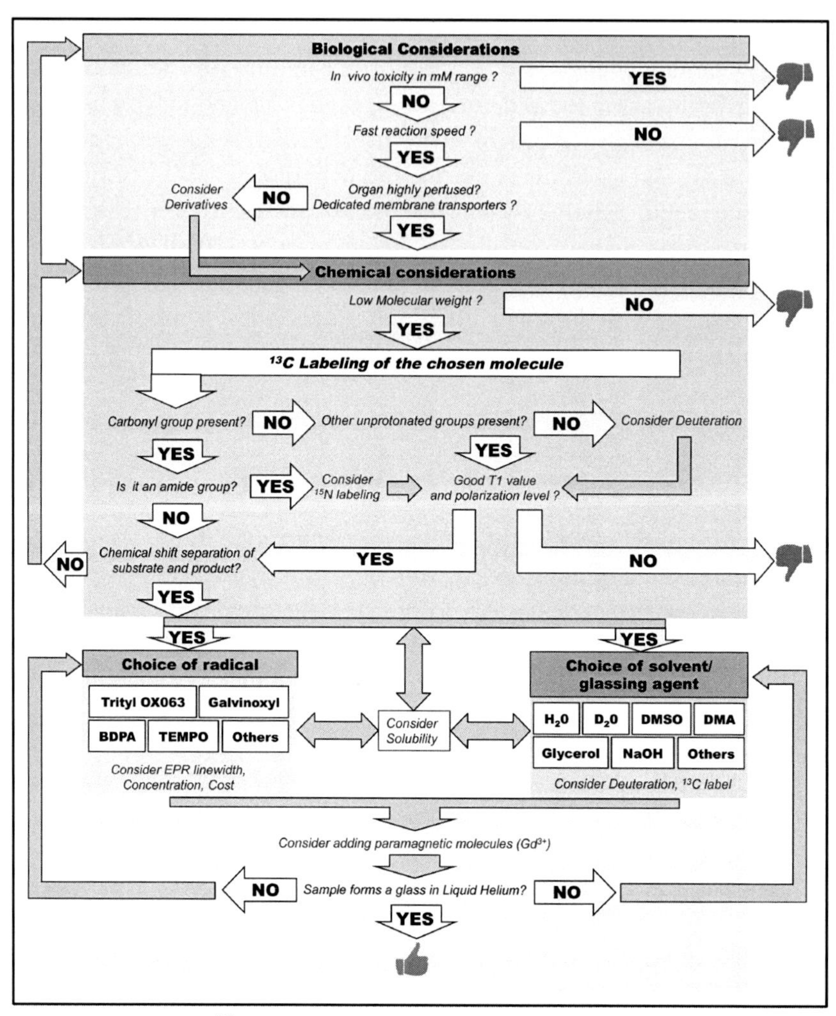

Figure 1 Designing a ^{13}C DNP probe for applications to the study of living systems: a brief guide. The figure summarizes Section 3 in a graphical manner. Briefly, when designing a ^{13}C DNP probe, one has to take into consideration biological parameters, such as reaction speed and delivery, as well as chemical parameters, such as ^{13}C label positioning, and deuteration and then come the choices of the radical and solvent/s for polarization, before final testing for low-temperature glassing. (See the color plate.)

3.1.2 Additional In Vivo *Considerations*

As opposed to positron emission tomography that requires the injection of a few nanomoles or less of a radioactive tracer, ^{13}C MRS of hyperpolarized DNP probes necessitates *in vivo* injections of solutions in the 10–150 m*M* range. As a consequence, toxicity needs to be ruled out when considering a new hyperpolarized agent for injection in living systems.

Another *in vivo* consideration is the delivery of the hyperpolarized probe to the organ of interest, which has to occur rapidly and well within the 3T1s of the probe. This is usually not an issue for highly perfused organs, such as kidney (Cunningham, Dominguez Viqueira, Hurd, & Chen, 2014; Reed et al., 2014; Schmidt et al., 2014; von Morze, Bok, et al., 2012; von Morze et al., 2011; Wiesinger et al., 2012; Xu et al., 2011) or highly angiogenic tumors (cf. review Kurhanewicz et al., 2011 and Section 2), but highlights the challenges of using DNP probes to study poorly perfused organs such as the pancreas (Grozinger, Grozinger, & Horger, 2014; Hansen, Nilsson, Gram, & Frokjaer, 2013; Nissan et al., 2014; Sugimoto et al., 2015).

In the case of the brain, despite high perfusion, the presence of the BBB is an additional factor that can drastically limit the delivery of hyperpolarized agent and needs to be considered (Larsen, Martin, & Byrne, 2014; Pardridge, 2012). A strategy that has been developed to overcome this issue is the use of metabolite derivatives, such as ethyl [1-^{13}C] pyruvate (Hurd et al., 2010) (cf. Section 2). However, using derivatives of endogenous metabolites adds an additional metabolic step in the chain of reactions (demethylation, deethylation, deesterification, etc.), reducing the likelihood of being able to monitor the metabolic reaction of interest within the lifetime of the hyperpolarization. It is also possible that nonphysiological transport and metabolism would be observed.

Once the hyperpolarized agent has reached the organ of interest, another important aspect is the transport of the hyperpolarized agent from the extracellular space to the cytoplasm. The presence of dedicated transporters seems to confer an advantage, both in terms of specific spatial delivery and in terms of speed (e.g., MCT1 in the case of [1-^{13}C] pyruvate) (Lodi et al., 2013).

3.1.3 Pool-Size Effect

Another interesting consideration in ^{13}C DNP-MR experiments is the so-called pool size effect, especially for rapid equilibrium reactions. It has been shown that the size of the pool of the product to be labeled (e.g., [1-^{13}C] lactate when [1-^{13}C] pyruvate is injected) affects the likelihood of its subsequent detection. This phenomenon was described for the pyruvate-to-lactate exchange both in cells (Day et al., 2007) and *in vivo* (Hurd et al., 2013) and implies that, in the case of rapid exchange reactions, the smaller the pool of the product, the less likely it is to be detected.

3.2 Chemical Considerations

Once the biological considerations have been addressed, the specific chemical design of the ^{13}C DNP probe needs to be considered. Most notably, T1 and P_{hp} values can be drastically improved by careful consideration of five main components of the ^{13}C probe and its preparation: (1) the molecular weight, (2) the location of the ^{13}C label, (3) the organic free radical allowing for spin polarization transfer, (4) the solvent, or solvent mixture, and (5) optionally, polarization enhancement molecules.

3.2.1 Molecular Weight

Small molecular weight probes are likely to be most useful for DNP-MR, primarily because the relaxation time T1 decreases with increasing molecular weight (due to an increase in correlation time τ_c). Additional considerations such as ease of synthesis and solubility can also favor a relatively low molecular weight (Keshari & Wilson, 2014; Wilson et al., 2009).

3.2.2 Labeling

3.2.2.1 Position of ^{13}C Label

Once the probe is chosen, the next choice to be made is the positioning of the ^{13}C label within the molecule. This choice is mostly driven by two considerations that need to be optimized simultaneously: (1) maximization of hyperpolarized T1 values for both substrate and product/s of interest, and (2) maximization of the chemical shift difference between substrate and product/s of interest.

3.2.2.1.1 Time Dimension: Maximization of T1 Values The positioning of the ^{13}C label is typically made with the goal to minimize the relaxation processes affecting the ^{13}C nuclei.

Among the relaxation processes, dipolar coupling is the most common. For ^{13}C DNP probes, dipolar coupling originates mainly from ^{1}H nuclei coupled to ^{13}C, and its effect is inversely related to the number of bonds between ^{13}C and ^{1}H. As a result, molecules in which ^{13}C nuclei are directly bound to ^{1}H are of limited use as DNP imaging agents, because the strong ^{1}H–^{13}C dipolar coupling considerably decreases the hyperpolarized T1. To address this issue, the ^{13}C label should be positioned in a chemical group free of direct ^{1}H bonds. A perfect example is the carbonyl carbon, and indeed most useful DNP probes to date have been ^{13}C-labeled at a carbonyl position (cf. Section 2). Such probes include carboxylic acids, ketones, and amides (Keshari & Wilson, 2014). Furthermore, carbonyls are abundant in cellular metabolites and, as such, are candidates of choice for ^{13}C labeling and DNP probes.

3.2.2.1.2 Spectral Dimension: Maximization of Chemical Shift Difference In addition to minimizing the relaxation processes, the choice for the positioning of the ^{13}C label will also affect detection of the metabolic reaction/s in the spectral dimension. Specifically, in order to detect the conversion of a hyperpolarized substrate to its product, the chemical shift difference between the ^{13}C resonances of substrate and product has to be large enough to resolve the two species. Let us say that, for example, one wants to hyperpolarize glutamine and monitor the conversion of hyperpolarized glutamine to glutamate. The glutamine molecule has two carbonyls that could be ^{13}C labeled: C1 ($\delta_{C1}^{Gln}=176.8$ ppm) and C5 ($\delta_{C5}^{Gln}=180.4$ ppm) (cf. Fig. 2). All studies of hyperpolarized glutamine to date have used glutamine labeled in the C5 position (Cabella et al., 2013; Canape et al., 2014; Chiavazza et al., 2013; Gallagher, Kettunen, Day, Lerche, et al., 2008; Qu et al., 2011). The reason is as follows. When considering only T1 and P_{hp}, both carbonyl carbons would be good candidates for hyperpolarization. However, the chemical shifts of the glutamate resonances are $\delta_{C1}^{Gln}=177.2$ ppm for C1 and $\delta_{C5}^{Gln}=183.9$ ppm for C5. As a

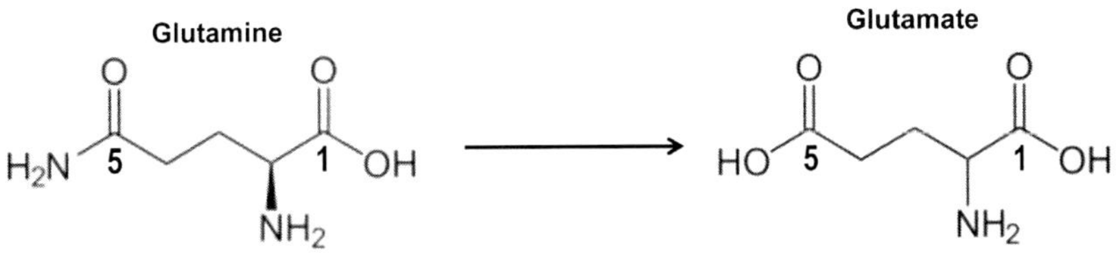

Chemical shift (ppm)	Δ (ppm)	Chemical shifts (ppm)
δ_{C1}^{Gln} =176.8 ppm	Δ_{C1}=0.4 ppm	δ_{C1}^{Glu} =177.2 ppm
δ_{C5}^{Gln} =180.4 ppm	Δ_{C5}=3.5 ppm	δ_{C5}^{Glu} =183.9 ppm

Figure 2 Schematic of the glutamine–glutamate conversion and corresponding chemical shift values. The molecular structure of glutamine and glutamate is illustrated, and the chemical shift values of the C1 and C5 carbons are reported for each molecule, as well as the chemical shift differences (Δ) for each carbon between glutamate and glutamine. The larger chemical difference (red/gray) illustrates the reason for labeling C5 when probing the metabolism of glutamine to glutamate.

consequence, the chemical shift difference between substrate and product would be of $\Delta_{C1}=0.4$ppm when labeling at the C1 position versus $\Delta_{C5}=3.5$ ppm when labeling at the C5 position (cf. Fig. 2). When developing a new DNP probe, these types of considerations need to be made *a priori*.

3.2.2.2 Deuteration and ^{15}N Labeling in Addition to ^{13}C Labeling

In some cases, unprotonated carbons are not present on the molecule of interest. In that case, a possible option to minimize dipolar coupling and enhance T1 is deuteration, as the nuclear spin of deuterium is 1 instead of 1/2 for ^{1}H. This strategy has proven useful to dramatically increase the T1 of several hyperpolarized molecules, such as [U-^{13}C, U-^{2}H] glucose (Allouche-Arnon et al., 2013; Rodrigues et al., 2014; Timm et al., 2011), [5-^{13}C, 4-$^{2}H_2$]-L-glutamine (Qu et al., 2011), and [2-^{13}C, 2-$^{2}H_4$] choline (Allouche-Arnon, Gamliel, et al., 2011; Allouche-Arnon, Lerche, Karlsson, Lenkinski, & Katz-Brull, 2011; Friesen-Waldner et al., 2014) (cf. Section 2). However, care should be taken when choosing the position of the deuterons. Depending on the distance between the ^{1}H and ^{13}C nuclei, replacing ^{1}H by ^{2}H can induce a splitting of the ^{13}C resonance of interest, thus decreasing signal intensity and chemical shift separations. It is also important to keep in mind that deuteration might also have a significant impact on metabolism and enzyme function due to the kinetic isotope effect (Meier, Jensen, et al., 2011).

If the DNP probe is ^{13}C-labeled on an amide group, another important phenomenon to remember is the type II scalar coupling relaxation process due to the presence of the fast-relaxing quadrupolar ^{14}N nucleus adjacent to the ^{13}C. This drastically shortens the T1 of the ^{13}C nuclei. This effect is higher at low field strength and can drastically decrease the liquid polarization during the transfer of the hyperpolarized solution from the polarizer to the MR magnet (Chiavazza et al., 2013). To limit this effect, one can replace ^{14}N by ^{15}N or use a magnetic field carrier during transport (Shang et al., 2015). It is interesting to note that, in addition to minimizing scalar coupling, ^{15}N labeling of amide groups can also dramatically increase the T2 (transverse relaxation rate constant) of the ^{13}C nucleus. This effect has been used to acquire angiographic images with submillimeter in-plane resolutions (Reed et al., 2014). On the other hand, despite the above-described advantages, ^{15}N labeling can induce significant splitting of the ^{13}C resonance/s of interest, thus reducing the *SNRs* and complicating spectral identification.

3.2.3 Radical

Once the choice of DNP probe is made, the choice of the radical needs to be considered. In order to decide which radical to select, one needs to consider several issues, including electron paramagnetic resonance (EPR) linewidth, solubility, and concentration, as described in detail below. The most successful radicals used for DNP to date are the trityl compound OX063 (tris[8-carboxyl-2,2,6,6-benzo(1,2-d:4,5-*d*)-bis(1,3)dithiole-4-yl] methyl sodium salt) (Ardenkjaer-Larsen, Macholl, & Jóhannesson, 2008; Lumata, Kovacs, et al., 2013), BDPA (1,3-bisdiphenylene-2-phenylallyl) (Lumata, Ratnakar, et al., 2011), the nitroxide TEMPO (2,2,6,6-tetramethylpiperidin-1-oxyl) (Montanari et al., 2007), and galvinoxyl (2,6-di-*tert*-butyl-α-(3,5-di-*tert*-butyl-4-oxo-2,5-cyclohexadien-1-ylidene)-*p*-tolyloxy) (Lumata, Merritt, Malloy, et al., 2013), although other radicals have also been developed (Gabellieri et al., 2010; Lumata, Merritt, Khemtong, et al., 2012; Macholl, Johannesson, & Ardenkjaer-Larsen, 2010; Paniagua et al., 2010). Their characteristics are summarized in Table 2.

3.2.3.1 EPR Linewidth

The first criterion for a useful organic radical for ^{13}C DNP is its EPR linewidth (Heckmann, Meyer, Radtke, Reicherz, & Goertz, 2006). Briefly, the use of radicals with a narrow linewidth, such as trityls (Lumata, Kovacs, et al., 2013) and BDPA (Lumata, Ratnakar, et al., 2011), allows for higher polarization of ^{13}C nuclei, whereas radicals with wider EPR linewidths, such as TEMPO, will enable polarization of both ^{13}C and ^{1}H nuclei that are present in their vicinity. However, it is important to note that, generally, the polarization levels obtained with trityls are approximately only two- to threefold higher than the ones obtained with TEMPO (Comment & Merritt, 2014).

For radicals with wide EPR linewidths, such as TEMPO, recent studies have shown that it is possible to take advantage of the significant ^{1}H polarization and use it to enhance ^{13}C polarization by applying ^{1}H–^{13}C cross-polarization schemes (Batel et al., 2014; Bornet et al., 2013; Shimon, Hovav, Feintuch, Goldfarb, & Vega, 2012).

3.2.3.2 Solubility

In order to achieve homogenous mixing of the DNP probe and radical cores, the choice of a solvent or solvent combination has to be optimized to maximize the solubility of both compounds. In terms of solubility, the nitroxide radical TEMPO is the most versatile, as it is highly soluble in both

Table 2 Common Radicals for ^{13}C DNP Probes

Radical	Molecular Structure	Solubility	EPR Linewidth (MHz) at 100 K
TEMPO		Protonated organic solvents	465[1]
		Unprotonated organic solvents	
		Water	
OX063		Water	115[2,3]
BDPA		Tetramethylene sulfone	62[4]
		Nonsoluble in water	
Galvinoxyl		Chloroform	250[5]
		Ethanol	
		Acetone	
		Hexane/cyclohexane	
		Pentane	
		Nonsoluble in water	

The four most common radicals used for dissolution DNP-MR are listed in Column 1. Their corresponding molecular structures, solubility characteristics, and EPR linewidths are reported in Columns 2, 3, and 4, respectively. References: 1, Montanari et al. (2007); 2, Lumata, Kovacs, Sherry, et al. (2013); 3, Ardenkjaer-Larsen, Macholl, et al. (2008); 4, Lumata, Ratnakar, et al. (2011); 5, Lumata, Merritt, Khemtong, et al. (2012); Lumata, Merritt, Malloy, et al. (2012).

apolar and polar organic solvents as well as in water, and as such can be used for almost any ^{13}C DNP probe preparation (Montanari et al., 2007). Trityl radicals are also very versatile in the sense that they can be formulated for increased solubility in several solvents (Reddy, Iwama, Halpern, & Rawal, 2002). The most commonly used radical, OX063, has the highest solubility in water, making it highly suitable for aqueous probe designs (cf. Table 2). The BDPA radical on the other hand is insoluble in water, but reasonably soluble in tetramethylene sulfone (sulfolane), a dipolar aprotic solvent (Lumata, Ratnakar, et al., 2011). As such, this radical is particularly suitable for hyperpolarization of hydrophobic compounds, even though it can be used to polarize water-soluble compounds such as [1-^{13}C] pyruvate or [1-^{13}C] urea (Lumata, Ratnakar, et al., 2011). Finally, the radical Galvinoxyl is not soluble in water, but is soluble in ethyl acetate, chloroform, and acetone (Lumata, Merritt, Malloy, et al., 2013). As a consequence, this radical might be suitable for nonhydrophilic molecules that present a high solubility in one of the above-mentioned solvents.

3.2.3.3 Concentration

Regarding the radical concentration, a ratio of 1 molecule of radical for 500–1000 ^{13}C nuclei has proven sufficient for good spin diffusion and relatively short DNP processes (build-up constant τ of 1–2 h). It is important to note that increasing the radical concentration above the above-mentioned value is generally detrimental to the liquid-state polarization level, as shown for [1-^{13}C] acetate (Flori et al., 2014). Furthermore, when increasing radical concentration, one has to be careful about the paramagnetic effect of the radical, which induces a decrease in the relaxation time T1 after dissolution. In some cases, toxicity of the radical is also an issue. Several approaches have therefore been proposed to safely remove the radical after dissolution. For the nitroxide water-soluble radicals, such as TEMPO, scavengers like vitamin C can be used to quench the radicals in solution after the dissolution (Miéville et al., 2010). For hydrophilic radicals such as BDPA or Galvinoxyl, the radical naturally precipitates in aqueous solutions post dissolution and can thus be completely and safely removed by rapid filtration (Lumata, Ratnakar, et al., 2011). For other radicals including trityls, state-of-the-art filtration methods using columns have been developed, especially for human clinical trials utilizing the hyperpolarized ^{13}C technology (Ardenkjaer-Larsen et al., 2011; Gajan et al., 2014; Nelson et al., 2013). In a recent study, an alternative elegant solution has been proposed: by exposing frozen [1-^{13}C] pyruvate to UV light, Eichhorn et al. were able to create a sufficiently large concentration of radicals to efficiently hyperpolarize [1-^{13}C] pyruvate using DNP (Eichhorn et al., 2013). These photo-induced radicals have the advantage of being transient, as they are readily thermally scavenged upon dissolution, thus removing paramagnetic effects and potential toxicity (Comment, 2014). Such a process could improve the feasibility and safety of future applications of hyperpolarized ^{13}C MR.

3.2.4 Solvent/Glassing Agent

3.2.4.1 Solubility of ^{13}C Probe and Radical

In order for the DNP process to take place, the mixture of ^{13}C probe and free radical has to be homogenous and form an amorphous glass at liquid helium temperature (<2 K). To achieve this, a solvent, or solvent mixture, is usually required. An interesting and widely used exception is [1-^{13}C] pyruvate, which acts as its own glassing agent and solvent for the radical. [1-^{13}C] pyruvate is thus used "neat" at a concentration of 14.1 *M* (Golman, Zandt, et al., 2006). In addition, [2-^{13}C] pyruvate (Hu et al., 2012; Josan et al., 2014; Josan, Park, et al., 2013; Park et al., 2013) and 2-keto[1-^{13}C] isocaproate (Butt et al., 2012; Karlsson et al., 2010) can also be used neat (cf. Section 2).

To choose the adequate solvent, one has to consider the respective solubilities of the ^{13}C probe as well as the chosen radical, as they both need to readily dissolve. The concentrations of the ^{13}C nuclei are typically in the 1–15 *M* range. In addition, as mentioned previously, a 1:500–1000 ratio between ^{13}C nuclei and radical is typically used, leading to a radical concentration of ~1–50 m*M*. Such concentrations allow for ^{13}C nuclei and free radical to be in "close proximity" to facilitate spin diffusion. Furthermore, high concentrations of the DNP probe also allow for postdissolution concentrations of 10–150 m*M*, which are, to date, necessary to observe physiological reactions *in vivo*. In some cases, solubility is increased by modulating parameters such pH and temperature, in order to reach the necessary concentration range.

3.2.4.2 Glassing Agents

As mentioned above, one of the required physical properties of a DNP sample mixture is that it forms an amorphous solid (glass matrix) at liquid helium temperature. In some cases, solvents can serve directly as glassing agents (e.g., pyruvate). In other cases, addition of a secondary cosolvent, such as glycerol, is necessary to facilitate the glassing (cf. Table 1). The glassing agent can also serve to dissolve the radical if it is not soluble in the main solvent (Lumata, Ratnakar, et al., 2011). In practical terms, in order to evaluate the glassing property of a newly designed DNP sample, one would typically place the DNP mixture in a benchtop liquid nitrogen Dewar and visually assess the glassing and homogeneity of the frozen sample.

3.2.4.3 Solvent Deuteration and ^{13}C Labeling

An additional parameter that can be used to increase polarization levels is deuteration of the solvent. For wide EPR linewidth polarizing agents such as the nitroxides TEMPO or galvinoxyl, deuteration can lead to a two- to threefold increase in polarization, achieving levels comparable to those obtained using trityls. However, for narrow EPR linewidth radicals, such as trityl or BDPA, deuteration led to a 40–50% *decrease* in maximal ^{13}C polarization (Kurdzesau et al., 2008; Lumata, Merritt, & Kovacs, 2013). ^{13}C labeling of the solvent of interest does not seem to have an effect on the polarization level of the DNP probe. However, one study reported that ^{13}C labeling of DMSO could decrease the build-up constant τ by a factor of 2, as compared to unlabeled DMSO, whereas maximum polarization remained unchanged (Lumata, Kovacs, Malloy, Sherry, & Merritt, 2011).

3.2.5 Polarization-Enhancing Molecules

Addition of Gadolinium (Gd^{3+})-based paramagnetic molecules, such as Gd-HP-DO3A or Gd-DOTA, is an additional optional strategy that can be used to increase the polarization level of ^{13}C-labeled compounds (Ardenkjaer-Larsen et al., 2008; Lumata, Merritt, Malloy, Sherry, & Kovacs, 2012). To date, the choice of the concentration of paramagnetic molecules remains fairly empirical. For [1-^{13}C] pyruvate, studies have shown that maximum polarization enhancement is reached at a concentration of Gd^{3+} of 4–5 m*M*. However, it is important to note that, as in the case of solvent deuteration, this approach is radical dependent. As reported by Lumata et al., Gd^{3+} doping leads to the best response in trityl OX063 [1-^{13}C] pyruvate preparations, with a 300% increase in the solid-state nuclear polarization. In BDPA- and TEMPO-doped samples, Gd^{3+} doping only led to a 5–20% increase in polarization level (Lumata, Merritt, Malloy, et al., 2012). Finally, it is important to note that doping with other paramagnetic nuclei, such Mn^{2+}, is also being considered as an alternative approach (Corzilius et al., 2011).

3.3 Characterization of the ^{13}C DNP Probe

Once the DNP sample preparation forms a glass at low temperature, final measurements have to be made to characterize the sample and confirm its utility.

3.3.1 DNP Microwave Frequency

First, one has to determine the microwave irradiation frequency that will provide maximal polarization enhancement, a procedure commonly referred to as "sweeping." To do so, a large volume (100–150 μL) of the prepared DNP probe is placed in the sample cup and introduced in the polarizer. For commercial systems, dedicated software can then be used to sweep over a large range of microwave frequencies and record the solid-state polarization of the DNP sample for each frequency. Once the sweep is performed, the frequency of maximum enhancement can be identified and will be used for any subsequent polarizations of this DNP sample preparation. It is important to note that the microwave frequency can vary from one polarizer to another and should thus be measured on each system.

3.3.2 T1 and Polarization Level

Acquisition methods to measure the relaxation time T1 and the polarization level P_{hp} of a DNP sample are described in detail in Section 5. Typically, polarization levels of more than 1% and T1 values of more than 10 s are needed to probe metabolism. And of course, the higher those values the better (cf. Section 2).

3.3.3 Dissolution Buffer Design

Once the DNP preparation is optimized and its final molarity is known, an important step is the preparation of the buffer for dissolution. If injected in perfused cells or *in vivo*, one has to ensure that the final hyperpolarized solution is physiologically isotonic with a pH in the physiological range. The dissolution buffer is therefore optimized for each injected concentration of hyperpolarized agents. For example, in the case of [1-^{13}C] pyruvate, a volume of 24 μL of 14.1 *M* pure [1-^{13}C] pyruvate is typically dissolved in 4.5 mL of an isotonic Tris-based buffer containing 80 m*M* NaOH resulting in a final isotonic solution of 80 m*M* [1-^{13}C] pyruvate at pH 7.5 (Ward et al., 2010).

4. CONDUCTING HYPERPOLARIZED EXPERIMENTS

In this section, we first describe the two main elements of a dissolution DNP-MR laboratory: the DNP polarizer and the MR system for ^{13}C data acquisitions. We then describe the practical aspects of how to perform dissolution DNP-MR experiments in solution, cells, organs, and *in vivo*. We hope that the reader will find this section useful in understanding the range of approaches that can be used to rapidly probe metabolism using DNP-MR.

4.1 Main Components of a DNP-MR Laboratory

4.1.1 DNP Equipment

Ardenkjaer-Larsen et al. published the design and proof-of-concept use of the first dissolution DNP system in 2003 (Ardenkjaer-Larsen et al., 2003). Since then, multiple commercial and custom-built systems have been developed.

4.1.1.1 Preclinical Commercial System: Hypersense™

To date, one of the most commonly used DNP polarizer for *in vitro* and preclinical *in vivo* studies is the Hypersense™ DNP polarizer (Oxford Instruments, Inc, Abingdon, United Kingdom). The main technical characteristics of this system are:

- An integrated 94 GHz microwave source with 0.5 GHz sweep, up to 100 mW
- A magnetic field strength of 3.35 T (actively shielded)
- A sample temperature of 1.2–1.6 K

The hypersense polarizer is versatile and not fully automated, which gives the user a lot of flexibility but also requires a certain level of training and expertise for usage. This system typically allows the polarization of up to 100 μL of a chosen DNP sample. Dissolution of the sample, once polarized, occurs into a volume of 3–10 mL of chosen aqueous solvent. These relatively small volumes are highly adequate for *in vitro*, cells, and preclinical *in vivo* studies. However, the hypersense polarizer only allows for polarization of one sample at a time, precluding several studies in quick succession and introducing the polarization time (build-up constant $\tau \sim 1$ h) as a minimum requirement between studies. The polarizer also uses large quantities of costly liquid cryogens (helium and nitrogen) to maintain low sample temperatures, but the installation of a helium recovery system can be considered to overcome this limitation.

4.1.1.2 Clinical Commercial System: SpinLab™

With the emergence of clinical studies using the hyperpolarized technology, a new commercial hyperpolarizer specifically designed for clinical use, the SpinLab™ (General Electrics, Milwaukee, USA) has been developed recently (Ardenkjaer-Larsen et al., 2011). Its main technical characteristics are:

- An integrated 140 GHz microwave source
- A magnetic field strength of 5 T
- A sample temperature of <1 K

When compared to the hypersense, the increase in magnetic field strength and decrease in temperature allows for a dramatic twofold improvement in the ^{13}C polarization level, at least for pyruvate (Hu et al., 2013). Furthermore, in contrast to the Hypersense™, the SpinLab™ is a fully automated system that allows for polarization of multiple samples all at the same time. Dissolution volumes are also larger (up to 100 mL), consistent with the clinical application. Additionally, low temperature is reached without any consumption of liquid cryogens, highly improving cost effectiveness. Whereas this design is highly suitable for human studies, and the higher polarization levels provide a significant advantage, the requirement for larger volumes can be a limiting factor for studies that involve optimization of a new and expensive agent aimed at preclinical studies.

4.1.1.3 Custom-Built Polarizers

Multiple groups have designed and developed custom-built polarizers to match their needs (Armstrong, Edwards, Wylde, Walker, & Han, 2010; Cheng, Capozzi, Takado, Balzan, & Comment, 2013; Comment, Rentsch, et al., 2008; Comment, van den Brandt, et al., 2008; Jannin et al., 2008; Jannin, Comment, & van der Klink, 2012; Johannesson, Macholl, & Ardenkjaer-Larsen, 2009; Lumata et al., 2015). In all cases, the design of these polarizers is based on the principle that the lower the temperature and the higher the field strength, the better the polarization. For further details, the authors refer the reader to the original publications (Armstrong et al., 2010; Cheng, Capozzi, et al., 2013; Jannin et al., 2008).

4.1.2 MR System

When planning a DNP experiment, the choice of the magnetic field strength of the MR system, as well as the location of the MR system relative to the hyperpolarizer, needs to be considered.

4.1.2.1 Magnetic Field Strength

In addition to the chemical properties discussed in Section 3, the T1 of the hyperpolarized nuclei intrinsically depends on the magnetic field at which the data acquisition is performed.

Dipolar coupling between ^{13}C and neighboring ^{1}H is usually the most dominant relaxation mechanism governing the T1. Dipolar coupling moderately decreases with increasing field strength and thus one would assume that a higher magnetic field would lead to increased T1 values. However, in ^{13}C carbonyls, the relaxation time T1 mainly depends on the chemical shift anisotropy (CSA) relaxation process. As CSA is proportional to B_0^2, its effect is more pronounced at high magnetic fields, leading to decreased T1 values for the most common carbonyl-based DNP probes currently used. In the case of hyperpolarized [1-^{13}C] pyruvate, for example, T1 is 67 s at 3 T, 48 s at 11.7 T, and 44 s at 14.1 T (Koelsch et al., 2013b; Wilson et al., 2010). As a consequence, whereas studies at high field strength usually benefit from increased spectral resolution and SNRs, decreased T1 values can present a dramatic drawback, at least for ^{13}C carbonyl probes and when trying to rapidly probe a metabolic reaction.

4.1.2.2 Position with Respect to DNP Polarizer

The positioning of the MR system relative to the DNP polarizer in the laboratory space is crucial. As a general rule, the polarizer should be placed as close as possible to the MR system, and if feasible in its fringe field. There are two main reasons for this: (1) the transport of the sample from the polarizer to the MR system should always be done as fast as possible to minimize polarization decay, and (2) low magnetic field values encountered by the hyperpolarized sample during transport can dramatically decrease its polarization. In the worst case, a point of zero field between the two systems would completely destroy the hyperpolarization of the solution.

4.2 *In Vitro* Experiments

4.2.1 Enzymes

Enzyme studies are simply performed by placing the enzyme that catalyzes a metabolic reaction of interest in an appropriate buffering solution inside a 5- or 10-mm NMR tube before mixing with the hyperpolarized agent in solution (Allouche-Arnon, Gamliel, et al., 2011; Barb, Hekmatyar, Glushka, & Prestegard, 2013; Keshari et al., 2009). Upon mixing, the conversion of hyperpolarized substrate to the product/s of interest is monitored by MRS (cf. Section 5). Such enzyme studies are often conducted right after designing a new DNP probe to provide initial insight into the feasibility of detecting metabolic conversion in subsequent cell and *in vivo* studies, wherein additional barriers to delivery are likely to be encountered.

4.2.2 Cell Lysates

Through the use of specific buffers, cells can be lysed and their content released, for further metabolic studies. To date, most studies have been performed on tumor cell lysates, generally to test a DNP probe (Chaumeil et al., 2013; Gallagher, Kettunen, Hu, et al., 2009; Kumagai et al., 2014). One study also reported the use of hyperpolarized [1-^{13}C] pyruvate in lysed human erythrocytes (Pages, Tan, & Kuchel, 2014).

Similar to an *in vitro* enzyme assay, lysates are placed in a 5- or 10-mm NMR tube, a hyperpolarized solution is rapidly injected, and the conversion of hyperpolarized substrate to the product of interest is monitored by MRS (cf. Section 5). This approach is valuable when the enzyme of interest is not readily available, or when there is a need to distinguish between hyperpolarized probe transport across a cell membrane and probe metabolism within the intracellular compartment (Lodi et al., 2013; Witney, Kettunen, & Brindle, 2011).

4.3 Live Cell Experiments

MRS studies of hyperpolarized substrates have enabled investigations of cellular metabolism in live cells. These use mainly two approaches: cell pellets (also known as cell suspensions) and perfused cell systems (also known as bioreactors). The advantages and limitations of these two methods are described below. However, it is important to note that, regardless of the type of approach, DNP-MR studies of live cells are no different from other MR studies and require a relatively high number of cells per experiment (10^7–10^8 cells). Such experiments are thus, to date, mainly performed on cancer cells or immortalized cell lines, which can be easily expanded in culture. Studies have also been reported in yeast and erythrocytes.

4.3.1 Cell Pellets/Cell Suspensions

The easiest and most common approach to study the metabolism of live, intact cells using DNP-MR is the use of cell pellets (Bohndiek et al., 2011; Canape et al., 2014; Day et al., 2007; Gallagher, Bohndiek, et al., 2011; Gallagher, Kettunen, Day, Lerche, et al., 2008; Hill, Jamin, et al., 2013; Hill, Orton, et al., 2013; Lin et al., 2014; Pages et al., 2013; Timm et al., 2014; Witney et al., 2011; Yang et al., 2014). Live cells in suspension (10^6–10^8 cells) are simply placed at the bottom of a 5- or 10-mm NMR tube, and, as for *in vitro* experiments, a hyperpolarized solution is rapidly injected, and the metabolic conversion of the polarized probe to its product is monitored by MRS (cf. Section 5). Cell pellet experiments are practical in the sense that, unlike bioreactors, they do not require a perfusion system. However, cell viability is limited, and so is the number of hyperpolarized injections that can be performed (usually only one hyperpolarized injection per pellet). This limitation can be particularly challenging if longitudinal metabolic information is desired (e.g., the effect of an intervention on metabolism).

4.3.2 Perfusion System, aka Bioreactor

4.3.2.1 Overview

MR-compatible bioreactors, or perfusion systems, have provided a platform for studying metabolic reactions for several decades (Brandes et al., 2010; Gillies, Liu, & Bhujwalla, 1994; Ronen, Rushkin, & Degani, 1991; Wolfe, Hsu, Reid, & Macdonald, 2002). They typically use an NMR tube to hold the cells and are connected via tubing to a medium reservoir and gas tank to provide a steady flow of cell culture medium and a constant atmosphere (typically 5% CO_2 in air). Cells in a bioreactor are therefore maintained under conditions that are similar to those of a tissue culture incubator and thus maintain viability during long periods of time (up to 72 h).

With the implementation of hyperpolarized ^{13}C MRS, such systems were adapted to address the specific needs of this new technique. Modifications were made to allow for rapid and frequent injections of hyperpolarized substrates while maintaining proper physiological pH and temperature (Harris et al., 2013; Keshari et al., 2010; Ward et al., 2010). This approach has proved useful for monitoring several metabolic reactions, including [1-^{13}C] pyruvate-to-[1-^{13}C] lactate conversion (Harris et al., 2009, 2013; Keshari et al., 2010; Keshari, Sriram, Koelsch, et al., 2013; Venkatesh et al., 2012; Ward et al., 2010), [1-^{13}C] α-ketoglutarate-to-[1-^{13}C] glutamate conversion (Chaumeil et al., 2014), or metabolism of [U-^{2}H, U^{13}C] glucose (Harris et al., 2013).

4.3.2.2 Cell Preparation

In a bioreactor, cells are placed in an NMR tube, and their metabolism studied while the perfusion is on. For this experiment to succeed, cells have to be immobilized within the "coil region" where the MR signal is acquired. This is achieved by immobilizing the cells either by encapsulation inside beads (Chandrasekaran, Seagle, Rice, Macdonald, & Gerber, 2006; Ronen et al., 1991) or by culture on microcarrier beads (Chaumeil et al., 2014), and the beads are then maintained with the coil region with a filter. The choice of the immobilization technique depends on several factors, the most important being cell adherence. For adherent cells, like most solid tumor cells, both approaches can be used. For nonadherent cells, like hematopoietic cells, encapsulation is the only option. Other considerations include substrate diffusion, which can be more limited in encapsulated cells, and detachment of cells, which is more likely to occur when cells are grown on beads.

4.3.2.3 Conducting a Bioreactor Experiment

Immobilized cells in the bioreactor are placed in a ^{13}C probe and typically maintained at 37 °C in an MR spectrometer. The perfusion is turned on and shimming performed. If feasible, a ^{31}P MR spectrum is acquired to confirm cell viability (Venkatesh et al., 2012; Ward et al., 2010).

After polarization of the DNP sample, dissolution is performed and, typically within 15 s, a defined volume of hyperpolarized agent (e.g., 3 mL for a 10-mm tube) is rapidly injected into the perfusion system (over 10–12 s), and acquisition of ^{13}C spectra is performed (cf. Section 5). It is important to note that, during injection and data acquisition, some groups briefly stop the perfusion (Chaumeil et al., 2014; Lodi et al., 2013; Venkatesh et al., 2012; Ward et al., 2010), while others do not (Keshari et al., 2010; Keshari, Sriram, Koelsch, et al., 2013; Keshari, Sriram, Van Criekinge, et al., 2013). This choice is operator specific but needs to be taken into consideration when analyzing and interpreting the data.

4.4 *Ex Vivo* Experiments

4.4.1 *Perfused Organs*

Several reports have demonstrated the application of DNP-MR to the study of perfused organs, especially perfused heart (Ball et al., 2013, 2014; Khemtong et al., 2014; Merritt et al., 2007; Moreno, Sabelhaus, Merritt, Sherry, & Malloy, 2010; Schroeder et al., 2009) and liver (Moreno et al., 2014). Perfused organs are highly controlled systems that can be studied independently of hormonal and/or neuronal influences that occur *in vivo*. Furthermore, unlike the *in vivo* setting, there is no metabolic contribution from adjacent tissues. The metabolism of perfused organs was evaluated using multiple DNP probes, including [1-^{13}C] pyruvate (Ball et al., 2013; Khemtong et al., 2014; Merritt et al., 2007; Moreno et al., 2010), [2-^{13}C] pyruvate (Schroeder et al., 2009), [1-^{13}C] butyrate (Ball et al., 2014), and [2-^{13}C] dihydroxyacetone (Moreno et al., 2014).

4.4.2 *Tissue Slices*

Perfusion systems can also be used to monitor the metabolism of live tissue samples. In a recent study, Keshari et al. showed that the conversion of hyperpolarized [1-^{13}C] pyruvate to [1-^{13}C] lactate could be detected in living primary human prostate slices and that its increase correlated with the presence of cancer (Keshari, Sriram, Van Criekinge, et al., 2013).

4.5 Preclinical *In Vivo* experiments

Conducting a preclinical *in vivo* DNP-MR experiment is, for the most part, identical to conducting any other *in vivo* preclinical study. However, because the goal of ^{13}C MRS/MRI of hyperpolarized probes is to assess metabolism, particular care should be taken to ascertain proper control of anesthesia and animal physiological parameters. Additionally, hyperpolarized *in vivo* experiments require intravenous injection of the hyperpolarized agent (cf. Section 5). These two points will be addressed below.

4.5.1 Control of Physiological Parameters

In most DNP-MR experiments, the animal is anesthetized using a volatile gas (e.g., isoflurane) and its physiological parameters (respiration rate, temperature, and sometimes heart rate) monitored throughout the experiment. This approach ensures animal well-being during imaging. However, in the case of DNP-MR studies, the monitoring of anesthesia and physiological vitals is even more critical, as these can affect metabolism and thus the hyperpolarized findings. In fact, anesthesia using isoflurane gas has been shown to modulate metabolism of hyperpolarized [1-^{13}C] pyruvate: the product-to-substrate ratios and apparent conversion rates were found to depend on isoflurane dose in the brain, although no dependence on isoflurane was observed in kidney or liver (Josan, Hurd, et al., 2013). As a consequence, care should be taken to maintain consistency of isoflurane anesthesia. Similarly, body temperature is an important player in the regulation of metabolic reactions (Sadler, Stratton, DeBerry, & Kolber, 2013; Zheng, Lee, & Golay, 2010) and this too should be monitored and maintained at a constant level.

4.5.2 Intravenous Injection of Hyperpolarized Agents

4.5.2.1 Catheter Placement

As mentioned, DNP-MR experiments require the placement of an intravenous catheter prior to the study. Tail veins are most commonly used in mice (Ward et al., 2010) and rats (Chaumeil et al., 2013, 2012; Chesnelong et al., 2014) because of the relative simplicity of the procedure. An alternative, but more technically challenging, placement is in the jugular vein, which can be used to increase the speed of delivery of the hyperpolarized agent (Kmiotek, Baimel, & Gill, 2012). In the case of primates, the saphenous vein is often use for catheter placement, given its accessibility (Chaumeil et al., 2009; Park, Larson, et al., 2014; Valette et al., 2008).

4.5.2.2 Controlled Injection: pH, Speed, Volume, and Frequency

Once the DNP sample has been polarized, dissolution in the appropriate buffer is performed (cf. Section 3). Within 15 s, a defined volume of hyperpolarized agent is rapidly injected intravenously, and acquisition of ^{13}C spectra performed (cf. Section 5).

Right before the hyperpolarized sample is injected, a quick pH measurement using a pH strip is recommended to ensure that the dissolution went properly. Once this verification is performed, the agent can be injected. For a rodent, the typical volume injected is as follows: for a healthy 30 g mouse,

200–300 μL; for a healthy 200 g rat, 1.5–2 mL. Lower volumes are typically used when studying pathological and/or symptomatic animals. Another crucial parameter is the speed of injection. The injection should be fairly rapid to preserve polarization, but the injection rate should be constant to ensure data reproducibility (a typical rate of injection is 12 s for a volume of 300 μL in mice). It is important to note that higher speed, while leading to faster delivery of the hyperpolarized agent, can also disrupt the integrity of the vein compromising the study. A final consideration is the frequency of injections in the context of animal well-being.

4.5.2.3 Toward Automation of the Injection

In an effort to minimize the variability between manual injections of hyperpolarized agents on preclinical animal models, Comment et al. have developed and implemented the use of a DNP polarizer directly coupled to a preclinical MR system (Cheng, Mishkovsky, et al., 2013; Comment, van den Brandt, et al., 2008; Comment et al., 2007). This system is based on a remotely controlled infusion especially designed to allow for fast transfer and controlled injection. In addition to enhancing the reproducibility of injection, this method also minimizes the polarization loss during transfer of the hyperpolarized solution from the polarizer to the MR system.

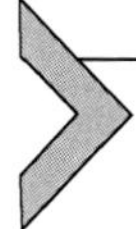

5. ACQUISITION STRATEGIES FOR HYPERPOLARIZED MR SPECTROSCOPY AND IMAGING

As previously mentioned, after dissolution, nuclear polarization decreases exponentially with a time constant T1, and this polarization is not renewable. MR data must thus be sampled quickly, while simultaneously taking into account the fact that each radiofrequency (RF) excitation applied during data acquisition in itself accelerates the decay of the hyperpolarized signal (Dafni & Ronen, 2010; Keshari & Wilson, 2014; Kurhanewicz et al., 2011). Additionally, when designing sequences for *in vivo* hyperpolarized studies, the rate of the conversion of the hyperpolarized agent into its metabolic product/s, as well as the vascular delivery of the hyperpolarized probe to the organ of interest, has to be carefully considered (cf. Section 3). In this section, we describe the different acquisition approaches that have developed to date and specifically optimized for MRS and imaging of hyperpolarized ^{13}C probes in different systems. We introduce the basic pulse sequences used, the strategies implemented to improve the spatial and temporal resolution, and recent developments.

5.1 *In Vitro* Hyperpolarized ^{13}C MR Studies

As mentioned, each DNP probe is characterized by its relaxation time T1 and its level of polarization P_{hp} (cf. Section 3). Here, we describe the MR measurements of T1 and polarization level *in vitro* (Allouche-Arnon, Lerche, et al., 2011; Ardenkjaer-Larsen et al., 2003; Ball et al., 2014; Harada, Kubo, Abe, Maezawa, & Otsuka, 2010), as well as the *in vitro* measurement of enzymatic reaction rates.

5.1.1 T1 Measurement

The hyperpolarized solution to be characterized is injected into a tube or syringe within 10–12 s after dissolution and then quickly transferred to an NMR spectrometer equipped with a ^{13}C probe.

Low flip angles ($\theta \sim 5$–$10°$) and short repetition times (TR $\sim$ 3 s) are used to maximize the SNRs. As shown in Fig. 3, dynamic ^{13}C spectra are recorded every TR. As mentioned above and illustrated in Fig. 3, both T1 and the acquisition pulse lead to a significant decrease in the amplitude of the longitudinal magnetization. For each RF excitation, the magnetization is tipped by an angle θ resulting in a decrease of the longitudinal component. Therefore, if not taken into consideration, fitting the data without correcting for the flip angle leads to an underestimation of T1. To determine the T1, the ^{13}C resonance of interest (peak height or amplitude, cf. Section 6) is therefore quantified at each time point, corrected for the flip angle and fitted to the following equation (Chiavazza et al., 2013; Zhao et al., 1996):

$$M_z(t) = M_0(t) sin(\theta)(cos(\theta))^{t/\mathrm{TR}} exp(-t/T_1) \quad (1)$$

5.1.2 Liquid-State Polarization Level

The measurement of the level of liquid-state polarization P_{hp} is often performed in combination with the T1 measurement. Specifically, to measure P_{hp}, one would first acquire the dynamic set of ^{13}C spectra required for T1 assessment, then wait until full decay of the hyperpolarized signal, and finally acquire a spectrum at thermal polarization. To acquire such a spectrum, a pulse-acquire sequence with a 90-degree flip angle and a repetition time TR greater than or equal to $10 \times$ T1 (allowing for full recovery of the longitudinal magnetization) is commonly used. Furthermore, it is important to note that a large number of averages are needed to acquire a high *SNR* spectrum, given the low polarization of the sample at thermal equilibrium. For instance, Keshari et al. measured the polarization level for hyperpolarized

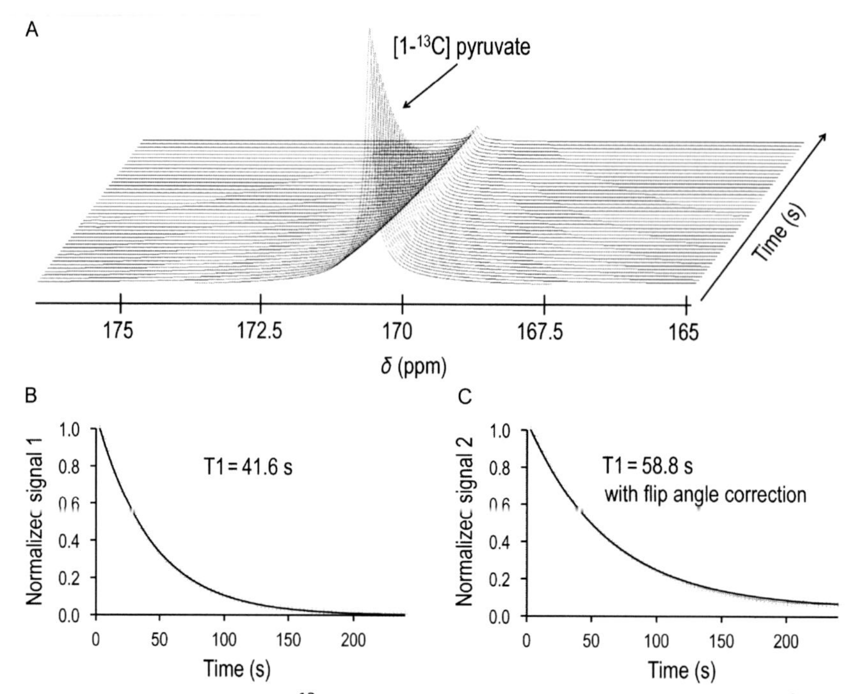

Figure 3 Measurement of [1-^{13}C] pyruvate relaxation time T1. (A) Stack plot of ^{13}C MR spectra of hyperpolarized [1-^{13}C] pyruvate in solution acquired at 11.7 T. (B) [1-^{13}C] pyruvate raw signal decay as a function of time (measured as the area under the peak) and normalized to the signal at the first time point. T1 was calculated and reported without correcting for the flip angle. (C) [1-^{13}C] pyruvate signal decay as a function of time (measured as the area under the peak), normalized to the signal at the first time point and corrected for the flip angle. T1 was calculated and reported using Eq. (1) (see Section 5).

[2-^{13}C] fructose using a nonlocalized pulse-acquire sequence with a 90-degree flip angle, repetition time TR = 3.5 s, an acquisition time of 0.5 s, and 9000 averages resulting in a total acquisition time of 10 h (Keshari et al., 2009). Both the hyperpolarized data at the first time point and the thermal data are then used to quantify the liquid-state polarization level using the equation (Chiavazza et al., 2013; Lumata, Jindal, et al., 2011):

$$P_{hp}(\%) = P_{th}\left(\frac{A_{hp}}{A_{th}}\right)\left(\frac{sin(\theta_{th})}{sin(\theta_{hp})}\right) \qquad (2)$$

where P is the polarization level and "th" and "hp" refer to the thermal and hyperpolarized polarizations, A is the NMR signal and θ is the flip angle (90° for θ_{th}). The liquid-state polarization level right after the dissolution can also be extrapolated by correcting P_{hp} for the loss of signal due to T1 relaxation during the transfer of the sample from the polarizer to the NMR spectrometer.

5.1.3 Enzymes and Cell Lysates

For enzymes and cell lysate studies (cf. Section 4), a simple pulse-acquire sequence with low flip angle (5–30°) and short TR (1–3 s) is used to acquire a dynamic set of ^{13}C spectra (50–100 transients) and monitor the metabolic conversion of a chosen DNP substrate to its product/s (Allouche-Arnon, Gamliel, et al., 2011; Chaumeil et al., 2013; Keshari et al., 2009; Kumagai et al., 2014). For instance, Allouche-Arnon et al. used a pulse-acquire sequence with a 20-degree flip angle, a repetition time of 2 s, and 50 transients to monitor the conversion of [1,1,2,2-D_4, 2-^{13}C] choline to [1,1,2,2-D_4, 2-^{13}C] acetylcholine.

5.2 Hyperpolarized ^{13}C MR Studies on Cells or Perfused Organs

5.2.1 Conventional Studies

Similar to *in vitro* studies, experiments in cell pellets, perfused cells, and perfused organs are commonly conducted on high-resolution NMR spectrometers. Following injection of the hyperpolarized solution (cf. Section 4), a set of sequential ^{13}C spectra are rapidly acquired to monitor the metabolic conversion of the DNP probe to its subsequent product/s. To do so, a simple pulse-acquire sequence with low flip angle (5–20°) and short repetition time TR (1–3 s) is generally used to maximize the SNR of all labeled metabolites (Cabella et al., 2013; Chaumeil et al., 2013; Day et al., 2007; Gallagher, Kettunen, et al., 2011; Harris et al., 2009, 2012; Keshari et al., 2010; Kumagai et al., 2014). It is important to note that knowledge of the T1 values of substrates and product/s can additionally be used to optimize the number of transients needed to acquire full dynamic datasets.

5.2.2 Diffusion and MAD-STEAM

Recently, the diffusion properties of hyperpolarized ^{13}C-labeled metabolites were evaluated using a diffusion-weighted pulsed-gradient spin-echo sequence in MCF-7 human breast cancer cells (Schilling et al., 2013) and using a diffusion-weighted double spin-echo sequence in UOK262 renal cancer carcinoma cells (Koelsch et al., 2013a). These studies provide new insight on the compartmentation of labeled metabolites by discriminating signals from the intra- and extracellular metabolites (Koelsch et al., 2013a; Schilling et al., 2013). Such investigations can improve our understanding of the delivery of hyperpolarized agents to the cell, the localization of metabolic conversions in intra- and extracellular compartments, and the transport of metabolites between the intracellular and extracellular spaces.

Another approach uses the MAD-STEAM sequence (metabolic activity decomposition stimulated-echo acquisition mode) (Larson, Kerr, Swisher, Pauly, & Vigneron, 2012). This approach is focused only on the enzymatic reaction and will be discussed in detail below in the context of *in vivo* studies where the method has mostly been applied.

5.3 *In Vivo* Hyperpolarized ^{13}C MR Studies

As opposed to *in vitro* or cell/organ studies, *in vivo* studies and sequence development for DNP-MR have been commonly performed at relatively low magnetic fields. The advantages of this are twofold: (1) as described in Sections 2–4, the T_1 of the most common DNP probes is higher at lower field strength and (2) the development of dedicated sequences on clinical systems facilitates clinical translation (Nelson et al., 2013). Whereas early sequences struggled between spatially resolved data with no dynamic information and poorly localized dynamic data, new strategies are now being developed to efficiently sample hyperpolarized ^{13}C data and thus maximize both spatial and temporal resolutions. Figure 4 gives an overview of the different sequences used for *in vivo* applications as well as their corresponding acquisition times and voxel sizes, for the reader's convenience.

5.3.1 *Early Sequences: Dynamic Time Courses and Static 2D CSI*

Early *in vivo* studies have employed simple spectroscopic imaging sequences such as nonlocalized, coil-localized, and slice-selective (slab-selective) pulse-acquire sequences. These approaches provide a high temporal resolution thanks to the use of short TR values. As a consequence, these sequences have enabled *in vivo* monitoring of the delivery and metabolic conversion of several injected hyperpolarized agents (Chen et al., 2009; Day et al., 2007; Gallagher, Kettunen, et al., 2011; Gallagher, Kettunen, Hu, et al., 2009; Golman, in 't Zandt, & Thaning, 2006; Golman, Zandt, et al., 2006; Park et al., 2013; Schroeder et al., 2008; Tyler, Schroeder, Cochlin, Clarke, & Radda, 2008). However, such sequences present an intrinsic lack of spatial information. The contributions of different physiological compartments, such as blood, normal, and disease tissues, can therefore not be resolved, and the use of the hyperpolarized data for localized diagnosis or response monitoring response is thus limited.

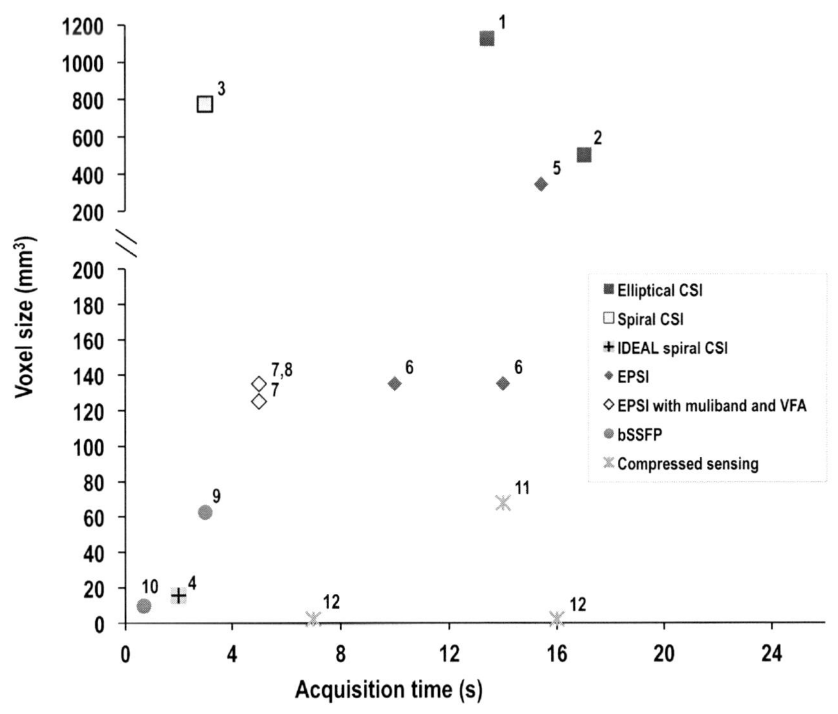

Figure 4 Voxel size and acquisition time for 2D slices. Summary of voxel sizes and acquisition times that can be achieved when acquiring a 2D slice using the different sequences found in the literature. References: 1, Golman et al. (2008); 2, Kohler et al. (2007); 3, Lau et al. (2010); 4, Wiesinger et al. (2012); 5, Cunningham et al. (2007); 6, Chen, Albers, et al. (2007); 7, Larson et al. (2010); 8, Larson et al. (2008); 9, von Morze et al. (2013); 10, Leupold et al. (2009); 11, Hu et al. (2008); 12, Hu et al. (2010). (See the color plate.)

Importantly however, the use of slab dynamic sequences can be useful to optimize the timing of other localized sequences, as it provides information on the kinetics of delivery and metabolism of the polarized agent (Chaumeil et al., 2013; Park et al., 2010). Indeed, to maximize SNR and spatial resolution, at the cost of temporal resolution, 2D chemical shift imaging sequences with centric phase encoding (CSI) at a single time point (typically that of maximum metabolic product based on slab dynamic data) have been used (Day et al., 2007; Golman, in 't Zandt, et al., 2006; Golman, Zandt, et al., 2006; Karlsson et al., 2010). The spatial resolution obtained with such 2D CSI sequences is limited by the number of phase encoding steps used to avoid excessive signal loss, but is typically between $2 \times 2 \times 6$ and $5 \times 5 \times 10$ mm^3.

5.3.2 *Elliptical, Spiral, Echo-Planar, and Steady-State Free Precession*

5.3.2.1 Elliptical Chemical Shift Imaging

A strategy that is frequently used to accelerate phase encoding is to cut the *k*-space corners using spherical or elliptical *k*-space encoding. Two studies performed in rats and pigs showed a 17% and 42% decrease in the total acquisition time with a spatial resolution of $5 \times 5 \times 20$ and $7.5 \times 7.5 \times 20$ mm^3, respectively (Golman et al., 2008; Kohler et al., 2007). However, this improvement is still not sufficient to perform a dynamic 2D CSI study, as the acquisition time is too long. Indeed, using such a sequence results in a signal drop of ~32% after the first time point (acquisition time ~ 13–17 s, flip angle ~5–15°), ~53% after the second time point, and ~85% after the fifth time point.

5.3.2.2 Spiral Chemical Shift Encoding

Fast metabolic imaging using sub-second undersample spiral chemical shift imaging (spCSI) provides better spatial and temporal resolution and allows for real-time metabolic imaging (Lau et al., 2010; Mayer, Levin, Hurd, Glover, & Spielman, 2006; Mayer et al., 2010, 2009; Wiesinger et al., 2012). In comparison to the conventional CSI, a decrease by a factor of 50 in the acquisition time can be achieved with the spCSI without compromising the SNR and spatial resolution (Mayer et al., 2010). This sequence, designed by an analytical algorithm (Adalsteinsson et al., 1998; Glover, 1999), allows simultaneous encoding of three *k*-space dimensions (k_x, k_y, k_f) using a series of spiral gradient waveforms in the *x*- and *y*-directions. Previous work has also shown the potential of this sequence to enable dynamic multislice imaging with a total acquisition time per slice of 375 ms and a spatial resolution per slice of $5 \times 5 \times 10$ mm^3 (Mayer et al., 2009). Recently, a higher spatial resolution ($1.25 \times 1.25 \times 10$ mm^3) and a decreased acquisition time (65 ms per slice) were demonstrated *in vivo* using the IDEAL spiral CSI for which an echo-time shifting in between excitation is added for the chemical shift encoding (Wiesinger et al., 2012).

5.3.2.3 Echo-Planar Spectroscopic Imaging

As mentioned, the decay of the hyperpolarized signal is accelerated by the application of each RF pulse. The double-spin-echo sequence using adiabatic refocusing and flyback or symmetric echo-planar readout gradient is insensitive to tip angle errors and allows encoding of the spectral and three spatial dimensions (Chaumeil et al., 2013; Chen, Hurd, et al., 2007; Cunningham et al., 2007; Larson et al., 2010; Yen et al., 2009). The readout consists in interleaving gradients with alternative polarity during data acquisition. Whereas the flyback echo-planar spectroscopic imaging (EPSI) uses only the echo generated by the positive gradient during data reconstruction and uses a shorter negative gradient to decrease the acquisition time, the symmetric EPSI uses the echo generated from both gradients to increase the SNR (Cunningham et al., 2005; Posse, Otazo, Dager, & Alger, 2013; Zeirhut et al., 2006). Chen et al. showed a 65% increase in SNRs using the symmetric EPSI at a cost of only 28% increase in acquisition time (Chen, Hurd, et al., 2007).

5.3.2.4 Steady-State Free Precession

Higher temporal and spatial resolutions can be achieved using a steady-state free precession (SSFP) sequence (Leupold et al., 2006; Reed et al., 2014; Svensson, Mansson, Johansson, Petersson, & Olsson, 2003; von Morze et al., 2011, 2013). In particular, fast imaging of DNP probes with T2 longer than TR can be performed. For instance, studies have used hyperpolarized ^{13}C urea (Reed et al., 2014; von Morze et al., 2011) (T2 ~ 300 ms *in vivo* at 3.4 T (von Morze et al., 2011)) or bis-1,1-(hydroxymethyl)-1-^{13}C-cyclopropane-d_8 (Johansson et al., 2004; Svensson et al., 2003) (T2 ~ 1.3 s *in vivo* at 2.35 T (Svensson et al., 2003)) for MR angiography and perfusion. Recently, one study showed the advantage of using urea labeled with ^{13}C and ^{15}N simultaneously to increase T2 (Reed et al., 2014) (cf. Section 3). In another study using the SSFP sequence, Leupold et al. showed an increase by a factor of 18.6 in temporal resolution and 1.6 in spatial resolution when comparing the multiecho balanced SSFP sequence to a conventional CSI sequence *in vivo* (Leupold, Mansson, Petersson, Hennig, & Wieben, 2009). A 3D version of this sequence was also implemented and applied *in vivo* (Perman et al., 2010). However, the acquisition time does not allow for acquisition of dynamic data (total acquisition time: 14.5 s and spatial resolution: 7 mm isotropic).

5.3.3 Variable Flip Angle and Multiband Excitation Pulses

Since each RF pulse excitation causes the hyperpolarized magnetization to irreversibly decay, low flip angles are generally used to preserve polarization (Day et al., 2007; Golman, in 't Zandt, et al., 2006; Kohler et al., 2007; Merritt et al., 2007, 2008; Schroeder et al., 2009, 2008). However, reducing the flip angle intrinsically results in a decrease in *SNRs*. Using the variable flip angle scheme to progressively increase the flip angle when sampling *k*-space compensates for the magnetization depletion and maintains a constant *SNR* during the acquisition (Nagashima, 2008; Zhao et al., 1996). The *n*th flip angle can be estimated from the T1, TR, and total number of excitation pulses (*N*) using the following formula (Nagashima, 2008; Xing, Reed, Pauly, Kerr, & Larson, 2013):

$$\theta_n = cos^{-1}\sqrt{\frac{(exp(-\mathrm{TR}/T_1))^2 - (exp(-\mathrm{TR}/T_1))^{2(N-n-1)}}{1-(exp(-\mathrm{TR}/T_1))^{2(N-n-1)}}} \quad (3)$$

Because of enzymatic reaction rates, the signal from metabolic product/s (e.g., [1-^{13}C] lactate) is commonly significantly lower than the signal of the injected DNP substrate (e.g., [1-^{13}C] pyruvate). With this in mind, previous studies have employed spectral–spatial radiofrequency pulses with multiple spectral bands to spectrally vary the flip angle (Kerr et al., 2008; Larson et al., 2010, 2008; Park, Larson, et al., 2014). During a dynamic acquisition, a low flip angle applied to the substrate signal can help retain its polarization and give more time for the product/s to be formed; in contrast, a larger flip angle applied to the resonances of the product/s maximizes their SNRs (Larson et al., 2010, 2008; Park, Larson, et al., 2014). This multiband technique, combined with the use of a variable flip angle scheme, showed an increase by a factor of ~3 for [1-^{13}C] lactate and [1-^{13}C] alanine, and a decrease by a factor of ~4 for [1-^{13}C] pyruvate in the SNR and signal duration in the mouse at 3 T compared to the signal obtained with 5- and 10-degree constant flip angle schemes (Larson et al., 2008). This sequence also enabled the *in vivo* detection of hyperpolarized [1-^{13}C] 2-hydroxyglutarate from [1-^{13}C] α-ketoglutarate in an orthotopic rat model of gliomas at clinical strength (Chaumeil et al., 2013). Finally, Xing et al. implemented a new scheme to independently control the flip angle of two metabolites by taking into account the differences in T1, the enzymatic reaction rate, and the number of excitation pulses (Xing et al., 2013). This improvement enhances the SNR at the latest time point and further extends the acquisition time window (Xing et al., 2013). Both schemes, multiband and variable flip angle, could also theoretically be applied *in vitro* and were used in one study performed with breast cancer cell cultures (Harris et al., 2013).

5.3.4 Compressed Sensing

Compressed sensing serves to enhance the spatial resolution (Hu et al., 2008) while achieving a drastic acceleration in imaging speed (Hu et al., 2010). The approach consists in randomly undersampling the *k*-space data and using a nonlinear reconstruction to generate the same image that would be obtained with a fully sampled *k*-space (Lustig, Donoho, & Pauly, 2007). The underlying signal must satisfy two conditions: the signal has to exhibit sparsity in the transform domain, and the aliasing, induced by the violation of the Nyquist theorem, must be incoherent in the same transform domain. While the second condition depends on the undersampling approach, hyperpolarized ^{13}C data satisfies the first criterion (Hu et al., 2010, 2008). Hu et al. implemented a 3D ^{13}C MRSI sequence using a double spin-echo scheme with phase encoding in *x*- and *y*-directions and a flyback echo-planar readout in the *z*-direction (Cunningham et al., 2007, 2005), a variable flip angle to limit the loss due to T1 relaxation (Zhao et al., 1996), a blipped scheme to randomly undersample the time domain (k_f), and one or two phase encoding directions (k_x, k_y) (Hu et al., 2008). First, a factor of 2 in spatial resolution was achieved in the mouse brain on a clinical 3 T system without increasing the acquisition time (spatial resolution: 0.25×0.5 cm and acquisition time: 14 s) by undersampling the k_f–k_x space (Hu et al., 2008). Then, fully sampling the central region and undersampling the rest of the k_x–k_y space achieved an acceleration factor of 3.37 (voxel size: 0.034 cm^3 in 16 s with the blipped scheme compared to a voxel size of 0.135 cm^3 in 14 s without the blipped scheme) (Hu et al., 2010). Finally, an even higher acceleration was achieved by equaling the blip areas to two or more phase encoding steps and consequently allowing undersampling the k_x–k_y space even further (voxel size: 0.034 cm^3 in 7 s).

5.3.5 Hyperpolarized Singlet State NMR

Hyperpolarized studies are intrinsically limited by the T1 of the DNP probes. Pyruvate labeled at the C1 position is the most common hyperpolarized agent to date due primarily to its long T1 (Dafni & Ronen, 2010; Gallagher, Kettunen, & Brindle, 2009) (cf. Section 2). For other agents, or in order to monitor slower reaction types, extending the lifetime of polarization would be critical. This may be achieved using a phenomenon called long-lived state (Carravetta & Levitt, 2004, 2005). Molecules with a pair of spin-1/2 nuclei can form either a triplet nuclear spin (i.e., three quantum states with a total spin $I=1$) or a singlet nuclear spin (i.e., a quantum state with a total spin $I=0$). The later is very interesting, as the singlet lifetime (T_S) is much longer than T1, since these spins are insensitive to dipolar interaction. However, singlet states cannot be directly observed using NMR. Therefore, the approach consists of transferring the magnetization from the triplet states to the singlet states, storing it in the singlet state to preserve the hyperpolarization, and recovering it at a later time. Different methods have been proposed to interconvert the spin in between the singlet and triplet states: applying a specific pulse sequence (Laustsen et al., 2012; Pileio, Carravetta, & Levitt, 2010; Vasos et al., 2009), using chemical reactions (Warren, Jenista, Branca, & Chen, 2009), or "filtering" the hyperpolarized spin in the singlet state after dissolution using a mu-metal cylinder (Marco-Rius et al., 2013; Tayler et al., 2012). Marco-Rius et al. performed the first *in vivo* study using this strategy in mice bearing an EL4 tumor with hyperpolarized [1,2-^{13}C] pyruvate at 7 T (Marco-Rius et al., 2013).

5.3.6 New Strategies

5.3.6.1 Suppressing Unwanted Resonances

Recently, a new approach was proposed to suppress the signal from unwanted contaminants that overlap with the signal of metabolic product/s of interest. von Morze et al. developed this method to suppress the resonances of lactide dimers originating from the polarization of [1-^{13}C] lactate, which were overlapping with the [1-^{13}C] alanine signal (von Morze, Larson, Shang, & Vigneron, 2014). To do so, a train of three 90-degree consecutive spectrally selective pulses centered on both dimer signals followed by crusher gradient were applied before the dynamic acquisition.

5.3.6.2 Focusing on the Enzymatic Reactions

In all the methods presented above, the signal from the hyperpolarized substrate and metabolic product/s are either monitored dynamically or at a single time point, and all provide an estimate of the metabolic reaction rate by measuring the ratio of the product/s to substrate or to noise. However, the metabolite signals acquired originate from fairly large volumes that generally include both tissue and vasculature. Therefore, the detected signal represents the weighted average of multiple signals, including signals from in-flowing metabolites originating in other tissues, metabolic conversion in the intra-cellular or extracellular compartments, etc. Recently, new methods have been proposed to overcome this complexity, including magnetization transfer (Kettunen et al., 2010), MAD-STEAM (Larson et al., 2012; Swisher et al., 2014), and diffusion-weighted MR (Koelsch et al., 2014; Larson et al., 2013; Sogaard, Schilling, Janich, Menzel, & Ardenkjaer-Larsen, 2014).

Magnetization transfer (Kettunen et al., 2010) and MAD-STEAM (Larson et al., 2012; Swisher et al., 2014) are two different acquisition strategies that allow suppression of the signal from in-flowing metabolites from tissues and blood, and metabolites traveling between the intra- and extracellular space. Therefore, only the metabolic conversion of interest is observed. Whereas the first strategy uses selective saturation pulses to measure the rate of exchange in between two spins populations (Alger & Shulman, 1984), the MAD-STEAM sequence labels the spins with dephasing and rephasing gradients and exploits the signal phase to measure the exchange and conversion during the mixing time (i.e., in between the second and third 90° of the STEAM sequence).

5.3.6.3 Differentiating Between Intra- and Extracellular Pools

Mapping the diffusion coefficients of hyperpolarized ^{13}C-labeled metabolites allows for localization of the metabolites in the intra- and/or extracellular spaces. Indeed, as compared to extracellular metabolites, metabolites diffusing inside the cells have lower diffusion coefficients, as the cell membrane restricts their range of motion (Koelsch et al., 2014; Larson et al., 2013; Sogaard et al., 2014). Different sequences were used such as diffusion-weighted STEAM (stimulated-echo acquisition mode) sequence (Larson et al., 2013), bipolar pulsed-gradient double spin echo with a variable flip angle scheme (Koelsch et al., 2014), and diffusion-weighted slice-selective pulsed-gradient spin-echo sequence (Sogaard et al., 2014) to measure the diffusion coefficient of [1-^{13}C] pyruvate, [1-^{13}C] lactate, and [1-^{13}C] alanine in normal and tumor model animals.

6. ANALYSIS OF HYPERPOLARIZED MR DATA

Absolute quantification provides valuable information on metabolite levels and on the changes associated with pathological conditions. However, to date, absolute quantification of hyperpolarized ^{13}C data has been a challenge due to the confounding factors that affect the signal intensity (i.e., longitudinal relaxation, polarization level, rate of injection, transport into the cell, metabolic conversion and exchange, etc.). Therefore, relative quantification is the most common approach, and assessment of product-to-substrate ratios (e.g., lactate-to-pyruvate) is frequently used. Besides absolute quantification, metabolic modeling of dynamic ^{13}C data is also an area of interest. This too is challenging, as most systems are underdetermined. Furthermore, due to the above-mentioned parameters influencing signal intensity, the models provide only an estimate of *pseudo* metabolic rates. In this section, we first describe the different approaches used to quantify and display substrates and product/s signals. We then report on the different models that are being developed to estimate metabolic pseudo-rate constants.

6.1 Relative Quantification

For any quantification approach, the signal of a ^{13}C-labeled metabolite is measured either as the amplitude (Cabella et al., 2013; Chaumeil et al., 2013; Larson et al., 2008, 2012; Mayer et al., 2010, 2009; Park et al., 2013) or as the peak height of its resonance (Kohler et al., 2007; Larson et al., 2010, 2008); the most common approach uses the amplitude as it is not dependent on shimming.

6.1.1 In Vitro*, Cells, and* Ex Vivo

The quantification approaches described in this section are valid for *in vitro* studies, studies of cell lysates, cell pellets, perfused cells, and perfused organs (cf. Section 4). First, the signals of the ^{13}C-labeled metabolites of interest are quantified using a dedicated software package (jMRUI (Naressi, Couturier, Castang, de Beer, & Graveron-Demilly, 2001), ACD/Spec Manager, Chenomx, MestReNova, etc.), and each signal is plotted as a function of time. The resulting time-courses represent a combination of delivery, metabolic conversion, and signal decay due to longitudinal relaxation (Ball et al., 2013; Cabella et al., 2013; Chaumeil et al., 2013; Day et al., 2007; Gallagher, Kettunen, et al., 2011; Harris et al., 2009; Harrison et al., 2012; Keshari et al., 2010; Kumagai et al., 2014; Merritt et al., 2008; Schroeder et al., 2009). In the majority of studies, the time-course of metabolic product/s is then normalized by dividing the signals at each time point by the maximum signal of the injected hyperpolarized substrate as well as by cell number or the amount of enzyme as appropriate (Dafni & Ronen, 2010; Gallagher, Kettunen, et al., 2011; Harris et al., 2009; Harrison et al., 2012; Schroeder et al., 2009; Venkatesh et al., 2012; Ward et al., 2010). Alternatively, the time-course of the ratio of product and substrate as a function of time can be calculated to provide additional insight into cellular processes (Merritt et al., 2008). In some studies, the area under the curve (AUC) for each metabolic time-course was evaluated (Cabella et al., 2013; Schroeder et al., 2009). The parameters obtained in this fashion can be used to compare different studies. However, this type of data does not provide absolute metabolic flux information.

6.1.2 In Vivo

6.1.2.1 Basic Approaches

For *in vivo* studies, the signal of substrates and product/s is quantified using dedicated software, such as SIVIC (Crane, Olson, & Nelson, 2013). The most basic approaches then involve normalization of the hyperpolarized

MR signals to noise (Chaumeil et al., 2013; Mishkovsky et al., 2012; Yen et al., 2009), to the liquid-state polarization (Larson et al., 2010; Mayer et al., 2010; Yen et al., 2009), to the signal intensity of an external ^{13}C-labeled reference phantom (Mayer et al., 2010), to the signal from contralateral control tissues in tumor studies (Dafni et al., 2010; Gallagher, Kettunen, Hu, et al., 2009), or to the pyruvate blood signal (Dafni & Ronen, 2010). Additionally, in most studies, the product-to-substrate ratio is also quantified. This ratio is a good indicator on the metabolic reaction and, importantly, is independent of the polarization level of the injected agent. As such, it provides a simple readout of metabolism (Chaumeil et al., 2013; Gallagher, Kettunen, et al., 2011; Lau et al., 2010; Mayer et al., 2009; Park et al., 2013). A more complex approach could use the arterial input function (AIF) to provide definitive flux data. However, measuring an AIF in DNP-MR studies remains challenging. This has been achieved in a few cases *in vivo* by using an implantable coil in the rat carotid (Marjanska et al., 2012), by quantifying the signal from the injected hyperpolarized agent from a catheter placed next to the coil (Kazan et al., 2013), or by extrapolation using a gamma-variate function (Kazan et al., 2013; von Morze et al., 2011).

6.1.2.2 Data Display

For nonlocalized, coil-localized, and localized 1D dynamic acquisitions, the time-course of metabolites is usually displayed (Cabella et al., 2013; Chaumeil et al., 2013; Day et al., 2007; Harris et al., 2009; Harrison et al., 2012; Keshari et al., 2010; Kohler et al., 2007; Larson et al., 2008; Lau et al., 2010; Li et al., 2013; Mayer et al., 2009; Wiesinger et al., 2012; Xu et al., 2011; Zierhut et al., 2010). For 2D and 3D single time point or dynamic hyperpolarized MR studies, heat maps can be generated and co-registered to high-resolution anatomic MRI (Chaumeil et al., 2014, 2013; Cunningham et al., 2007; Day et al., 2007; Golman et al., 2008; Hu et al., 2010; Kohler et al., 2007; Larson et al., 2010; Lau et al., 2010; Mayer et al., 2010, 2009; Wiesinger et al., 2012; Xu et al., 2011; Yen et al., 2009). Heat maps are visually pleasing but simply correspond to an interpolated color-coded graphical display of the spectroscopic data.

6.2 Quantification Using Modeling

Modeling the time-dependent variation of the hyperpolarized metabolite signals is expected to provide a more quantitative and robust estimate of physiological parameters, such as transport rates and enzymatic kinetic constants. However, modeling hyperpolarized data is complex. As an example,

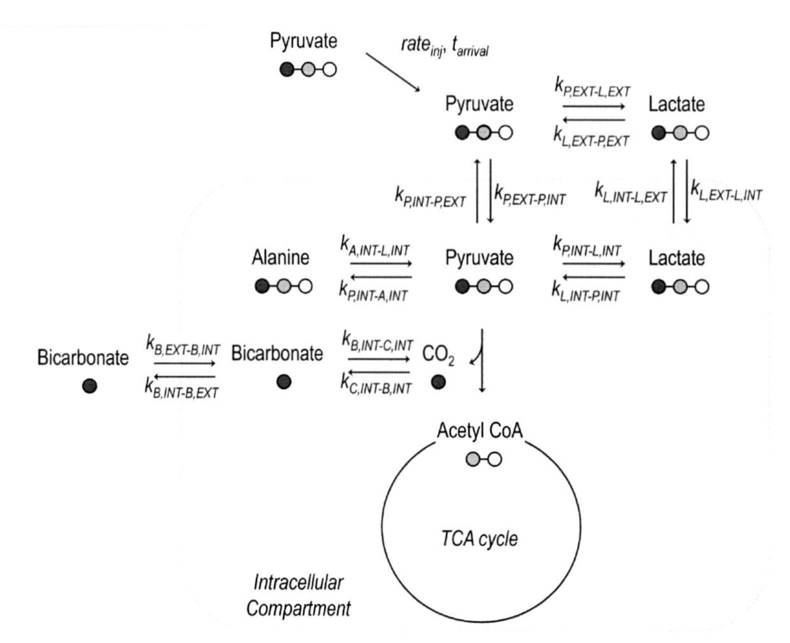

Figure 5 Schematic of pyruvate uptake and metabolism. Illustration of parameters that need to be included in modeling pyruvate delivery and conversion. These include the rate of injection ($rate_{inj}$), the delivery of the hyperpolarized substrate ($t_{arrival}$), its transport across membranes (e.g., $k_{P,EXP-P,INT}$ is the rate for pyruvate transport from the extra- to intracellular space and $k_{P,INT-P,EXT}$ the reverse) and its conversion through multiple enzymatic reactions. Here, we depict the conversion of pyruvate into lactate, alanine, bicarbonate, and acetyl-CoA (e.g., $k_{P,INT-L,INT}$ is the rate for the pyruvate (P)-to-lactate (L) conversion in the intracellular space). Note: C1 position: green/dark; C2 position: blue/light. (See the color plate.)

Fig. 5 illustrates the metabolism of hyperpolarized [1-^{13}C] pyruvate. In an *in vivo* setting, to properly model pyruvate data, one should take into account the rate of injection ($rate_{inj}$), vascular delivery of the hyperpolarized substrate ($t_{arrival}$), its transport across membranes (e.g., $k_{P,EXP-P,INT}$ being the rate for pyruvate transport from the extra- to intracellular space and $k_{P,INT-P,EXT}$ the opposite pyruvate transport), and its conversion via enzymatic reaction/s (e.g., $k_{P,INT-L,INT}$ being the rate for pyruvate (P)-to-lactate (L) conversion in the intracellular space). In addition to these biological steps, proper modeling of hyperpolarized data must take into account the intrinsic properties of the hyperpolarized signals, namely their exponential decay due to T1 and the effect of the RF pulses on their intensity. To date, different approaches have been proposed to model the hyperpolarized datasets obtained in enzyme experiments, cells lysates, live cells, perfused organs, and *in vivo*. Although these models can theoretically be applied to any enzymatic reaction, the following sections will focus on the modeling of the [1-^{13}C] pyruvate-to-[1-^{13}C] lactate conversion mediated by LDH-A.

6.2.1 Modified Bloch Equations

The most common approach to model the kinetics of hyperpolarized [1-^{13}C] pyruvate and [1-^{13}C] lactate is the two- or three-site exchange model based on the modified Bloch equations (Day et al., 2007; Harris et al., 2009; Harrison et al., 2012; Hill, Orton, et al., 2013).

6.2.1.1 Description of the Model

The forward ($k_{P\text{–}L}$) and backward ($k_{L\text{–}P}$) reaction rates, longitudinal relaxation rate constants ($\rho_P = 1/T_{1,P}$ and $\rho_L = 1/T_{1,L}$), and repetition of RF excitation with a flip angle (θ) during TR are included in these equations, as follows (Harris et al., 2009):

$$\frac{dM_P}{dt} = -k_{P-L}M_P + k_{L-P}M_L - \rho_P M_P - (1 - cos(\theta))^{1/TR} M_P \quad (4)$$

$$\frac{dM_L}{dt} = -k_{L-P}M_L + k_{P-L}M_P - \rho_L M_L - (1 - cos(\theta))^{1/TR} M_L \quad (5)$$

where M is the longitudinal magnetization. However, in this model, the injection rate, the mean arrival time of the labeled pyruvate to the organ of interest, and the reperfusion of all labeled metabolites are neglected.

6.2.1.2 Cell Studies

Several studies have used the modified Bloch equations to model data obtained in perfused cells and organs, as well as in cell pellets (Day et al., 2007; Harris et al., 2009; Harrison et al., 2012; Ward et al., 2010). Simplifying this model by neglecting the backward reaction ($k_{L\text{–}P} = 0$), Harris et al. modeled the pyruvate-to-lactate conversion in T47D human breast cancer cells culture (Harris et al., 2009). The combination of ^{13}C and ^{31}P MRS allowed them to measure a pseudo metabolic rate constant ($k_{P\text{–}L}$) per cell (Neeman, Rushkin, Kadouri, & Degani, 1988). Additionally, they measured $k_{P\text{–}L}$ for different initial concentrations of hyperpolarized pyruvate in order to perform a Michaelis–Menten kinetic analysis and study the transport of pyruvate from the extra- to the intracellular pool and its conversion. Finally, Harrison et al. proposed and compared six different models (three-site models including the extracellular lactate pool and two-site models excluding the extracellular lactate) to study the pyruvate-to-lactate conversion in SF188-derived glioblastoma cells (Harrison et al., 2012). Although the three-site models provide insight into the transport of lactate from the intra- to the extracellular space, results showed no strong differences in the pyruvate-to-lactate flux (corresponding to the product of $k_{\mathrm{P-L}}$ and the initial pyruvate concentration) among the six models (Harrison et al., 2012), suggesting that less complex models, as generally used in the literature, should be sufficient for estimating the initial pyruvate-to-lactate flux.

6.2.1.3 *In Vivo* Studies

In an *in vivo* setting, modeling the hyperpolarized metabolite kinetics using the Bloch equations is most commonly performed on data with high temporal resolution and low spatial resolution, especially slab dynamic, non-localized, or coil-localized data (Bastiaansen, Cheng, Mishkovsky, Comment, & Gruetter, 2011; Bohndiek et al., 2010; Day et al., 2007). For example, in a study on EL4 mouse lymphoma, Day et al. used a two-site exchange bidirectional model to estimate the pseudo-rate constants of pyruvate-to-lactate conversion by LDH-A, as well as the longitudinal relaxation rate constants of both metabolites (Day et al., 2007). The signal loss due to repetitive RF pulse excitation was not considered since a low flip angle (5°) was used to acquire the hyperpolarized data.

Finally, as mentioned in Section 5, the use of magnetization transfer (Kettunen et al., 2010) and MAD-STEAM (Larson et al., 2012) acquisition schemes lead to simplification of the data modeling by suppressing the signal from metabolites that are not implicated in the metabolic conversion. Therefore, the time-course can be fitted using the modified Bloch equation and the backward, forward, and longitudinal relaxation constant rates can be estimated.

6.2.2 *Piecewise Equations*

Another modeling approach consists in using piecewise equations to model the changes in labeled pyruvate magnetization and lactate as a function of time (Keshari et al., 2010; Merritt et al., 2007; Zierhut et al., 2010):

$$M_{\mathrm{P}}(t) = \begin{cases} \dfrac{\mathrm{rate_{inj}}}{k_{\mathrm{P}}}\left(1 - e^{-k_{\mathrm{P}}(t - t_{\mathrm{arrival}})}\right), t_{\mathrm{arrival}} \leq t \leq t_{\mathrm{end}} \\ M_{\mathrm{P}}(t_{\mathrm{end}})e^{-k_{\mathrm{P}}(t - t_{\mathrm{end}})}, t \geq t_{\mathrm{end}} \end{cases} \quad (6)$$

$$M_P(t) = \begin{cases} \frac{k_{P-L}\text{rate}_{\text{inj}}}{k_P - k_L}\left(\frac{1 - e^{-k_L(t-t_{\text{arrival}})}}{k_L} - \frac{1 - e^{-k_P(t-t_{\text{arrival}})}}{k_P}\right), t_{\text{arrival}} \le t \le t_{\text{end}} \\ \frac{M_P(t_{\text{end}})k_{P-L}}{k_P - k_L}\left(e^{-k_L(t-t_{\text{end}})} - e^{-k_P(t-t_{\text{end}})}\right) + M_L(t_{\text{end}})e^{-k_L(t-t_{\text{end}})}, t \ge t_{\text{end}} \end{cases} \quad (7)$$

k_P and k_L account for the pyruvate and lactate signal decay rate constants due to the longitudinal relaxation and flip angles. The backward enzymatic reaction (k_{L-P}) and delivery of pyruvate to the intracellular space are neglected here. This model, combined with a Michaelis–Menten analysis, was used *in vivo* to study the pyruvate-to-lactate conversion in the normal rat kidney, in the prostate of normal and TRAMP mice (Zierhut et al., 2010), and in rat hepatoma cell cultures (Keshari et al., 2010).

6.2.3 Mass Balance Model

A three-compartment non-steady-state brain/blood/body metabolic model using dynamic mass balance equations was proposed to study the pyruvate-to-lactate reaction in the rat brain (Henry et al., 2006; Marjanska et al., 2010). Within each compartment, the exchange between pyruvate and lactate depends on the LDH enzyme. The exchange between compartments is described by Michaelis–Menten kinetics. Additionally, the model takes into account the conversion of pyruvate into bicarbonate in the body and brain. This model allows fitting the rate of pyruvate transport through the BBB, the pyruvate-to-lactate, and pyruvate-to-bicarbonate conversion rates in the brain.

6.2.4 Ratio Modeling

Recently, Li et al. proposed a ratio modeling method to study the pyruvate–lactate interconversion in a transgenic and tumor xenograft mouse models (Li et al., 2013). The model, derived from the modified Bloch equations for a two-pool exchange, is defined by two parameters, R_{LP} and q (Li et al., 2013):

$$R_{LP}(t) = \frac{r(1 + R_{LP}(t_0)) + (R_{LP}(t_0) - r)e^{-s(t-t_0)}}{1 + R_{LP}(t_0) + (r - R_{LP}(t_0))e^{-s(t-t_0)}} \quad (8)$$

$$q = -k_{LP}R_{LP}(t) + k_{PL} \quad (9)$$

where $R_{LP}(t_0) = M_L(t_0)/M_P(t)$, $r = k_{PL}/k_{LP}$, and $s = k_{PL} + k_{LP}$. Fitting the linear range of the time-course of $q(t)$ versus $R_{LP}(t)$ yields the forward and backward rate constants.

6.2.5 Gamma-Variate Analysis

The hyperpolarized kinetics can be empirically assessed using a gamma-variate analysis (Chaumeil et al., 2014) defined by the following equation (Lupo et al., 2010):

$$y(t) = \frac{AUC(t - t_{\text{arrival}})^{\alpha} e^{-(t - t_{\text{arrival}})/\beta}}{\beta^{\alpha+1}\Gamma(\alpha+1)} \tag{10}$$

where the AUC is the area under the dynamic curve, t_{arrival} is the bolus arrival time, α is the skewness of the function shape, and β defines the time scaling. This model is well adapted to describe complex systems with multiple enzymes involved. It uses the known kinetics and no assumptions are made when fitting the data. So whereas the data does not reflect specific metabolic fluxes, it provides a robust metric of metabolism.

7. SUMMARY AND CONCLUSION

DNP-hyperpolarized agents in combination with ^{13}C MRS provide a noninvasive nonradioactive method for detecting metabolic reactions in real time. This field has developed dramatically over the past decade, culminating with a recent clinical trial that demonstrates the value of this approach for probing pyruvate metabolism in patients. With only about a decade of research, metabolic studies using hyperpolarized compounds are still in their infancy and further applications are likely to emerge in the years to come. No doubt the near future will see the translation of additional hyperpolarized probes and improvements in data acquisition and analysis resulting in high-resolution quantitative metabolic flux information. Upon further development, DNP-MR is likely to enhance our ability to monitor metabolism as a readout of health and disease, both in the laboratory and in the clinic.

REFERENCES

Adalsteinsson, E., Irarrazabal, P., Topp, S., Meyer, C., Macovski, A., & Spielman, D. M. (1998). Volumetric spectroscopic imaging with spiral-based k-space trajectories. *Magnetic Resonance in Medicine*, *39*(6), 889–898.

Albers, M. J., Bok, R., Chen, A. P., Cunningham, C. H., Zierhut, M. L., Zhang, V. Y., et al. (2008). Hyperpolarized 13C lactate, pyruvate, and alanine: Noninvasive biomarkers for prostate cancer detection and grading. *Cancer Research*, *68*(20), 8607–8615. http://dx.doi.org/10.1158/0008-5472.CAN-08-0749.

Alger, J. R., & Shulman, R. G. (1984). NMR studies of enzymatic rates in vitro and in vivo by magnetization transfer. *Quarterly Reviews of Biophysics*, *17*(1), 83–124.

Allouche-Arnon, H., Gamliel, A., Barzilay, C. M., Nalbandian, R., Gomori, J. M., Karlsson, M., et al. (2011). A hyperpolarized choline molecular probe for monitoring

acetylcholine synthesis. *Contrast Media & Molecular Imaging*, *6*(3), 139–147. http://dx.doi.org/10.1002/cmmi.418.
Allouche-Arnon, H., Lerche, M. H., Karlsson, M., Lenkinski, R. E., & Katz-Brull, R. (2011). Deuteration of a molecular probe for DNP hyperpolarization—A new approach and validation for choline chloride. *Contrast Media & Molecular Imaging*, *6*(6), 499–506. http://dx.doi.org/10.1002/cmmi.452.
Allouche-Arnon, H., Wade, T., Waldner, L. F., Miller, V. N., Gomori, J. M., Katz-Brull, R., et al. (2013). In vivo magnetic resonance imaging of glucose—Initial experience. *Contrast Media & Molecular Imaging*, *8*(1), 72–82. http://dx.doi.org/10.1002/cmmi.1497.
Ardenkjaer-Larsen, J. H., Fridlund, B., Gram, A., Hansson, G., Hansson, L., Lerche, M. H., et al. (2003). Increase in signal-to-noise ratio of >10,000 times in liquid-state NMR. *Proceedings of the National Academy of Sciences of the United States of America*, *100*(18), 10158–10163. http://dx.doi.org/10.1073/pnas.1733835100.
Ardenkjaer-Larsen, J. H., Leach, A. M., Clarke, N., Urbahn, J., Anderson, D., & Skloss, T. W. (2011). Dynamic nuclear polarization polarizer for sterile use intent. *NMR in Biomedicine*, *24*(8), 927–932. http://dx.doi.org/10.1002/nbm.1682.
Ardenkjaer-Larsen, J. H., Macholl, S., & Jóhannesson, H. (2008). Dynamic nuclear polarization with trityls at 1.2 K. *Applied Magnetic Resonance*, *34*(3–4), 509–522. http://dx.doi.org/10.1007/s00723-008-0134-4.
Armstrong, B. D., Edwards, D. T., Wylde, R. J., Walker, S. A., & Han, S. (2010). A 200 GHz dynamic nuclear polarization spectrometer. *Physical Chemistry Chemical Physics*, *12*(22), 5920–5926. http://dx.doi.org/10.1039/c002290j.
Bahrami, N., Swisher, C. L., Von Morze, C., Vigneron, D. B., & Larson, P. E. (2014). Kinetic and perfusion modeling of hyperpolarized (13)C pyruvate and urea in cancer with arbitrary RF flip angles. *Quantitative Imaging in Medicine and Surgery*, *4*(1), 24–32. http://dx.doi.org/10.3978/j.issn.2223-4292.2014.02.02.
Ball, D. R., Cruickshank, R., Carr, C. A., Stuckey, D. J., Lee, P., Clarke, K., et al. (2013). Metabolic imaging of acute and chronic infarction in the perfused rat heart using hyperpolarised [1-13C]pyruvate. *NMR in Biomedicine*, *26*(11), 1441–1450. http://dx.doi.org/10.1002/nbm.2972.
Ball, D. R., Rowlands, B., Dodd, M. S., Le Page, L., Ball, V., Carr, C. A., et al. (2014). Hyperpolarized butyrate: A metabolic probe of short chain fatty acid metabolism in the heart. *Magnetic Resonance in Medicine*, *71*(5), 1663–1669. http://dx.doi.org/10.1002/mrm.24849.
Barb, A. W., Hekmatyar, S. K., Glushka, J. N., & Prestegard, J. H. (2013). Probing alanine transaminase catalysis with hyperpolarized 13CD3-pyruvate. *Journal of Magnetic Resonance*, *228*, 59–65. http://dx.doi.org/10.1016/j.jmr.2012.12.013.
Bastiaansen, J. A., Cheng, T., Mishkovsky, M., Comment, A., & Gruetter, R. (2011). Study of acetyl carnitine kinetics in skeletal muscle *in vivo* using hyperpolarized 1-13C acetate. *Proceedings on International Society for Magnetic Resonance in Medicine*, *19*, 3536.
Bastiaansen, J. A., Yoshihara, H. A., Takado, Y., Gruetter, R., & Comment, A. (2014). Hyperpolarized 13C lactate as a substrate for in vivo metabolic studies in skeletal muscle. *Metabolomics*, *10*, 986–994.
Batel, M., Dapp, A., Hunkeler, A., Meier, B. H., Kozerke, S., & Ernst, M. (2014). Cross-polarization for dissolution dynamic nuclear polarization. *Physical Chemistry Chemical Physics*, *16*(39), 21407–21416. http://dx.doi.org/10.1039/c4cp02696a.
Bohndiek, S. E., Kettunen, M. I., Hu, D. E., & Brindle, K. M. (2012). Hyperpolarized (13)C spectroscopy detects early changes in tumor vasculature and metabolism after VEGF neutralization. *Cancer Research*, *72*(4), 854–864. http://dx.doi.org/10.1158/0008-5472.CAN-11-2795.
Bohndiek, S. E., Kettunen, M. I., Hu, D. E., Kennedy, B. W., Boren, J., Gallagher, F. A., et al. (2011). Hyperpolarized [1-13C]-ascorbic and dehydroascorbic acid: Vitamin C as a

probe for imaging redox status in vivo. *Journal of the American Chemical Society*, *133*(30), 11795–11801. http://dx.doi.org/10.1021/ja2045925.

Bohndiek, S. E., Kettunen, M. I., Hu, D. E., Witney, T. H., Kennedy, B. W., Gallagher, F. A., et al. (2010). Detection of tumor response to a vascular disrupting agent by hyperpolarized 13C magnetic resonance spectroscopy. *Molecular Cancer Therapeutics*, *9*(12), 3278–3288. http://dx.doi.org/10.1158/1535-7163.MCT-10-0706.

Bornet, A., Melzi, R., Linde, P., Angel, J., Hautle, P., van den Brandt, B., et al. (2013). Boosting dissolution dynamic nuclear polarization by cross polarization. *The Journal of Physical Chemistry Letters*, *4*(1), 111–114. http://dx.doi.org/10.1021/jz301781t.

Brandes, A. H., Ward, C. S., & Ronen, S. M. (2010). 17-Allyamino-17-demethoxygeldanamycin treatment results in a magnetic resonance spectroscopy-detectable elevation in choline-containing metabolites associated with increased expression of choline transporter SLC44A1 and phospholipase A2. *Breast Cancer Research*, *12*(5), R84. http://dx.doi.org/10.1186/bcr2729.

Brindle, K. M., Bohndiek, S. E., Gallagher, F. A., & Kettunen, M. I. (2011). Tumor imaging using hyperpolarized 13C magnetic resonance spectroscopy. *Magnetic Resonance in Medicine*, *66*(2), 505–519. http://dx.doi.org/10.1002/mrm.22999.

Butt, S. A., Sogaard, L. V., Magnusson, P. O., Lauritzen, M. H., Laustsen, C., Akeson, P., et al. (2012). Imaging cerebral 2-ketoisocaproate metabolism with hyperpolarized (13)C magnetic resonance spectroscopic imaging. *Journal of Cerebral Blood Flow and Metabolism*, *32*(8), 1508–1514. http://dx.doi.org/10.1038/jcbfm.2012.34.

Cabella, C., Karlsson, M., Canape, C., Catanzaro, G., Colombo Serra, S., Miragoli, L., et al. (2013). In vivo and in vitro liver cancer metabolism observed with hyperpolarized [5-(13)C]glutamine. *Journal of Magnetic Resonance*, *232*, 45–52. http://dx.doi.org/10.1016/j.jmr.2013.04.010.

Canape, C., Catanzaro, G., Terreno, E., Karlsson, M., Lerche, M. H., & Jensen, P. R. (2014). Probing treatment response of glutaminolytic prostate cancer cells to natural drugs with hyperpolarized [5-13 C]glutamine. *Magnetic Resonance in Medicine*, *73*(6), 2296–2305. http://dx.doi.org/10.1002/mrm.25360.

Carravetta, M., & Levitt, M. H. (2004). Long-lived nuclear spin states in high-field solution NMR. *Journal of the American Chemical Society*, *126*(20), 6228–6229. http://dx.doi.org/10.1021/ja0490931.

Carravetta, M., & Levitt, M. H. (2005). Theory of long-lived nuclear spin states in solution nuclear magnetic resonance. I. Singlet states in low magnetic field. *The Journal of Chemical Physics*, *122*(21), 214505. http://dx.doi.org/10.1063/1.1893983.

Carver, T., & Slichter, C. (1953). Polarization of nuclear spins in metals. *Physical Review*, *92*(1), 212–213. http://dx.doi.org/10.1103/PhysRev.92.212.2.

Chandrasekaran, P., Seagle, C., Rice, L., Macdonald, J., & Gerber, D. A. (2006). Functional analysis of encapsulated hepatic progenitor cells. *Tissue Engineering*, *12*(7), 2001–2008. http://dx.doi.org/10.1089/ten.2006.12.2001.

Chaumeil, M. M., Larson, P. E., Woods, S. M., Cai, L., Eriksson, P., Robinson, A. E., et al. (2014). Hyperpolarized [1–13C] glutamate: A metabolic imaging biomarker of IDH1 mutational status in glioma. *Cancer Research*, *74*(16), 4247–4257. http://dx.doi.org/10.1158/0008-5472.CAN-14-0680.

Chaumeil, M. M., Larson, P. E., Yoshihara, H. A., Danforth, O. M., Vigneron, D. B., Nelson, S. J., et al. (2013). Non-invasive in vivo assessment of IDH1 mutational status in glioma. *Nature Communications*, *4*, 2429. http://dx.doi.org/10.1038/ncomms3429.

Chaumeil, M. M., Ozawa, T., Park, I., Scott, K., James, C. D., Nelson, S. J., et al. (2012). Hyperpolarized 13C MR spectroscopic imaging can be used to monitor Everolimus treatment in vivo in an orthotopic rodent model of glioblastoma. *NeuroImage*, *59*(1), 193–201. http://dx.doi.org/10.1016/j.neuroimage.2011.07.034.

Chaumeil, M. M., Valette, J., Guillermier, M., Brouillet, E., Boumezbeur, F., Herard, A. S., et al. (2009). Multimodal neuroimaging provides a highly consistent picture of energy metabolism, validating 31P MRS for measuring brain ATP synthesis. *Proceedings of the National Academy of Sciences of the United States of America, 106*(10), 3988–3993. http://dx.doi.org/10.1073/pnas.0806516106.

Chen, A. P., Albers, M. J., Cunningham, C. H., Kohler, S. J., Yen, Y. F., Hurd, R. E., et al. (2007). Hyperpolarized C-13 spectroscopic imaging of the TRAMP mouse at 3T-initial experience. *Magnetic Resonance in Medicine, 58*(6), 1099–1106. http://dx.doi.org/10.1002/mrm.21256.

Chen, A. P., Hurd, R. E., Cunningham, C. H., Albers, M. J., Zierhut, M. L., Yen, Y. F., et al. (2007). Symmetric echo acquisition of hyperpolarized C-13 MRSI data in the TRAMP mouse at 3T. In *Proceedings of the 15th annual meeting of ISMRM, Berlin* (p. 538).

Chen, A. P., Hurd, R. E., Schroeder, M. A., Lau, A. Z., Gu, Y. P., Lam, W. W., et al. (2012). Simultaneous investigation of cardiac pyruvate dehydrogenase flux, Krebs cycle metabolism and pH, using hyperpolarized [1,2-(13)C2]pyruvate in vivo. *NMR in Biomedicine, 25*(2), 305–311. http://dx.doi.org/10.1002/nbm.1749.

Chen, A. P., Kurhanewicz, J., Bok, R., Xu, D., Joun, D., Zhang, V., et al. (2008). Feasibility of using hyperpolarized [1-13C]lactate as a substrate for in vivo metabolic 13C MRSI studies. *Magnetic Resonance Imaging, 26*(6), 721–726. http://dx.doi.org/10.1016/j.mri.2008.01.002.

Chen, A. P., Tropp, J., Hurd, R. E., Van Criekinge, M., Carvajal, L. G., Xu, D., et al. (2009). In vivo hyperpolarized 13C MR spectroscopic imaging with 1H decoupling. *Journal of Magnetic Resonance, 197*(1), 100–106. http://dx.doi.org/10.1016/j.jmr.2008.12.004.

Cheng, T., Capozzi, A., Takado, Y., Balzan, R., & Comment, A. (2013). Over 35% liquid-state 13C polarization obtained via dissolution dynamic nuclear polarization at 7 T and 1 K using ubiquitous nitroxyl radicals. *Physical Chemistry Chemical Physics, 15*(48), 20819–20822. http://dx.doi.org/10.1039/c3cp53022a.

Cheng, T., Mishkovsky, M., Bastiaansen, J. A., Ouari, O., Hautle, P., Tordo, P., et al. (2013). Automated transfer and injection of hyperpolarized molecules with polarization measurement prior to in vivo NMR. *NMR in Biomedicine, 26*(11), 1582–1588. http://dx.doi.org/10.1002/nbm.2993.

Chesnelong, C., Chaumeil, M. M., Blough, M. D., Al-Najjar, M., Stechishin, O. D., Chan, J. A., et al. (2014). Lactate dehydrogenase A silencing in IDH mutant gliomas. *Neuro-Oncology, 16*(5), 686–695. http://dx.doi.org/10.1093/neuonc/not243.

Chiavazza, E., Kubala, E., Gringeri, C. V., Duwel, S., Durst, M., Schulte, R. F., et al. (2013). Earth's magnetic field enabled scalar coupling relaxation of 13C nuclei bound to fast-relaxing quadrupolar 14N in amide groups. *Journal of Magnetic Resonance, 227*, 35–38. http://dx.doi.org/10.1016/j.jmr.2012.11.016.

Clatworthy, M. R., Kettunen, M. I., Hu, D. E., Mathews, R. J., Witney, T. H., Kennedy, B. W., et al. (2012). Magnetic resonance imaging with hyperpolarized [1,4-(13)C2]fumarate allows detection of early renal acute tubular necrosis. *Proceedings of the National Academy of Sciences of the United States of America, 109*(33), 13374–13379. http://dx.doi.org/10.1073/pnas.1205539109.

Comment, A. (2013). Hyperpolarization: Concepts, techniques and applications. In L. Garrido & N. Beckmann (Eds.), *New developments in NMR: Vol. 2. New applications of NMR in drug discovery and development* (pp. 252–272). Cambridge, UK: The Royal Society of Chemistry. http://dx.doi.org/10.1039/9781849737661-00252.

Comment, A. (2014). The benefits of not using exogenous substances to prepare substrates for hyperpolarized MRI. *Imaging in Medicine, 6*(1), 1–3. http://dx.doi.org/10.2217/iim.13.75.

Comment, A., & Merritt, M. E. (2014). Hyperpolarized magnetic resonance as a sensitive detector of metabolic function. *Biochemistry, 53*(47), 7333–7357. http://dx.doi.org/10.1021/bi501225t.

Comment, A., Rentsch, J., Kurdzesau, F., Jannin, S., Uffmann, K., van Heeswijk, R. B., et al. (2008). Producing over 100 ml of highly concentrated hyperpolarized solution by means of dissolution DNP. *Journal of Magnetic Resonance*, *194*(1), 152–155. http://dx.doi.org/10.1016/j.jmr.2008.06.003.

Comment, A., van den Brandt, B., Uffmann, K., Kurdzesau, F., Jannin, S., Konter, J. A., et al. (2007). Design and performance of a DNP prepolarizer coupled to a rodent MRI scanner. *Concepts in Magnetic Resonance Part B: Magnetic Resonance Engineering*, *31B*(4), 255–269. http://dx.doi.org/10.1002/cmr.b.20099.

Comment, A., van den Brandt, B., Uffmann, K., Kurdzesau, F., Jannin, S., Konter, J. A., et al. (2008). Principles of operation of a DNP prepolarizer coupled to a rodent MRI scanner. *Applied Magnetic Resonance*, *34*(3–4), 313–319. http://dx.doi.org/10.1007/s00723-008-0119-3.

Corzilius, B., Smith, A. A., Barnes, A. B., Luchinat, C., Bertini, I., & Griffin, R. G. (2011). High-field dynamic nuclear polarization with high-spin transition metal ions. *Journal of the American Chemical Society*, *133*(15), 5648–5651. http://dx.doi.org/10.1021/ja1109002.

Crane, J. C., Olson, M. P., & Nelson, S. J. (2013). SIVIC: Open-source, standards-based software for DICOM MR spectroscopy workflows. *International Journal of Biomedical Imaging*, *2013*, 169526. http://dx.doi.org/10.1155/2013/169526.

Cunningham, C. H., Chen, A. P., Albers, M. J., Kurhanewicz, J., Hurd, R. E., Yen, Y. F., et al. (2007). Double spin-echo sequence for rapid spectroscopic imaging of hyperpolarized 13C. *Journal of Magnetic Resonance*, *187*(2), 357–362. http://dx.doi.org/10.1016/j.jmr.2007.05.014.

Cunningham, C. H., Dominguez Viqueira, W., Hurd, R. E., & Chen, A. P. (2014). Frequency correction method for improved spatial correlation of hyperpolarized 13C metabolites and anatomy. *NMR in Biomedicine*, *27*(2), 212–218. http://dx.doi.org/10.1002/nbm.3055.

Cunningham, C. H., Vigneron, D. B., Chen, A. P., Xu, D., Nelson, S. J., Hurd, R. E., et al. (2005). Design of flyback echo-planar readout gradients for magnetic resonance spectroscopic imaging. *Magnetic Resonance in Medicine*, *54*(5), 1286–1289. http://dx.doi.org/10.1002/mrm.20663.

Dafni, H., Larson, P. E., Hu, S., Yoshihara, H. A., Ward, C. S., Venkatesh, H. S., et al. (2010). Hyperpolarized 13C spectroscopic imaging informs on hypoxia-inducible factor-1 and myc activity downstream of platelet-derived growth factor receptor. *Cancer Research*, *70*(19), 7400–7410. http://dx.doi.org/10.1158/0008-5472.CAN-10-0883.

Dafni, H., & Ronen, S. M. (2010). Dynamic nuclear polarization in metabolic imaging of metastasis: Common sense, hypersense and compressed sensing. *Cancer Biomarkers*, 7, 1–11.

Day, S. E., Kettunen, M. I., Gallagher, F. A., Hu, D. E., Lerche, M., Wolber, J., et al. (2007). Detecting tumor response to treatment using hyperpolarized 13C magnetic resonance imaging and spectroscopy. *Nature Medicine*, *13*(11), 1382–1387. http://dx.doi.org/10.1038/nm1650.

DeBerardinis, R. J., Mancuso, A., Daikhin, E., Nissim, I., Yudkoff, M., Wehrli, S., et al. (2007). Beyond aerobic glycolysis: Transformed cells can engage in glutamine metabolism that exceeds the requirement for protein and nucleotide synthesis. *Proceedings of the National Academy of Sciences of the United States of America*, *104*(49), 19345–19350. http://dx.doi.org/10.1073/pnas.0709747104.

De Graaf, R. A. (2007). In vivo *NMR spectroscopy: Principles and techniques* (2nd ed.). Chichester, West Sussex, England; Hoboken, NJ: John Wiley & Sons.

Eichhorn, T. R., Takado, Y., Salameh, N., Capozzi, A., Cheng, T., Hyacinthe, J. N., et al. (2013). Hyperpolarization without persistent radicals for in vivo real-time metabolic imaging. *Proceedings of the National Academy of Sciences of the United States of America*, *110*(45), 18064–18069. http://dx.doi.org/10.1073/pnas.1314928110.

Flori, A., Liserani, M., Frijia, F., Giovannetti, G., Lionetti, V., Casieri, V., et al. (2014). Real-time cardiac metabolism assessed with hyperpolarized [1-^{13}C]acetate in a large-animal model. *Contrast Media & Molecular Imaging, 119*(10), 1885–1893. http://dx.doi.org/10.1002/cmmi.1618.

Friesen-Waldner, L. J., Wade, T. P., Thind, K., Chen, A. P., Gomori, J. M., Sosna, J., et al. (2014). Hyperpolarized choline as an MR imaging molecular probe: Feasibility of in vivo imaging in a rat model. *Journal of Magnetic Resonance Imaging, 41*(4), 917–923. http://dx.doi.org/10.1002/jmri.24659.

Gabellieri, C., Mugnaini, V., Paniagua, J. C., Roques, N., Oliveros, M., Feliz, M., et al. (2010). Dynamic nuclear polarization with polychlorotriphenylmethyl radicals: Supramolecular polarization-transfer effects. *Angewandte Chemie (International Ed. in English), 49*(19), 3360–3362. http://dx.doi.org/10.1002/anie.201000031.

Gajan, D., Bornet, A., Vuichoud, B., Milani, J., Melzi, R., van Kalkeren, H. A., et al. (2014). Hybrid polarizing solids for pure hyperpolarized liquids through dissolution dynamic nuclear polarization. *Proceedings of the National Academy of Sciences of the United States of America, 111*(41), 14693–14697. http://dx.doi.org/10.1073/pnas.1407730111.

Gallagher, F. A., Bohndiek, S. E., Kettunen, M. I., Lewis, D. Y., Soloviev, D., & Brindle, K. M. (2011). Hyperpolarized 13C MRI and PET: In vivo tumor biochemistry. *Journal of Nuclear Medicine, 52*(9), 1333–1336. http://dx.doi.org/10.2967/jnumed.110.085258.

Gallagher, F. A., Kettunen, M. I., & Brindle, K. M. (2009). Biomedical applications of hyperpolarized 13C magnetic resonance imaging. *Progress in Nuclear Magnetic Resonance Spectroscopy, 55*, 285–295.

Gallagher, F. A., Kettunen, M. I., Day, S. E., Hu, D. E., Ardenkjaer-Larsen, J. H., Zandt, Ri, et al. (2008). Magnetic resonance imaging of pH in vivo using hyperpolarized 13C-labelled bicarbonate. *Nature, 453*(7197), 940–943. http://dx.doi.org/10.1038/nature07017.

Gallagher, F. A., Kettunen, M. I., Day, S. E., Hu, D. E., Karlsson, M., Gisselsson, A., et al. (2011). Detection of tumor glutamate metabolism in vivo using (13)C magnetic resonance spectroscopy and hyperpolarized [1-(13)C]glutamate. *Magnetic Resonance in Medicine, 66*(1), 18–23. http://dx.doi.org/10.1002/mrm.22851.

Gallagher, F. A., Kettunen, M. I., Day, S. E., Lerche, M., & Brindle, K. M. (2008). 13C MR spectroscopy measurements of glutaminase activity in human hepatocellular carcinoma cells using hyperpolarized 13C-labeled glutamine. *Magnetic Resonance in Medicine, 60*(2), 253–257. http://dx.doi.org/10.1002/mrm.21650.

Gallagher, F. A., Kettunen, M. I., Hu, D. E., Jensen, P. R., Zandt, R. I., Karlsson, M., et al. (2009). Production of hyperpolarized [1,4-13C2]malate from [1,4-13C2]fumarate is a marker of cell necrosis and treatment response in tumors. *Proceedings of the National Academy of Sciences of the United States of America, 106*(47), 19801–19806. http://dx.doi.org/10.1073/pnas.0911447106.

Gillies, R. J., Liu, Z., & Bhujwalla, Z. (1994). 31P-MRS measurements of extracellular pH of tumors using 3-aminopropylphosphonate. *The American Journal of Physiology, 267*(1 Pt. 1), C195–C203.

Gillies, R. J., & Morse, D. L. (2005). In vivo magnetic resonance spectroscopy in cancer. *Annual Review of Biomedical Engineering*, 7, 287–326. http://dx.doi.org/10.1146/annurev.bioeng.7.060804.100411.

Gillies, R. J., Robey, I., & Gatenby, R. A. (2008). Causes and consequences of increased glucose metabolism of cancers. *Journal of Nuclear Medicine, 49*(Suppl. 2), 24S–42S. http://dx.doi.org/10.2967/jnumed.107.047258.

Glover, G. H. (1999). Simple analytic spiral K-space algorithm. *Magnetic Resonance in Medicine, 42*(2), 412–415.

Golman, K., in 't Zandt, R., & Thaning, M. (2006). Real-time metabolic imaging. *Proceedings of the National Academy of Sciences of the United States of America, 103*(30), 11270–11275. http://dx.doi.org/10.1073/pnas.0601319103.

Golman, K., Olsson, L. E., Axelsson, O., Mansson, S., Karlsson, M., & Petersson, J. S. (2003). Molecular imaging using hyperpolarized 13C. *The British Journal of Radiology, 76*(Spec No. 2), S118–S127.

Golman, K., Petersson, J. S., Magnusson, P., Johansson, E., Akeson, P., Chai, C. M., et al. (2008). Cardiac metabolism measured noninvasively by hyperpolarized 13C MRI. *Magnetic Resonance in Medicine, 59*(5), 1005–1013. http://dx.doi.org/10.1002/mrm.21460.

Golman, K., Zandt, R. I., Lerche, M., Pehrson, R., & Ardenkjaer-Larsen, J. H. (2006). Metabolic imaging by hyperpolarized 13C magnetic resonance imaging for in vivo tumor diagnosis. *Cancer Research, 66*(22), 10855–10860. http://dx.doi.org/10.1158/0008-5472.CAN-06-2564.

Grozinger, G., Grozinger, A., & Horger, M. (2014). The role of volume perfusion CT in the diagnosis of pathologies of the pancreas. *Röfo, 186*(12), 1082–1093. http://dx.doi.org/10.1055/s-0034-1384876.

Gruetter, R., Adriany, G., Choi, I. Y., Henry, P. G., Lei, H., & Oz, G. (2003). Localized in vivo 13C NMR spectroscopy of the brain. *NMR in Biomedicine, 16*(6–7), 313–338. http://dx.doi.org/10.1002/nbm.841.

Hansen, T. M., Nilsson, M., Gram, M., & Frokjaer, J. B. (2013). Morphological and functional evaluation of chronic pancreatitis with magnetic resonance imaging. *World Journal of Gastroenterology, 19*(42), 7241–7246. http://dx.doi.org/10.3748/wjg.v19.i42.7241.

Harada, M., Kubo, H., Abe, T., Maezawa, H., & Otsuka, H. (2010). Selection of endogenous 13C substrates for observation of intracellular metabolism using the dynamic nuclear polarization technique. *Japanese Journal of Radiology, 28*(2), 173–179. http://dx.doi.org/10.1007/s11604-009-0390-8.

Harris, T., Degani, H., & Frydman, L. (2013). Hyperpolarized 13C NMR studies of glucose metabolism in living breast cancer cell cultures. *NMR in Biomedicine, 26*(12), 1831–1843. http://dx.doi.org/10.1002/nbm.3024.

Harris, T., Eliyahu, G., Frydman, L., & Degani, H. (2009). Kinetics of hyperpolarized 13C1-pyruvate transport and metabolism in living human breast cancer cells. *Proceedings of the National Academy of Sciences of the United States of America, 106*(43), 18131–18136. http://dx.doi.org/10.1073/pnas.0909049106.

Harrison, C., Yang, C., Jindal, A., DeBerardinis, R. J., Hooshyar, M. A., Merritt, M., et al. (2012). Comparison of kinetic models for analysis of pyruvate-to-lactate exchange by hyperpolarized 13 C NMR. *NMR in Biomedicine, 25*(11), 1286–1294. http://dx.doi.org/10.1002/nbm.2801.

Heckmann, J., Meyer, W., Radtke, E., Reicherz, G., & Goertz, S. (2006). Electron spin resonance and its implication on the maximum nuclear polarization of deuterated solid target materials. *Physical Review B, 74*(13), 13418–1/9. http://dx.doi.org/10.1103/PhysRevB.74.134418.

Henry, P. G., Adriany, G., Deelchand, D., Gruetter, R., Marjanska, M., Oz, G., et al. (2006). In vivo 13C NMR spectroscopy and metabolic modeling in the brain: A practical perspective. *Magnetic Resonance Imaging, 24*(4), 527–539. http://dx.doi.org/10.1016/j.mri.2006.01.003.

Hill, D. K., Jamin, Y., Orton, M. R., Tardif, N., Parkes, H. G., Robinson, S. P., et al. (2013). ^{1}H NMR and hyperpolarized ^{13}C NMR assays of pyruvate-lactate: A comparative study. *NMR in Biomedicine, 26*(10), 1321–1325. http://dx.doi.org/10.1002/nbm.2957.

Hill, D. K., Orton, M. R., Mariotti, E., Boult, J. K., Panek, R., Jafar, M., et al. (2013). Model free approach to kinetic analysis of real-time hyperpolarized 13C magnetic resonance spectroscopy data. *PLoS One. 8*(9). http://dx.doi.org/10.1371/journal.pone.0071996, e71996.

Hu, S., Balakrishnan, A., Bok, R. A., Anderton, B., Larson, P. E., Nelson, S. J., et al. (2011). 13C-pyruvate imaging reveals alterations in glycolysis that precede c-Myc-induced tumor formation and regression. *Cell Metabolism*, *14*(1), 131–142. http://dx.doi.org/10.1016/j.cmet.2011.04.012.

Hu, S., Larson, P. E. Z., VanCriekinge, M., Leach, A. M., Park, I., Leon, C., et al. (2013). Rapid sequential injections of hyperpolarized [1-13C]pyruvate in vivo using a sub-kelvin, multi-sample DNP polarizer. *Magnetic Resonance Imaging*, *31*(4), 490–496. http://dx.doi.org/10.1016/j.mri.2012.09.002.

Hu, S., Lustig, M., Balakrishnan, A., Larson, P. E., Bok, R., Kurhanewicz, J., et al. (2010). 3D compressed sensing for highly accelerated hyperpolarized (13)C MRSI with in vivo applications to transgenic mouse models of cancer. *Magnetic Resonance in Medicine*, *63*(2), 312–321. http://dx.doi.org/10.1002/mrm.22233.

Hu, S., Lustig, M., Chen, A. P., Crane, J., Kerr, A., Kelley, D. A., et al. (2008). Compressed sensing for resolution enhancement of hyperpolarized 13C flyback 3D-MRSI. *Journal of Magnetic Resonance*, *192*(2), 258–264. http://dx.doi.org/10.1016/j.jmr.2008.03.003.

Hu, S., Yoshihara, H. A., Bok, R., Zhou, J., Zhu, M., Kurhanewicz, J., et al. (2012). Use of hyperpolarized [1-13C]pyruvate and [2-13C]pyruvate to probe the effects of the anticancer agent dichloroacetate on mitochondrial metabolism in vivo in the normal rat. *Magnetic Resonance Imaging*, *30*(10), 1367–1372. http://dx.doi.org/10.1016/j.mri.2012.05.012.

Hu, S., Zhu, M., Yoshihara, H. A., Wilson, D. M., Keshari, K. R., Shin, P., et al. (2011). In vivo measurement of normal rat intracellular pyruvate and lactate levels after injection of hyperpolarized [1-(13)C]alanine. *Magnetic Resonance Imaging*, *29*(8), 1035–1040. http://dx.doi.org/10.1016/j.mri.2011.07.001.

Hurd, R. E., Spielman, D., Josan, S., Yen, Y. F., Pfefferbaum, A., & Mayer, D. (2013). Exchange-linked dissolution agents in dissolution-DNP (13)C metabolic imaging. *Magnetic Resonance in Medicine*, *70*(4), 936–942. http://dx.doi.org/10.1002/mrm.24544.

Hurd, R. E., Yen, Y. F., Mayer, D., Chen, A., Wilson, D., Kohler, S., et al. (2010). Metabolic imaging in the anesthetized rat brain using hyperpolarized [1-13C] pyruvate and [1-13C] ethyl pyruvate. *Magnetic Resonance in Medicine*, *63*(5), 1137–1143. http://dx.doi.org/10.1002/mrm.22364.

Jannin, S., Comment, A., Kurdzesau, F., Konter, J. A., Hautle, P., van den Brandt, B., et al. (2008). A 140 GHz prepolarizer for dissolution dynamic nuclear polarization. *The Journal of Chemical Physics*, *128*(24), 241102. http://dx.doi.org/10.1063/1.2951994.

Jannin, S., Comment, A., & van der Klink, J. J. (2012). Dynamic nuclear polarization by thermal mixing under partial saturation. *Applied Magnetic Resonance*, *43*(1–2), 59–68. http://dx.doi.org/10.1007/s00723-012-0363-4.

Jensen, P. R., Meier, S., Ardenkjaer-Larsen, J. H., Duus, J. O., Karlsson, M., & Lerche, M. H. (2009). Detection of low-populated reaction intermediates with hyperpolarized NMR. *Chemical Communications: Cambridge, England*, (34), 5168–5170. http://dx.doi.org/10.1039/b910626j.

Jensen, P. R., Peitersen, T., Karlsson, M., In 't Zandt, R., Gisselsson, A., Hansson, G., et al. (2009). Tissue-specific short chain fatty acid metabolism and slow metabolic recovery after ischemia from hyperpolarized NMR in vivo. *The Journal of Biological Chemistry*, *284*(52), 36077–36082. http://dx.doi.org/10.1074/jbc.M109.066407.

Johannesson, H., Macholl, S., & Ardenkjaer-Larsen, J. H. (2009). Dynamic nuclear polarization of [1-13C]pyruvic acid at 4.6 tesla. *Journal of Magnetic Resonance*, *197*(2), 167–175. http://dx.doi.org/10.1016/j.jmr.2008.12.016.

Johansson, E., Mansson, S., Wirestam, R., Svensson, J., Petersson, J. S., Golman, K., et al. (2004). Cerebral perfusion assessment by bolus tracking using hyperpolarized 13C. *Magnetic Resonance in Medicine*, *51*, 464–472.

Josan, S., Hurd, R., Billingsley, K., Senadheera, L., Park, J. M., Yen, Y. F., et al. (2013). Effects of isoflurane anesthesia on hyperpolarized (13)C metabolic measurements in rat brain. *Magnetic Resonance in Medicine, 70*(4), 1117–1124. http://dx.doi.org/10.1002/mrm.24532.

Josan, S., Hurd, R., Park, J. M., Yen, Y. F., Watkins, R., Pfefferbaum, A., et al. (2014). Dynamic metabolic imaging of hyperpolarized [2-(13) C]pyruvate using spiral chemical shift imaging with alternating spectral band excitation. *Magnetic Resonance in Medicine, 71*(6), 2051–2058. http://dx.doi.org/10.1002/mrm.24871.

Josan, S., Park, J. M., Hurd, R., Yen, Y. F., Pfefferbaum, A., Spielman, D., et al. (2013). In vivo investigation of cardiac metabolism in the rat using MRS of hyperpolarized [1-13C] and [2-13C]pyruvate. *NMR in Biomedicine, 26*(12), 1680–1687. http://dx.doi.org/10.1002/nbm.3003.

Karlsson, M., Jensen, P. R., in 't Zandt, R., Gisselsson, A., Hansson, G., Duus, J. O., et al. (2010). Imaging of branched chain amino acid metabolism in tumors with hyperpolarized 13C ketoisocaproate. *International Journal of Cancer, 127*(3), 729–736. http://dx.doi.org/10.1002/ijc.25072.

Kazan, S. M., Reynolds, S., Kennerley, A., Wholey, E., Bluff, J. E., Berwick, J., et al. (2013). Kinetic modeling of hyperpolarized (13)C pyruvate metabolism in tumors using a measured arterial input function. *Magnetic Resonance in Medicine, 70*(4), 943–953. http://dx.doi.org/10.1002/mrm.24546.

Kerr, A. B., Larson, P. E., Lustig, M. S., Cunningham, C. H., Chen, A. P., Kurhanewicz, J., et al. (2008). Multiband spectral-spatial design for high-field and hyperpolarized C-13 applications. In *Proceedings of the 16th annual meeting of ISMRM, Toronto* (p. 226).

Keshari, K. R., Kurhanewicz, J., Bok, R., Larson, P. E., Vigneron, D. B., & Wilson, D. M. (2011). Hyperpolarized 13C dehydroascorbate as an endogenous redox sensor for in vivo metabolic imaging. *Proceedings of the National Academy of Sciences of the United States of America, 108*(46), 18606–18611. http://dx.doi.org/10.1073/pnas.1106920108.

Keshari, K. R., Kurhanewicz, J., Jeffries, R. E., Wilson, D. M., Dewar, B. J., Van Criekinge, M., et al. (2010). Hyperpolarized (13)C spectroscopy and an NMR-compatible bioreactor system for the investigation of real-time cellular metabolism. *Magnetic Resonance in Medicine, 63*(2), 322–329. http://dx.doi.org/10.1002/mrm.22225.

Keshari, K. R., Sai, V., Wang, Z. J., Vanbrocklin, H. F., Kurhanewicz, J., & Wilson, D. M. (2013). Hyperpolarized [1-13C]dehydroascorbate MR spectroscopy in a murine model of prostate cancer: Comparison with 18F-FDG PET. *Journal of Nuclear Medicine, 54*(6), 922–928. http://dx.doi.org/10.2967/jnumed.112.115402.

Keshari, K. R., Sriram, R., Koelsch, B. L., Van Criekinge, M., Wilson, D. M., Kurhanewicz, J., et al. (2013). Hyperpolarized 13C-pyruvate magnetic resonance reveals rapid lactate export in metastatic renal cell carcinomas. *Cancer Research, 73*(2), 529–538. http://dx.doi.org/10.1158/0008-5472.CAN-12-3461.

Keshari, K. R., Sriram, R., Van Criekinge, M., Wilson, D. M., Wang, Z. J., Vigneron, D. B., et al. (2013). Metabolic reprogramming and validation of hyperpolarized 13C lactate as a prostate cancer biomarker using a human prostate tissue slice culture bioreactor. *Prostate, 73*(11), 1171–1181. http://dx.doi.org/10.1002/pros.22665.

Keshari, K. R., & Wilson, D. A. (2014). Chemistry and biochemistry of 13C hyperpolarized magnetic resonance using dynamic nuclear polarization. *Chemical Society Reviews, 43*, 1627–1659.

Keshari, K. R., Wilson, D. M., Chen, A. P., Bok, R., Larson, P. E., Hu, S., et al. (2009). Hyperpolarized [2-13C]-fructose: A hemiketal DNP substrate for in vivo metabolic imaging. *Journal of the American Chemical Society, 131*(48), 17591–17596. http://dx.doi.org/10.1021/ja9049355.

Keshari, K. R., Wilson, D. M., Sai, V., Bok, R., Jen, K. Y., Larson, P., et al. (2015). Non-invasive in vivo imaging of diabetes-induced renal oxidative stress and response to therapy using hyperpolarized 13C dehydroascorbate magnetic resonance. *Diabetes, 64*(2), 344–352. http://dx.doi.org/10.2337/db13-1829.

Kettunen, M. I., Hu, D. E., Witney, T. H., McLaughlin, R., Gallagher, F. A., Bohndiek, S. E., et al. (2010). Magnetization transfer measurements of exchange between hyperpolarized [1-13C]pyruvate and [1-13C]lactate in a murine lymphoma. *Magnetic Resonance in Medicine, 63*(4), 872–880. http://dx.doi.org/10.1002/mrm.22276.

Khemtong, C., Carpenter, N. R., Lumata, L. L., Merritt, M. E., Moreno, K. X., Kovacs, Z., et al. (2014). Hyperpolarized ^{13}C NMR detects rapid drug-induced changes in cardiac metabolism. *Magnetic Resonance in Medicine*. http://dx.doi.org/10.1002/mrm.25419.

Kmiotek, E. K., Baimel, C., & Gill, K. J. (2012). Methods for intravenous self administration in a mouse model. *Journal of Visualized Experiments*, (70), e3739. http://dx.doi.org/10.3791/3739.

Koelsch, B. L., Keshari, K. R., Peeters, T. H., Larson, P. E., Wilson, D. A., & Kurhanewicz, J. (2013a). Complete separation of extra- and intracellular hyperpolarized 13C metabolite signal with diffusion weighted MR. *Proceedings on International Society for Magnetic Resonance in Medicine, 21*, 567.

Koelsch, B. L., Keshari, K. R., Peeters, T. H., Larson, P. E., Wilson, D. M., & Kurhanewicz, J. (2013b). Diffusion MR of hyperpolarized 13C molecules in solution. *Analyst, 138*(4), 1011–1014. http://dx.doi.org/10.1039/c2an36715g.

Koelsch, B. L., Reed, G. D., Keshari, K. R., Chaumeil, M. M., Bok, R., Ronen, S. M., et al. (2014). Rapid in vivo apparent diffusion coefficient mapping of hyperpolarized C metabolites. *Magnetic Resonance in Medicine*. http://dx.doi.org/10.1002/mrm.25422.

Kohler, S. J., Yen, Y., Wolber, J., Chen, A. P., Albers, M. J., Bok, R., et al. (2007). In vivo 13 carbon metabolic imaging at 3T with hyperpolarized 13C-1-pyruvate. *Magnetic Resonance in Medicine, 58*(1), 65–69. http://dx.doi.org/10.1002/mrm.21253.

Kumagai, K., Akakabe, M., Tsuda, M., Tsuda, M., Fukushi, E., Kawabata, J., et al. (2014). Observation of glycolytic metabolites in tumor cell lysate by using hyperpolarization of deuterated glucose. *Biological & Pharmaceutical Bulletin, 37*(8), 1416–1421.

Kurdzesau, F., van den Brandt, B., Comment, A., Hautle, P., Jannin, S., van der Klink, J. J., et al. (2008). Dynamic nuclear polarization of small labelled molecules in frozen water–alcohol solutions. *Journal of Physics D: Applied Physics, 41*(15), 155506. http://dx.doi.org/10.1088/0022-3727/41/15/155506.

Kurhanewicz, J., Vigneron, D. B., Brindle, K., Chekmenev, E. Y., Comment, A., Cunningham, C. H., et al. (2011). Analysis of cancer metabolism by imaging hyperpolarized nuclei: Prospects for translation to clinical research. *Neoplasia, 13*(2), 81–97.

Larsen, J. M., Martin, D. R., & Byrne, M. E. (2014). Recent advances in delivery through the blood-brain barrier. *Current Topics in Medicinal Chemistry, 14*(9), 1148–1160.

Larson, P. E., Bok, R., Kerr, A. B., Lustig, M., Hu, S., Chen, A. P., et al. (2010). Investigation of tumor hyperpolarized [1-13C]-pyruvate dynamics using time-resolved multiband RF excitation echo-planar MRSI. *Magnetic Resonance in Medicine, 63*(3), 582–591. http://dx.doi.org/10.1002/mrm.22264.

Larson, P. E., Hurd, R. E., Kerr, A. B., Pauly, J. M., Bok, R. A., Kurhanewicz, J., et al. (2013). Perfusion and diffusion sensitive 13C stimulated-echo MRSI for metabolic imaging of cancer. *Magnetic Resonance Imaging, 31*(5), 635–642. http://dx.doi.org/10.1016/j.mri.2012.10.020.

Larson, P. E., Kerr, A. B., Chen, A. P., Lustig, M. S., Zierhut, M. L., Hu, S., et al. (2008). Multiband excitation pulses for hyperpolarized 13C dynamic chemical-shift imaging. *Journal of Magnetic Resonance, 194*(1), 121–127. http://dx.doi.org/10.1016/j.jmr.2008.06.010.

Larson, P. E., Kerr, A. B., Swisher, C. L., Pauly, J. M., & Vigneron, D. B. (2012). A rapid method for direct detection of metabolic conversion and magnetization exchange with application to hyperpolarized substrates. *Journal of Magnetic Resonance, 225*, 71–80. http://dx.doi.org/10.1016/j.jmr.2012.09.014.

Lau, A. Z., Chen, A. P., Ghugre, N. R., Ramanan, V., Lam, W. W., Connelly, K. A., et al. (2010). Rapid multislice imaging of hyperpolarized 13C pyruvate and bicarbonate in the heart. *Magnetic Resonance in Medicine, 64*(5), 1323–1331. http://dx.doi.org/10.1002/mrm.22525.

Laustsen, C., Ostergaard, J. A., Lauritzen, M. H., Norregaard, R., Bowen, S., Sogaard, L. V., et al. (2013). Assessment of early diabetic renal changes with hyperpolarized [1-(13) C] pyruvate. *Diabetes/Metabolism Research and Reviews, 29*(2), 125–129. http://dx.doi.org/10.1002/dmrr.2370.

Laustsen, C., Pileio, G., Tayler, M. C., Brown, L. J., Brown, R. C., Levitt, M. H., et al. (2012). Hyperpolarized singlet NMR on a small animal imaging system. *Magnetic Resonance in Medicine, 68*(4), 1262–1265. http://dx.doi.org/10.1002/mrm.24430.

Lee, P., Leong, W., Tan, T., Lim, M., Han, W., & Radda, G. K. (2013). In vivo hyperpolarized carbon-13 magnetic resonance spectroscopy reveals increased pyruvate carboxylase flux in an insulin-resistant mouse model. *Hepatology, 57*(2), 515–524. http://dx.doi.org/10.1002/hep.26028.

Leftin, A., Degani, H., & Frydman, L. (2013). In vivo magnetic resonance of hyperpolarized [(13)C1]pyruvate: Metabolic dynamics in stimulated muscle. *American Journal of Physiology. Endocrinology and Metabolism, 305*(9), E1165–E1171. http://dx.doi.org/10.1152/ajpendo.00296.2013.

Leftin, A., Roussel, T., & Frydman, L. (2014). Hyperpolarized functional magnetic resonance of murine skeletal muscle enabled by multiple tracer-paradigm synchronizations. *PLoS One, 9*(4), e96399. http://dx.doi.org/10.1371/journal.pone.0096399.

Leupold, J., Mansson, S., Petersson, J. S., Hennig, J., & Wieben, O. (2009). Fast multiecho balanced SSFP metabolite mapping of (1)H and hyperpolarized (13)C compounds. *Magma, 22*(4), 251–256. http://dx.doi.org/10.1007/s10334-009-0169-z.

Leupold, J., Wieben, O., Mansson, S., Speck, O., Scheffler, K., Petersson, J. S., et al. (2006). Fast chemical shift mapping with multiecho balanced SSFP. *Magma, 19*(5), 267–273. http://dx.doi.org/10.1007/s10334-006-0056-9.

Li, L. Z., Kadlececk, S., Xu, H. N., Daye, D., Pullinger, B., Profka, H., et al. (2013). Ratiometric analysis in hyperpolarized NMR (I): Test of the two-site exchange model and the quantification of reaction rate constants. *NMR in Biomedicine, 26*(10), 1308–1320. http://dx.doi.org/10.1002/nbm.2953.

Lin, G., Andrejeva, G., Wong Te Fong, A. C., Hill, D. K., Orton, M. R., Parkes, H. G., et al. (2014). Reduced Warburg effect in cancer cells undergoing autophagy: Steady-state 1H-MRS and real-time hyperpolarized 13C-MRS studies. *PLoS One, 9*(3), e92645. http://dx.doi.org/10.1371/journal.pone.0092645.

Lodi, A., Woods, S. M., & Ronen, S. M. (2013). Treatment with the MEK inhibitor U0126 induces decreased hyperpolarized pyruvate to lactate conversion in breast, but not prostate, cancer cells. *NMR in Biomedicine, 26*(3), 299–306. http://dx.doi.org/10.1002/nbm.2848.

Lumata, L., Jindal, A. K., Merritt, M. E., Malloy, C. R., Sherry, A. D., & Kovacs, Z. (2011). DNP by thermal mixing under optimized conditions yields >60,000-fold enhancement of 89Y NMR signal. *Journal of the American Chemical Society, 133*(22), 8673–8680. http://dx.doi.org/10.1021/ja201880y.

Lumata, L., Kovacs, Z., Malloy, C., Sherry, A. D., & Merritt, M. (2011). The effect of13C enrichment in the glassing matrix on dynamic nuclear polarization of [1-13C]pyruvate. *Physics in Medicine and Biology, 56*(5), N85–N92. http://dx.doi.org/10.1088/0031-9155/56/5/n01.

Lumata, L., Kovacs, Z., Sherry, A. D., Malloy, C., Hill, S., van Tol, J., et al. (2013). Electron spin resonance studies of trityl OX063 at a concentration optimal for

DNP. *Physical Chemistry Chemical Physics*, *15*(24), 9800–9807. http://dx.doi.org/10.1039/c3cp50186h.

Lumata, L. L., Martin, R., Jindal, A. K., Kovacs, Z., Conradi, M. S., & Merritt, M. E. (2015). Development and performance of a 129-GHz dynamic nuclear polarizer in an ultra-wide bore superconducting magnet. *Magma*, *28*(2), 195–205. http://dx.doi.org/10.1007/s10334-014-0455-2.

Lumata, L., Merritt, M., Khemtong, C., Ratnakar, S. J., van Tol, J., Yu, L., et al. (2012). The efficiency of DPPH as a polarising agent for DNP-NMR spectroscopy. *RSC Advances*, *2*(33), 12812–12817. http://dx.doi.org/10.1039/C2RA21853D.

Lumata, L., Merritt, M. E., & Kovacs, Z. (2013). Influence of deuteration in the glassing matrix on 13C dynamic nuclear polarization. *Physical Chemistry Chemical Physics*, *15*(19), 7032–7035. http://dx.doi.org/10.1039/c3cp50750e.

Lumata, L., Merritt, M. E., Malloy, C. R., Sherry, A. D., & Kovacs, Z. (2012). Impact of Gd3+on DNP of [1-13C]pyruvate doped with trityl OX063, BDPA, or 4-Oxo-TEMPO. *The Journal of Physical Chemistry A*, *116*(21), 5129–5138. http://dx.doi.org/10.1021/jp302399f.

Lumata, L. L., Merritt, M. E., Malloy, C. R., Sherry, A. D., van Tol, J., Song, L., et al. (2013). Dissolution DNP-NMR spectroscopy using galvinoxyl as a polarizing agent. *Journal of Magnetic Resonance*, *227*, 14–19. http://dx.doi.org/10.1016/j.jmr.2012.11.006.

Lumata, L., Ratnakar, S. J., Jindal, A., Merritt, M., Comment, A., Malloy, C., et al. (2011). BDPA: An efficient polarizing agent for fast dissolution dynamic nuclear polarization NMR spectroscopy. *Chemistry*, *17*(39), 10825–10827. http://dx.doi.org/10.1002/chem.201102037.

Lupo, J. M., Chen, A. P., Zierhut, M. L., Bok, R. A., Cunningham, C. H., Kurhanewicz, J., et al. (2010). Analysis of hyperpolarized dynamic 13C lactate imaging in a transgenic mouse model of prostate cancer. *Magnetic Resonance Imaging*, *28*(2), 153–162. http://dx.doi.org/10.1016/j.mri.2009.07.007.

Lustig, M., Donoho, D., & Pauly, J. M. (2007). Sparse MRI: The application of compressed sensing for rapid MR imaging. *Magnetic Resonance in Medicine*, *58*(6), 1182–1195. http://dx.doi.org/10.1002/mrm.21391.

Macholl, S., Johannesson, H., & Ardenkjaer-Larsen, J. H. (2010). Trityl biradicals and 13C dynamic nuclear polarization. *Physical Chemistry Chemical Physics*, *12*(22), 5804–5817. http://dx.doi.org/10.1039/c002699a.

MacKenzie, J. D., Yen, Y. F., Mayer, D., Tropp, J. S., Hurd, R. E., & Spielman, D. M. (2011). Detection of inflammatory arthritis by using hyperpolarized 13C-pyruvate with MR imaging and spectroscopy. *Radiology*, *259*(2), 414–420. http://dx.doi.org/10.1148/radiol.10101921.

Marco-Rius, I., Tayler, M. C., Kettunen, M. I., Larkin, T. J., Timm, K. N., Serrao, E. M., et al. (2013). Hyperpolarized singlet lifetimes of pyruvate in human blood and in the mouse. *NMR in Biomedicine*, *26*(12), 1696–1704. http://dx.doi.org/10.1002/nbm.3005.

Marjanska, M., Iltis, I., Shestov, A. A., Deelchand, D. K., Nelson, C., Ugurbil, K., et al. (2010). In vivo 13C spectroscopy in the rat brain using hyperpolarized [1-(13)C] pyruvate and [2-(13)C]pyruvate. *Journal of Magnetic Resonance*, *206*(2), 210–218. http://dx.doi.org/10.1016/j.jmr.2010.07.006.

Marjanska, M., Teisseyre, T. Z., Halpern-Manners, N. W., Zhang, Y., Iltis, I., Bajaj, V., et al. (2012). Measurement of arterial input function in hyperpolarized 13C studies. *Applied Magnetic Resonance*, *43*, 289–297.

Mayer, D., Levin, Y. S., Hurd, R. E., Glover, G. H., & Spielman, D. M. (2006). Fast metabolic imaging of systems with sparse spectra: Application for hyperpolarized 13C imaging. *Magnetic Resonance in Medicine*, *56*(4), 932–937. http://dx.doi.org/10.1002/mrm.21025.

Mayer, D., Yen, Y. F., Josan, S., Park, J. M., Pfefferbaum, A., Hurd, R. E., et al. (2012). Application of hyperpolarized [1-(1)(3)C]lactate for the in vivo investigation of cardiac metabolism. *NMR in Biomedicine*, *25*(10), 1119–1124. http://dx.doi.org/10.1002/nbm.2778.

Mayer, D., Yen, Y. F., Levin, Y. S., Tropp, J., Pfefferbaum, A., Hurd, R. E., et al. (2010). In vivo application of sub-second spiral chemical shift imaging (CSI) to hyperpolarized 13C metabolic imaging: Comparison with phase-encoded CSI. *Journal of Magnetic Resonance*, *204*(2), 340–345. http://dx.doi.org/10.1016/j.jmr.2010.03.005.

Mayer, D., Yen, Y. F., Tropp, J., Pfefferbaum, A., Hurd, R. E., & Spielman, D. M. (2009). Application of subsecond spiral chemical shift imaging to real-time multislice metabolic imaging of the rat in vivo after injection of hyperpolarized 13C1-pyruvate. *Magnetic Resonance in Medicine*, *62*(3), 557–564. http://dx.doi.org/10.1002/mrm.22041.

Meier, S., Jensen, P. R., & Duus, J. O. (2011). Real-time detection of central carbon metabolism in living Escherichia coli and its response to perturbations. *FEBS Letters*, *585*(19), 3133–3138. http://dx.doi.org/10.1016/j.febslet.2011.08.049.

Meier, S., Jensen, P. R., & Duus, J. O. (2012). Direct observation of metabolic differences in living Escherichia coli strains K-12 and BL21. *Chembiochem*, *13*(2), 308–310. http://dx.doi.org/10.1002/cbic.201100654.

Meier, S., Karlsson, M., Jensen, P. R., Lerche, M. H., & Duus, J. O. (2011). Metabolic pathway visualization in living yeast by DNP-NMR. *Molecular BioSystems*, 7(10), 2834–2836. http://dx.doi.org/10.1039/c1mb05202k.

Merritt, M. E., Harrison, C., Sherry, A. D., Malloy, C. R., & Burgess, S. C. (2011). Flux through hepatic pyruvate carboxylase and phosphoenolpyruvate carboxykinase detected by hyperpolarized 13C magnetic resonance. *Proceedings of the National Academy of Sciences of the United States of America*, *108*(47), 19084–19089. http://dx.doi.org/10.1073/pnas.1111247108.

Merritt, M. E., Harrison, C., Storey, C., Jeffrey, F. M., Sherry, A. D., & Malloy, C. R. (2007). Hyperpolarized 13C allows a direct measure of flux through a single enzyme-catalyzed step by NMR. *Proceedings of the National Academy of Sciences of the United States of America*, *104*(50), 19773–19777. http://dx.doi.org/10.1073/pnas.0706235104.

Merritt, M. E., Harrison, C., Storey, C., Sherry, A. D., & Malloy, C. R. (2008). Inhibition of carbohydrate oxidation during the first minute of reperfusion after brief ischemia: NMR detection of hyperpolarized 13CO2 and H13CO3. *Magnetic Resonance in Medicine*, *60*(5), 1029–1036. http://dx.doi.org/10.1002/mrm.21760.

Miéville, P., Ahuja, P., Sarkar, R., Jannin, Sami, Vasos, P. R., Gerber-Lemaire, S., et al. (2010). Scavenging free radicals to preserve enhancement and extend relaxation times in NMR using dynamic nuclear polarization. *Angewandte Chemie (International Ed. in English)*, *49*(35), 6182–6185. http://dx.doi.org/10.1002/anie.201000934.

Mishkovsky, M., Comment, A., & Gruetter, R. (2012). In vivo detection of brain Krebs cycle intermediate by hyperpolarized magnetic resonance. *Journal of Cerebral Blood Flow and Metabolism*, *32*(12), 2108–2113. http://dx.doi.org/10.1038/jcbfm.2012.136.

Montanari, F., Quici, S., Henry-Riyad, H., Tidwell, T., Studer, A., & Vogler, T. (2007). 2,2,6,6-Tetramethylpiperidin-1-oxyl. In e-EROS (Ed.), *Encyclopedia of reagents for organic synthesis* (pp. 1–9). Hoboken, New Jersey: John Wiley and Son, Ltd. http://dx.doi.org/10.1002/047084289X.rt069.pub3.

Moreno, K. X., Sabelhaus, S. M., Merritt, M. E., Sherry, A. D., & Malloy, C. R. (2010). Competition of pyruvate with physiological substrates for oxidation by the heart: Implications for studies with hyperpolarized [1-13C]pyruvate. *American Journal of Physiology. Heart and Circulatory Physiology*, *298*(5), H1556–H1564. http://dx.doi.org/10.1152/ajpheart.00656.2009.

Moreno, K. X., Satapati, S., DeBerardinis, R. J., Burgess, S. C., Malloy, C. R., & Merritt, M. E. (2014). Real-time detection of hepatic gluconeogenic and glycogenolytic states using hyperpolarized [2-13C]dihydroxyacetone. *The Journal of Biological Chemistry, 289*(52), 35859–35867. http://dx.doi.org/10.1074/jbc.M114.613265.

Nagashima, K. (2008). Optimum pulse flip angles for multi-scan acquisition of hyperpolarized NMR and MRI. *Journal of Magnetic Resonance, 190*(2), 183–188. http://dx.doi.org/10.1016/j.jmr.2007.10.011.

Naressi, A., Couturier, C., Castang, I., de Beer, R., & Graveron-Demilly, D. (2001). Java-based graphical user interface for MRUI, a software package for quantitation of in vivo/medical magnetic resonance spectroscopy signals. *Computers in Biology and Medicine, 31*(4), 269–286.

Neeman, M., Rushkin, E., Kadouri, A., & Degani, H. (1988). Adaptation of culture methods for NMR studies of anchorage-dependent cells. *Magnetic Resonance in Medicine*, 7(2), 236–242.

Nelson, S. J., Kurhanewicz, J., Vigneron, D. B., Larson, P. E., Harzstark, A. L., Ferrone, M., et al. (2013). Metabolic imaging of patients with prostate cancer using hyperpolarized [1-(1)(3)C]pyruvate. *Science Translational Medicine, 5*(198), 198ra108. http://dx.doi.org/10.1126/scitranslmed.3006070.

Nissan, N., Golan, T., Furman-Haran, E., Apter, S., Inbar, Y., Ariche, A., et al. (2014). Diffusion tensor magnetic resonance imaging of the pancreas. *PLoS One, 9*(12), e115783. http://dx.doi.org/10.1371/journal.pone.0115783.

Overhauser, Albert. (1953). Polarization of nuclei in metals. *Physical Review, 92*(2), 411–415. http://dx.doi.org/10.1103/PhysRev.92.411.

Pages, G., Puckeridge, M., Liangfeng, G., Tan, Y. L., Jacob, C., Garland, M., et al. (2013). Transmembrane exchange of hyperpolarized 13C-urea in human erythrocytes: Subminute timescale kinetic analysis. *Biophysical Journal, 105*(9), 1956–1966. http://dx.doi.org/10.1016/j.bpj.2013.09.034.

Pages, G., Tan, Y. L., & Kuchel, P. W. (2014). Hyperpolarized [1, (13)C]pyruvate in lysed human erythrocytes: Effects of co-substrate supply on reaction time courses. *NMR in Biomedicine, 27*(10), 1203–1210. http://dx.doi.org/10.1002/nbm.3176.

Paniagua, J. C., Mugnaini, V., Gabellieri, C., Feliz, M., Roques, N., Veciana, J., et al. (2010). Polychlorinated trityl radicals for dynamic nuclear polarization: The role of chlorine nuclei. *Physical Chemistry Chemical Physics, 12*(22), 5824–5829. http://dx.doi.org/10.1039/c003291n.

Pardridge, W. M. (2012). Drug transport across the blood-brain barrier. *Journal of Cerebral Blood Flow and Metabolism, 32*(11), 1959–1972. http://dx.doi.org/10.1038/jcbfm.2012.126.

Park, J. M., Josan, S., Grafendorfer, T., Yen, Y. F., Hurd, R. E., Spielman, D. M., et al. (2013). Measuring mitochondrial metabolism in rat brain in vivo using MR spectroscopy of hyperpolarized [2-(1)(3)C]pyruvate. *NMR in Biomedicine, 26*(10), 1197–1203. http://dx.doi.org/10.1002/nbm.2935.

Park, I., Larson, P. E., Tropp, J. L., Carvajal, L., Reed, G., Bok, R., et al. (2014). Dynamic hyperpolarized carbon-13 MR metabolic imaging of nonhuman primate brain. *Magnetic Resonance in Medicine, 71*(1), 19–25. http://dx.doi.org/10.1002/mrm.25003.

Park, I., Larson, P. E., Zierhut, M. L., Hu, S., Bok, R., Ozawa, T., et al. (2010). Hyperpolarized 13C magnetic resonance metabolic imaging: Application to brain tumors. *Neuro-Oncology, 12*(2), 133–144. http://dx.doi.org/10.1093/neuonc/nop043.

Park, I., Mukherjee, J., Ito, M., Chaumeil, M. M., Jalbert, L. E., Gaensler, K., et al. (2014). Changes in pyruvate metabolism detected by magnetic resonance imaging are linked to DNA damage and serve as a sensor of temozolomide response in glioblastoma cells. *Cancer Research, 74*(23), 7115–7124. http://dx.doi.org/10.1158/0008-5472.CAN-14-0849.

Patrick, P. S., Kettunen, M. I., Tee, S. S., Rodrigues, T. B., Serrao, E., Timm, K. N., et al. (2015). Detection of transgene expression using hyperpolarized C urea and diffusion-weighted magnetic resonance spectroscopy. *Magnetic Resonance in Medicine, 73*(4), 1401–1406. http://dx.doi.org/10.1002/mrm.25254.

Perman, W. H., Bhattacharya, P., Leupold, J., Lin, A. P., Harris, K. C., Norton, V. A., et al. (2010). Fast volumetric spatial-spectral MR imaging of hyperpolarized 13C-labeled compounds using multiple echo 3D bSSFP. *Magnetic Resonance Imaging, 28*(4), 459–465. http://dx.doi.org/10.1016/j.mri.2009.12.003.

Pileio, G., Carravetta, M., & Levitt, M. H. (2010). Storage of nuclear magnetization as long-lived singlet order in low magnetic field. *Proceedings of the National Academy of Sciences of the United States of America, 107*(40), 17135–17139. http://dx.doi.org/10.1073/pnas.1010570107.

Posse, S., Otazo, R., Dager, S. R., & Alger, J. (2013). MR spectroscopic imaging: Principles and recent advances. *Journal of Magnetic Resonance Imaging, 37*(6), 1301–1325. http://dx.doi.org/10.1002/jmri.23945.

Qu, W., Zha, Z., Lieberman, B. P., Mancuso, A., Stetz, M., Rizzi, R., et al. (2011). Facile synthesis [5-(13)C-4-(2)H(2)]-L-glutamine for hyperpolarized MRS imaging of cancer cell metabolism. *Academic Radiology, 18*(8), 932–939. http://dx.doi.org/10.1016/j.acra.2011.05.002.

Reddy, T. J., Iwama, T., Halpern, H. J., & Rawal, V. H. (2002). General synthesis of persistent trityl radicals for EPR imaging of biological systems. *The Journal of Organic Chemistry, 67*(14), 4635–4639.

Reed, G. D., von Morze, C., Bok, R., Koelsch, B. L., Van Criekinge, M., Smith, K. J., et al. (2014). High resolution (13)C MRI with hyperpolarized urea: In vivo T(2) mapping and (15)N labeling effects. *IEEE Transactions on Medical Imaging, 33*(2), 362–371. http://dx.doi.org/10.1109/TMI.2013.2285120.

Rodrigues, T. B., Serrao, E. M., Kennedy, B. W., Hu, D. E., Kettunen, M. I., & Brindle, K. M. (2014). Magnetic resonance imaging of tumor glycolysis using hyperpolarized 13C-labeled glucose. *Nature Medicine, 20*(1), 93–97. http://dx.doi.org/10.1038/nm.3416.

Ronen, S. M., Rushkin, E., & Degani, H. (1991). Lipid metabolism in T47D human breast cancer cells: 31P and 13C-NMR studies of choline and ethanolamine uptake. *Biochimica et Biophysica Acta, 1095*(1), 5–16.

Sadler, K. E., Stratton, J. M., DeBerry, J. J., & Kolber, B. J. (2013). Optimization of a pain model: Effects of body temperature and anesthesia on bladder nociception in mice. *PLoS One, 8*(11), e79617. http://dx.doi.org/10.1371/journal.pone.0079617.

Schilling, F., Duwel, S., Kollisch, U., Durst, M., Schulte, R. F., Glaser, S. J., et al. (2013). Diffusion of hyperpolarized (13) C-metabolites in tumor cell spheroids using real-time NMR spectroscopy. *NMR in Biomedicine, 26*(5), 557–568. http://dx.doi.org/10.1002/nbm.2892.

Schmidt, R., Laustsen, C., Dumez, J. N., Kettunen, M. I., Serrao, E. M., Marco-Rius, I., et al. (2014). In vivo single-shot 13C spectroscopic imaging of hyperpolarized metabolites by spatiotemporal encoding. *Journal of Magnetic Resonance, 240*, 8–15. http://dx.doi.org/10.1016/j.jmr.2013.12.013.

Scholz, D. J., Janich, M. A., Kollisch, U., Schulte, R. F., Ardenkjaer-Larsen, J. H., Frank, A., et al. (2015). Quantified pH imaging with hyperpolarized C-bicarbonate. *Magnetic Resonance in Medicine, 73*(6), 2274–2282. http://dx.doi.org/10.1002/mrm.25357.

Schroeder, M. A., Atherton, H. J., Ball, D. R., Cole, M. A., Heather, L. C., Griffin, J. L., et al. (2009). Real-time assessment of Krebs cycle metabolism using hyperpolarized 13C magnetic resonance spectroscopy. *The FASEB Journal, 23*(8), 2529–2538. http://dx.doi.org/10.1096/fj.09-129171.

Schroeder, M. A., Atherton, H. J., Dodd, M. S., Lee, P., Cochlin, L. E., Radda, G. K., et al. (2012). The cycling of acetyl-coenzyme A through acetylcarnitine buffers cardiac substrate supply: A hyperpolarized 13C magnetic resonance study. *Circulation. Cardiovascular Imaging, 5*(2), 201–209. http://dx.doi.org/10.1161/CIRCIMAGING.111.969451.

Schroeder, M. A., Cochlin, L. E., Heather, L. C., Clarke, K., Radda, G. K., & Tyler, D. J. (2008). In vivo assessment of pyruvate dehydrogenase flux in the heart using hyperpolarized carbon-13 magnetic resonance. *Proceedings of the National Academy of Sciences of the United States of America*, *105*(33), 12051–12056. http://dx.doi.org/10.1073/pnas.0805953105.

Schroeder, M. A., Lau, A. Z., Chen, A. P., Gu, Y., Nagendran, J., Barry, J., et al. (2013). Hyperpolarized (13)C magnetic resonance reveals early- and late-onset changes to in vivo pyruvate metabolism in the failing heart. *European Journal of Heart Failure*, *15*(2), 130–140. http://dx.doi.org/10.1093/eurjhf/hfs192.

Shang, H., Skloss, T., von Morze, C., Carvajal, L., Van Criekinge, M., Milshteyn, E., et al. (2015 Mar 11). Handheld electromagnet carrier for transfer of hyperpolarized carbon-13 samples. *Magnetic Resonance in Medicine*. http://dx.doi.org/10.1002/mrm.25657. [Epub ahead of print].

Shimon, D., Hovav, Y., Feintuch, A., Goldfarb, D., & Vega, S. (2012). Dynamic nuclear polarization in the solid state: A transition between the cross effect and the solid effect. *Physical Chemistry Chemical Physics*, *14*(16), 5729–5743. http://dx.doi.org/10.1039/c2cp23915a.

Sogaard, L. V., Schilling, F., Janich, M. A., Menzel, M. I., & Ardenkjaer-Larsen, J. H. (2014). In vivo measurement of apparent diffusion coefficients of hyperpolarized (1)(3)C-labeled metabolites. *NMR in Biomedicine*, *27*(5), 561–569. http://dx.doi.org/10.1002/nbm.3093.

Sugimoto, M., Takahashi, S., Kobayashi, T., Kojima, M., Gotohda, N., Satake, M., et al. (2015). Pancreatic perfusion data and post-pancreaticoduodenectomy outcomes. *The Journal of Surgical Research*, *194*(2), 441–449. http://dx.doi.org/10.1016/j.jss.2014.11.046.

Svensson, J., Mansson, S., Johansson, E., Petersson, J. S., & Olsson, L. E. (2003). Hyperpolarized 13C MR angiography using trueFISP. *Magnetic Resonance in Medicine*, *50*(2), 256–262. http://dx.doi.org/10.1002/mrm.10530.

Swisher, C. L., Larson, P. E., Kruttwig, K., Kerr, A. B., Hu, S., Bok, R. A., et al. (2014). Quantitative measurement of cancer metabolism using stimulated echo hyperpolarized carbon-13 MRS. *Magnetic Resonance in Medicine*, *71*(1), 1–11. http://dx.doi.org/10.1002/mrm.24634.

Tayler, M. C., Marco-Rius, I., Kettunen, M. I., Brindle, K. M., Levitt, M. H., & Pileio, G. (2012). Direct enhancement of nuclear singlet order by dynamic nuclear polarization. *Journal of the American Chemical Society*, *134*(18), 7668–7671. http://dx.doi.org/10.1021/ja302814e.

Thind, K., Chen, A., Friesen-Waldner, L., Ouriadov, A., Scholl, T. J., Fox, M., et al. (2013). Detection of radiation-induced lung injury using hyperpolarized (13) C magnetic resonance spectroscopy and imaging. *Magnetic Resonance in Medicine*, *70*(3), 601–609. http://dx.doi.org/10.1002/mrm.24525.

Thind, K., Jensen, M. D., Hegarty, E., Chen, A. P., Lim, H., Martinez-Santiesteban, F., et al. (2014). Mapping metabolic changes associated with early radiation induced lung injury post conformal radiotherapy using hyperpolarized (1)(3)C-pyruvate magnetic resonance spectroscopic imaging. *Radiotherapy and Oncology*, *110*(2), 317–322. http://dx.doi.org/10.1016/j.radonc.2013.11.016.

Timm, K. N., Hartl, J., Keller, M. A., Hu, D., Kettunen, M. I., Rodrigues, T. B., et al. (2014). Hyperpolarized [U-2 H, U-13 C]glucose reports on glycolytic and pentose phosphate pathway activity in EL4 tumors and glycolytic activity in yeast cells. *Magnetic Resonance in Medicine*. http://dx.doi.org/10.1002/mrm.25561.

Tyler, D. J., Schroeder, M. A., Cochlin, L. E., Clarke, K., & Radda, G. K. (2008). Application of hyperpolarized magnetic resonance in the study of cardiac metabolism. *Applied Magnetic Resonance*, *34*, 523–531.

Valette, J., Chaumeil, M., Guillermier, M., Bloch, G., Hantraye, P., & Lebon, V. (2008). Diffusion-weighted NMR spectroscopy allows probing of 13C labeling of glutamate inside distinct metabolic compartments in the brain. *Magnetic Resonance in Medicine, 60*(2), 306–311. http://dx.doi.org/10.1002/mrm.21661.

Vander Heiden, M. G., Cantley, L. C., & Thompson, C. B. (2009). Understanding the Warburg effect: The metabolic requirements of cell proliferation. *Science, 324*(5930), 1029–1033. http://dx.doi.org/10.1126/science.1160809.

Vasos, P. R., Comment, A., Sarkar, R., Ahuja, P., Jannin, S., Ansermet, J. P., et al. (2009). Long-lived states to sustain hyperpolarized magnetization. *Proceedings of the National Academy of Sciences of the United States of America, 106*(44), 18469–18473. http://dx.doi.org/10.1073/pnas.0908123106.

Venkatesh, H. S., Chaumeil, M. M., Ward, C. S., Haas-Kogan, D. A., James, C. D., & Ronen, S. M. (2012). Reduced phosphocholine and hyperpolarized lactate provide magnetic resonance biomarkers of PI3K/Akt/mTOR inhibition in glioblastoma. *Neuro-Oncology, 14*(3), 315–325. http://dx.doi.org/10.1093/neuonc/nor209.

von Morze, C., Bok, R. A., Reed, G. D., Ardenkjaer-Larsen, J. H., Kurhanewicz, J., & Vigneron, D. B. (2014a). Simultaneous multiagent hyperpolarized (13)C perfusion imaging. *Magnetic Resonance in Medicine, 72*(6), 1599–1609. http://dx.doi.org/10.1002/mrm.25071.

von Morze, C., Bok, R. A., Reed, G. D., Ardenkjaer-Larsen, J. H., Kurhanewicz, J., & Vigneron, D. B. (2014b). Simultaneous multiagent hyperpolarized13C perfusion imaging. *Magnetic Resonance in Medicine, 72*(6), 1599–1609. http://dx.doi.org/10.1002/mrm.25071.

von Morze, C., Bok, R. A., Sands, J. M., Kurhanewicz, J., & Vigneron, D. B. (2012). Monitoring urea transport in rat kidney in vivo using hyperpolarized ^{13}C magnetic resonance imaging. *American Journal of Physiology. Renal Physiology, 302*(12), F1658–F1662. http://dx.doi.org/10.1152/ajprenal.00640.2011.

von Morze, C., Larson, P. E., Hu, S., Keshari, K., Wilson, D. M., Ardenkjaer-Larsen, J. H., et al. (2011). Imaging of blood flow using hyperpolarized [(13)C]urea in preclinical cancer models. *Journal of Magnetic Resonance Imaging, 33*(3), 692–697. http://dx.doi.org/10.1002/jmri.22484.

von Morze, C., Larson, P. E., Hu, S., Yoshihara, H. A., Bok, R. A., Goga, A., et al. (2012). Investigating tumor perfusion and metabolism using multiple hyperpolarized (13)C compounds: HP001, pyruvate and urea. *Magnetic Resonance Imaging, 30*(3), 305–311. http://dx.doi.org/10.1016/j.mri.2011.09.026.

von Morze, C., Larson, P. E., Shang, H., & Vigneron, D. B. (2014). Suppression of unwanted resonances in hyperpolarized MR studies with neat [1-13C]lactic acid. *Proceedings on International Society for Magnetic Resonance in Medicine, 22*, 2792.

von Morze, C., Sukumar, S., Reed, G. D., Larson, P. E., Bok, R. A., Kurhanewicz, J., et al. (2013). Frequency-specific SSFP for hyperpolarized ^{13}C metabolic imaging at 14.1 T. *Magnetic Resonance Imaging, 31*(2), 163–170. http://dx.doi.org/10.1016/j.mri.2012.06.037.

Warburg, O. (1956a). On respiratory impairment in cancer cells. *Science, 124*(3215), 269–270.

Warburg, O. (1956b). On the origin of cancer cells. *Science, 123*(3191), 309–314.

Ward, C. S., Venkatesh, H. S., Chaumeil, M. M., Brandes, A. H., Vancriekinge, M., Dafni, H., et al. (2010). Noninvasive detection of target modulation following phosphatidylinositol 3-kinase inhibition using hyperpolarized 13C magnetic resonance spectroscopy. *Cancer Research, 70*(4), 1296–1305. http://dx.doi.org/10.1158/0008-5472.CAN-09-2251.

Warren, W. S., Jenista, E., Branca, R. T., & Chen, X. (2009). Increasing hyperpolarized spin lifetimes through true singlet eigenstates. *Science, 323*(5922), 1711–1714. http://dx.doi.org/10.1126/science.1167693.

Wiesinger, F., Weidl, E., Menzel, M. I., Janich, M. A., Khegai, O., Glaser, S. J., et al. (2012). IDEAL spiral CSI for dynamic metabolic MR imaging of hyperpolarized [1-13C]pyruvate. *Magnetic Resonance in Medicine, 68*(1), 8–16. http://dx.doi.org/10.1002/mrm.23212.

Wijnen, J. P., Van der Graaf, M., Scheenen, T. W., Klomp, D. W., de Galan, B. E., Idema, A. J., et al. (2010). In vivo 13C magnetic resonance spectroscopy of a human brain tumor after application of 13C-1-enriched glucose. *Magnetic Resonance Imaging, 28*(5), 690–697. http://dx.doi.org/10.1016/j.mri.2010.03.006.

Wilson, D. M., Hurd, R. E., Keshari, K., Van Criekinge, M., Chen, A. P., Nelson, S. J., et al. (2009). Generation of hyperpolarized substrates by secondary labeling with [1,1-13C] acetic anhydride. *Proceedings of the National Academy of Sciences of the United States of America, 106*(14), 5503–5507. http://dx.doi.org/10.1073/pnas.0810190106.

Wilson, D. M., Keshari, K. R., Larson, P. E., Chen, A. P., Hu, S., Van Criekinge, M., et al. (2010). Multi-compound polarization by DNP allows simultaneous assessment of multiple enzymatic activities in vivo. *Journal of Magnetic Resonance, 205*(1), 141–147. http://dx.doi.org/10.1016/j.jmr.2010.04.012.

Wise, D. R., & Thompson, C. B. (2010). Glutamine addiction: A new therapeutic target in cancer. *Trends in Biochemical Sciences, 35*(8), 427–433. http://dx.doi.org/10.1016/j.tibs.2010.05.003.

Witney, T. H., Kettunen, M. I., & Brindle, K. M. (2011). Kinetic modeling of hyperpolarized 13C label exchange between pyruvate and lactate in tumor cells. *The Journal of Biological Chemistry, 286*(28), 24572–24580. http://dx.doi.org/10.1074/jbc.M111.237727.

Witney, T. H., Kettunen, M. I., Hu, D. E., Gallagher, F. A., Bohndiek, S. E., Napolitano, R., et al. (2010). Detecting treatment response in a model of human breast adenocarcinoma using hyperpolarised [1-13C]pyruvate and [1,4-13C2]fumarate. *British Journal of Cancer, 103*(9), 1400–1406. http://dx.doi.org/10.1038/sj.bjc.6605945.

Wolfe, S. P., Hsu, E., Reid, L. M., & Macdonald, J. M. (2002). A novel multi-coaxial hollow fiber bioreactor for adherent cell types. Part 1: hydrodynamic studies. *Biotechnology and Bioengineering*, 77(1), 83–90.

Xing, Y., Reed, G. D., Pauly, J. M., Kerr, A. B., & Larson, P. E. (2013). Optimal variable flip angle schemes for dynamic acquisition of exchanging hyperpolarized substrates. *Journal of Magnetic Resonance, 234*, 75–81. http://dx.doi.org/10.1016/j.jmr.2013.06.003.

Xu, T., Mayer, D., Gu, M., Yen, Y. F., Josan, S., Tropp, J., et al. (2011). Quantification of in vivo metabolic kinetics of hyperpolarized pyruvate in rat kidneys using dynamic 13C MRSI. *NMR in Biomedicine, 24*(8), 997–1005. http://dx.doi.org/10.1002/nbm.1719.

Yang, C., Harrison, C., Jin, E. S., Chuang, D. T., Sherry, A. D., Malloy, C. R., et al. (2014). Simultaneous steady-state and dynamic 13C NMR can differentiate alternative routes of pyruvate metabolism in living cancer cells. *The Journal of Biological Chemistry, 289*(9), 6212–6224. http://dx.doi.org/10.1074/jbc.M113.543637.

Yen, Y. F., Kohler, S. J., Chen, A. P., Tropp, J., Bok, R., Wolber, J., et al. (2009). Imaging considerations for in vivo 13C metabolic mapping using hyperpolarized 13C-pyruvate. *Magnetic Resonance in Medicine, 62*(1), 1–10. http://dx.doi.org/10.1002/mrm.21987.

Zeirhut, M. L., Ozturk, E., Chen, A. P., Pels, P., Cunningham, C. H., Vigneron, D. B., et al. (2006). Spectroscopic imaging of 1H at 3T: Comparing SNR between traditional phase encoding and echo-planar techniques in the human brain. In *Proceedings of the 14th annual meeting of ISMRM, Seattle* (p. 65).

Zhao, L., Mulkern, R., Tseng, C. H., Williamson, D., Patz, S., Kraft, R., et al. (1996). Gradient-echo imaging considerations for hyperpolarized 129Xe MR. *Journal of Magnetic Resonance. Series B, 113*(2), 179–183.

Zheng, B., Lee, P. T., & Golay, X. (2010). High-sensitivity cerebral perfusion mapping in mice by kbGRASE-FAIR at 9.4 T. *NMR in Biomedicine, 23*(9), 1061–1070. http://dx.doi.org/10.1002/nbm.1533.

Zierhut, M. L., Yen, Y. F., Chen, A. P., Bok, R., Albers, M. J., Zhang, V., et al. (2010). Kinetic modeling of hyperpolarized 13C1-pyruvate metabolism in normal rats and TRAMP mice. *Journal of Magnetic Resonance, 202*(1), 85–92. http://dx.doi.org/10.1016/j.jmr.2009.10.003.

CHAPTER TWO

Hyperpolarized ^{13}C Magnetic Resonance and Its Use in Metabolic Assessment of Cultured Cells and Perfused Organs

Lloyd Lumata*[,1], Chendong Yang[†], Mukundan Ragavan[‡], Nicholas Carpenter[‡], Ralph J. DeBerardinis[†,1], Matthew E. Merritt[‡,1]

*Department of Physics, University of Texas at Dallas, Richardson, Texas, USA

[†]Children's Medical Center Research Institute, University of Texas Southwestern Medical Center, Dallas, Texas, USA

[‡]Advanced Imaging Research Center, University of Texas Southwestern Medical Center, Dallas, Texas, USA

[1]Corresponding authors: e-mail address: lloyd.lumata@utdallas.edu; ralph.deberardinis@utsouthwestern.edu; matthew.merritt@utsouthwestern.edu

Contents

Methods in Enzymology, Volume 561
ISSN 0076-6879
http://dx.doi.org/10.1016/bs.mie.2015.04.006

Abstract

Diseased tissue is often characterized by abnormalities in intermediary metabolism. Observing these alterations *in situ* may lead to an improved understanding of pathological processes and novel ways to monitor these processes noninvasively in human patients. Although ^{13}C is a stable isotope safe for use in animal models of disease as well as human subjects, its utility as a metabolic tracer has largely been limited to *ex vivo* analyses employing analytical techniques like mass spectrometry or nuclear magnetic resonance spectroscopy. Neither of these techniques is suitable for noninvasive metabolic monitoring, and the low abundance and poor gyromagnetic ratio of conventional ^{13}C make it a poor nucleus for imaging. However, the recent advent of hyperpolarization methods, particularly dynamic nuclear polarization (DNP), makes it possible to enhance the spin polarization state of ^{13}C by many orders of magnitude, resulting in a temporary amplification of the signal sufficient for monitoring kinetics of enzyme-catalyzed reactions in living tissue through magnetic resonance spectroscopy or magnetic resonance imaging. Here, we review DNP techniques to monitor metabolism in cultured cells, perfused hearts, and perfused livers, focusing on our experiences with hyperpolarized [1-^{13}C]pyruvate. We present detailed approaches to optimize the DNP procedure, streamline biological sample preparation, and maximize detection of specific metabolic activities. We also discuss practical aspects in the choice of metabolic substrates for hyperpolarization studies and outline some of the current technical and conceptual challenges in the field, including efforts to use hyperpolarization to quantify metabolic rates *in vivo*.

1. INTRODUCTION: IMPORTANCE OF DEVELOPING METHODS TO OBSERVE METABOLIC FLUX IN DISEASE STATES

Metabolism is at the root of essentially all physiological processes (DeBerardinis & Thompson, 2012). The production and expenditure of energy, storage and breakdown of macromolecules, disposal of waste, and many other processes are subserved by thousands of enzymatic reactions at work in human cells. Recent work has produced insights that greatly extend the influence of metabolism and metabolites to include such seemingly disparate processes as signal transduction, posttranslational modification of proteins, and epigenetic effects on gene expression (Choudhary, Weinert, Nishida, Verdin, & Mann, 2014; Kaelin & McKnight, 2013; Ward & Thompson, 2012). These observations further emphasize the principle that metabolism is inexorably intertwined with cellular function and tissue homeostasis. In short, normal tissue function cannot occur unless metabolism is properly regulated.

For most of the past century, metabolism research has been dominated by studies in organs like the liver, one of whose major functions is to maintain metabolic homeostasis for the entire body, and the skeletal muscle, heart, and brain, whose normal function involves energetically demanding processes. However, many other tissues are equally dependent on a broad complement of metabolic activities. As an example, an emerging theme in cell biology research is the importance of acute metabolic changes for enabling physiological cell growth and proliferation (Metallo & Vander Heiden, 2013; Plas & Thompson, 2005). Because growth and proliferation are viewed as metabolically demanding in terms of the need for energy and macromolecular synthesis, there has been interest in understanding how the signals that stimulate punctuated bursts of cell proliferation engage the metabolic network to satisfy these demands. T-cell activation is one of the many processes now known to require several specific metabolic activities to support cell growth and proliferation (Gerriets et al., 2015).

Given the many links between metabolism and cellular function, it is unsurprising that most diseases feature abnormal metabolism at the cellular level (DeBerardinis & Thompson, 2012). Of the hundreds of monogenic diseases caused by mutations in single enzymes (the so-called inborn errors of metabolism), a high fraction affects the liver, heart, muscle, and brain. Many other monogenic metabolic diseases involve poorly defined effects on growth, either because of systemic metabolic imbalances or perhaps effects on specific populations of cells whose function is required for normal growth. Importantly, common diseases also involve altered metabolism. Among the most common causes of death in the United States, heart disease, cancer, stroke, and diabetes can all be viewed as involving altered metabolism at the level of the cell and/or organ. Thus, we need better tools to understand metabolic regulation in diseased tissues. Preferably, some of these tools will support metabolic analysis in intact tissue.

Metabolic dysregulation is a prominent and clinically relevant feature of cancer biology. As early as the 1920s, Warburg demonstrated that malignant cells have a propensity to take up excess amounts of glucose and convert it into lactate, even when oxygen availability was sufficient to oxidize glucose completely to carbon dioxide (Warburg, 1956b). The successful use of ^{18}fluoro-2-deoxyglucose (FDG) as a radiotracer for positron emission tomography (PET) studies in cancer patients has validated the clinical importance of glucose uptake in human tumors (Gallamini, Zwarthoed, & Borra, 2014). FDG-PET is commonly used to image the distribution of malignant tissue and to monitor the effects of therapy. Other imaging techniques have also been employed to observe altered metabolic states in tumors. Proton magnetic resonance spectroscopy (^{1}H MRS) enables the detection and quantitation of abundant metabolites, some of which have prognostic or diagnostic value. To date, this technique has been used most extensively in brain tumors and other central nervous system diseases (Oz et al., 2014). An example of a recent technical development is the use of ^{1}H MRS to detect 2-hydroxyglutarate, an "oncometabolite" produced by mutant forms of the metabolic enzymes isocitrate dehydrogenase-1 and -2 in brain tumors (Andronesi et al., 2012; Choi et al., 2012; Elkhaled et al., 2012; Pope et al., 2012). After a considerable amount of research over the past decade, metabolic reprogramming as a consequence of tumorigenic mutations in oncogenes or tumor suppressor genes is now considered to be one of the major biological hallmarks of cancer (Hanahan & Weinberg, 2011).

Stable isotope tracers like ^{13}C are widely used to investigate metabolism in living systems. Transfer of ^{13}C from a parent molecule (e.g., glucose) into downstream metabolites reports the activity of metabolic pathways, providing an important complement to measurements of steady-state metabolite levels. Because ^{13}C is a nonradioactive tracer that is effectively monitored using analytical techniques like nuclear magnetic resonance or mass spectrometry, administering ^{13}C to animals or human subjects is safe. A typical experiment involves the administration of one or more isotope-labeled nutrients to the subject, followed by periodic or endpoint acquisition of tissues of interest (e.g., blood, urine, tumor), extraction of informative metabolites, and analysis of isotope enrichment patterns to infer metabolic pathway activity. This type of approach has been used successfully in mice and humans to probe metabolic alterations in cancer and other metabolic disorders (Busch, Neese, Awada, Hayes, & Hellerstein, 2007; Maher et al., 2012; Marin-Valencia et al., 2012; Sunny, Parks, Browning, & Burgess, 2011; Ying et al., 2012; Yuneva et al., 2012).

Each of these methods has only a limited ability to report on the metabolic network. Stable isotope approaches are invasive and destructive; that is, they require tissue sampling and metabolite extraction in order to obtain information about metabolism. In addition, metabolic activity is inferred from ^{13}C enrichment rather than by observing metabolism directly through an imaging technique. On the other hand, while PET supports direct, noninvasive detection of a labeled metabolic probe, the resulting information is limited to anatomic localization of probe uptake and accumulation, with essentially no information about downstream metabolic activities. ^{1}H MRS quantifies metabolite pools but provides no information about flux and is somewhat limited by the narrow chemical shift range of the ^{1}H spectrum.

Methods to image ^{13}C would be powerful tools to probe disease-associated metabolic changes because they would enable direct visualization of metabolic activity and in principle could be repeated multiple times in the same subject. The wide chemical shift range of the ^{13}C NMR spectrum means that a very large number of individual carbon positions from intermediary metabolites could be monitored simultaneously. The major challenges associated with carbon imaging involve the low *in vivo* abundance and poor gyromagnetic ratio of the ^{13}C nucleus; these factors, combined with the low level of spin polarization, result in a very small signal detectable by magnetic resonance. However, using a process termed hyperpolarization, the spin polarization can be greatly enhanced, thereby enhancing the magnetic resonance signal (Ardenkjaer-Larsen et al., 2003; Kurhanewicz et al., 2011). Hyperpolarization techniques temporarily redistribute the population of energy levels of ^{13}C nuclei into a nonequilibrium state. This state is highly unstable, with a T_1 varying with the chemical environment of the nucleus but typically lasting less than 60 s. However, the signal gain during this period of time can routinely exceed 10,000-fold, leading to a massive if transient increase in MR signal (Ardenkjaer-Larsen et al., 2003). This gain provides sufficient signal to introduce hyperpolarized materials into a biological system, observe the labeled nucleus, and—if metabolism of the substrate is rapid enough—observe real-time transfer of the hyperpolarized nucleus to new molecules through enzyme-catalyzed metabolic reactions.

A large and growing number of metabolites from central carbon metabolism have been hyperpolarized for real-time metabolic studies (Gallagher, Kettunen, Day, Hu, et al., 2008; Gallagher, Kettunen, Day, Lerche, & Brindle, 2008; Gallagher et al., 2009; Golman, Zandt, Lerche, Pehrson, & Ardenkjaer-Larsen, 2006; Moreno et al., 2014; Rodrigues et al., 2014). However, [1-^{13}C]pyruvate has received the most attention to date. Carbon-1 of pyruvate has an unusually long T_1 (approximately 45 s), providing ample opportunity to observe the transfer of this carbon to metabolites of pyruvate. Furthermore, pyruvate is positioned at the interface between anaerobic and aerobic metabolism, a key metabolic node. Pyruvate can be reduced to lactate, transaminated to alanine, carboxylated to oxaloacetate, or decarboxylated to acetyl-CoA, and in the right tissue, all of these reactions can occur rapidly enough to observe by NMR (Fig. 1).

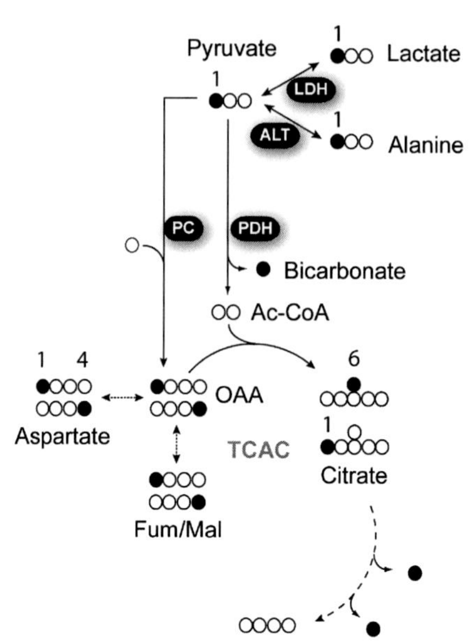

Figure 1 Pyruvate is a substrate for several reactions that may be monitored by hyperpolarized ^{13}C. The strengths of [1-^{13}C]pyruvate as a probe for hyperpolarization experiments are demonstrated in this illustration. The ^{13}C nuclei are in black, with nonenriched carbons in white. Pyruvate exchanges rapidly with the large lactate and alanine pools via the enzymes lactate dehydrogenase (LDH) and alanine aminotransferase (ALT), respectively. These exchange reactions result in the production of [1-^{13}C]lactate and [1-^{13}C]alanine, both of which are readily visible in hyperpolarization experiments in many systems. Depending on the tissue type, it may also be possible to observe flux through pyruvate dehydrogenase (PDH), which releases ^{13}C as bicarbonate and produces unlabeled acetyl-CoA (Ac-CoA). Ac-CoA may then enter the TCA cycle (TCAC). Pyruvate may also be carboxylated by pyruvate carboxylase (PC), generating labeled oxaloacetate (OAA). OAA may be transaminated to aspartate, condensed with Ac-CoA to produce citrate, or equilibrate with other 4-carbon TCAC intermediates (e.g., fumarate and malate, Fum/Mal). Thus, carbon-1 of pyruvate is positioned to probe multiple aspects of aerobic and anaerobic metabolism simultaneously.

Hyperpolarized [1-^{13}C]pyruvate has been used extensively in mouse models of cancer because the rapid interconversion of pyruvate and lactate is readily detected by NMR and has been demonstrated to differentiate benign from malignant tissue (Golman et al., 2006; Hu et al., 2011). Hyperpolarized [1-^{13}C]pyruvate has also been used in human studies to detect malignant tissue in the prostate, where its integration with magnetic resonance imaging has great potential as a clinical tool to diagnose and monitor cancer in patients (Nelson et al., 2013).

In addition to its use for *in vivo* studies, hyperpolarized [1-^{13}C]pyruvate has been used in a number of other applications, including cell culture studies and perfused organs (Harrison et al., 2012; Merritt, Harrison, Sherry, Malloy, & Burgess, 2011; Merritt et al., 2007; Moreno et al., 2014; Yang, Harrison, et al., 2014; Yang, Ko, et al., 2014). These studies capitalize on the extremely fine temporal resolution afforded by hyperpolarization (up to one data point per second) to directly observe, and in some cases quantify, flux through one or more metabolic reactions. Here, we discuss our experiences using hyperpolarization as an analytical approach to understand metabolism in *ex vivo* systems. We describe methods to achieve high levels of polarization in [1-^{13}C]pyruvate using dynamic nuclear polarization (DNP), maximize the efficiency and information yield of cell culture experiments, and analyze metabolism in the perfused heart and liver. We also discuss future directions and challenges for hyperpolarization, including roles for substrates other than pyruvate, and prospects for deriving quantitative flux information from *in vivo* experiments.

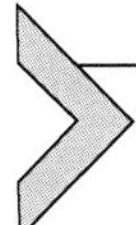

2. HYPERPOLARIZATION METHODS AND SAMPLE PREPARATION

2.1 Hyperpolarization via Dissolution DNP

Among different techniques of resolving the problem of low sensitivity of NMR signals, hyperpolarization via dissolution DNP is regarded as one of the most versatile and effective for increasing the amplitude of NMR signals (Ardenkjaer-Larsen et al., 2003). The underlying principle of DNP is the transfer of high electron spin polarization to the nuclear spins via microwaves, hence creating a larger spin population difference in the Zeeman energy levels of the system, resulting in significant enhancements of NMR signal. Originally used in particle and nuclear physics, the microwave-driven polarization transfer process in nonconducting solids occurs at cryogenic temperatures close to 1 K and at magnetic fields greater than 1 T (Abragam & Goldman, 1978). The NMR signal-amplifying power of DNP has recently been harnessed for biomedical and metabolic research via rapid dissolution, which converts the frozen polarized ^{13}C sample at cryogenic temperature into hyperpolarized ^{13}C liquid solution at physiologically tolerable temperatures (Ardenkjaer-Larsen et al., 2003). In this case, the lifetime of the hyperpolarized ^{13}C NMR signal is dictated by the ^{13}C spin–lattice relaxation time T_1.

2.2 ^{13}C DNP Sample Components and Preparation

Optimized DNP sample preparation is crucial to the success of the hyperpolarized ^{13}C NMR or MRI experiments because it can significantly affect the maximum ^{13}C NMR signal enhancement levels. Typically, a ^{13}C DNP sample is a solution with 10–100 μL volume consisting of:

(a) ^{13}C-*enriched substrates*. With the superb ^{13}C NMR signal sensitivity afforded by dissolution DNP, real-time assessment of cellular metabolism is now feasible. Some of the commonly used ^{13}C-enriched biomolecules for metabolic research via hyperpolarized ^{13}C NMR are ^{13}C glucose, ^{13}C pyruvate, ^{13}C acetate, ^{13}C glutamine, ^{13}C fumarate, ^{13}C ascorbic acid, and numerous others (Keshari & Wilson, 2014). Most of these ^{13}C-enriched compounds are in powder form and thus need to be dissolved in a glassing matrix prior to DNP. In special cases, neat liquids such as ^{13}C pyruvic acid are self-glassing and thus, one only needs to add and mix trace amounts of free radicals to complete the DNP sample preparation. As mentioned above, [1-^{13}C]pyruvate is the most commonly used agent, in part because the position of ^{13}C on the carbonyl is associated with a fairly long T_1 relaxation time. The long T_1 of this carbon is related to the fact that its nearest intramolecular contact with protons is with pyruvate's methyl group (carbon 3). Long ^{13}C T_1s are highly advantageous for DNP because they translate to a long lifetime of the hyperpolarized ^{13}C NMR signal and therefore provide greater opportunity to observe metabolic activities. The intrinsic biochemistry of intermediary metabolites drastically limits the number of compounds suitable for hyperpolarization studies because the presence of protons directly bonded to the carbon nucleus reduces the T_1 to 2–10 s. A list of some of the commonly used ^{13}C-enriched compounds and their corresponding liquid-state T_1 relaxation times is given in Table 1.

(b) *Glassing solvents*. The choice of glassing solvents or matrix depends mainly on the criterion of maximum solubility of the ^{13}C substrates. Equally important is the requirement that these solutions form amorphous or noncrystalline solids when subjected to cryogenic temperatures (Kurdzesau et al., 2008; Lumata, Jindal, et al., 2011). Amorphous conditions are also required to evenly distribute the ^{13}C nuclear spins and the free radical electrons across the sample volumes. Common examples of glassing matrices used for ^{13}C DNP are composed of two solvents mixed in 1:1 vol/vol ratio such as glycerol/water,

Table 1 List of Some of the ^{13}C-Enriched Compounds Discussed in This Text and Their Corresponding Liquid-State ^{13}C Spin–Lattice T_1 Relaxation Times at 9.4 T and Ambient Temperature

^{13}C-Enriched Compound	^{13}C T_1 (s)	References
[1-^{13}C]pyruvate	42	Lumata, Ratnakar, et al. (2011)
[2-^{13}C]dihydroxyacetone	32	Moreno et al. (2014)
[U-^{13}C, U-^{2}H]glucose	10	Rodrigues et al. (2014)
[1,4-$^{13}C_2$]fumarate	24	Gallagher et al. (2009)
[1-^{13}C]glutamine	25	Jensen, Karlsson, Meier, Duus, and Lerche (2009)
[1-^{13}C]ascorbic acid	16	Bohndiek et al. (2011)
[1-^{13}C]alanine	29	Jensen et al. (2009)

DMSO/water, ethanol/water, methanol/water, sulfolane/DMSO, and ethyl acetate/DMSO among others. The volume components may be adjusted when higher solubility of a hydrophilic ^{13}C substrate is needed; for example, increasing the ratio of water:glycerol to 4:1 has been used to accommodate some substrates (Lumata, Jindal, et al., 2011). Concentrations of ^{13}C-enriched biomolecules in glassing matrices are typically on the order of a few moles per liter.

(c) *Free radicals.* The next step is the addition of trace amounts of free radicals to the ^{13}C substrate solutions. The following stable, organic free radicals were proven to be effective polarizing agents for dissolution DNP (Lumata, Merritt, Khemtong, et al., 2012; Lumata, Merritt, Malloy, et al., 2013; Lumata, Ratnakar, et al., 2011) as well as their experimentally determined optimal concentrations for DNP: tris{8-carboxyl-2,2,6,6-benzo(1,2-d:4,5-d)-bis(1,3)dithiole-4-yl} methyl sodium salt (trityl OX063, 15 m*M*); 1,3-bisdiphenylene-2-phenylallyl (BDPA, 20 or 40 m*M*); 2,6-di-*tert*-butyl-α-(3,5-di-*tert*-butyl-4-oxo-2,5-cyclohexadien-1-ylidene)-*p*-tolyloxy (galvinoxyl, 20 m*M*); 2,2-diphenyl-1-picrylhydrazyl (DPPH, 40 m*M*); and 4-Oxo-2,2,6,6-tetramethyl-1-piperidinyloxy (4-oxo-TEMPO, 30–50 m*M*). Of these five free radicals, the most commonly used are the water-soluble polarizing agents trityl OX063 and TEMPO, since most ^{13}C-enriched substrates used in DNP are hydrophilic. The other three free radicals—BDPA, galvinoxyl, and DPPH—can be dissolved in special solvents such as sulfolane/DMSO and sulfolane/ethyl acetate. Generally, the narrow electron spin resonance (ESR) linewidth free radicals trityl OX063 and BDPA were shown to be better polarizing agents for DNP of low-gamma nuclei such as ^{13}C spins (Lumata et al., 2015; Lumata, Merritt, Malloy, Sherry, & Kovacs, 2012a). This is due to the fact that their ESR linewidths match the ^{13}C Larmor frequencies, leading to more efficient hyperpolarization.

A typical sample preparation procedure is outlined below:

- Weigh out the desired amounts of ^{13}C substrate and free radical using analytical balance. Use 1.5-mL microcentrifuge tubes as containers.
- Prepare the glassing solvent solution. Dissolve the ^{13}C substrate in the glassing solvents using a vortex mixer.
- Transfer the ^{13}C substrate solution into the tube containing the free radical. Use the vortex mixer to dissolve the free radical in solution. In some cases, a sonicator bath may be used to completely dissolve the substrate and/or free radical (Lumata, Merritt, Hashami, Ratnakar, & Kovacs, 2012; Lumata, Ratnakar, et al., 2011).

2.3 Other DNP Sample Optimization Tips

2.3.1 Gadolinium Doping

Inclusion of trace amounts (1–2 m*M*) of Gd^{3+} complexes such as Gd-DOTA into the ^{13}C DNP sample was shown to increase the ^{13}C DNP-enhanced polarization level by as much as 300% in the commercial 3.35 T HyperSense polarizer (Lumata, Jindal, et al., 2011; Lumata, Merritt, Malloy, Sherry, & Kovacs, 2012b). However, this sample preparation practice is so far only recommended for ^{13}C DNP samples that are doped with trityl OX063 free radical. Inclusion of the Gd-complex in ^{13}C samples doped with the other free radicals led to negligible increases in ^{13}C polarization levels. In addition to Gd^{3+} complexes, other lanthanides such as Holmium-DOTA have shown similar beneficial effects on trityl OX063-doped ^{13}C samples. This DNP-enhancing phenomenon is ascribed to the shortening the T_1 relaxation time of the free radical electrons, leading to a more efficient DNP (Lumata, Kovacs, et al., 2013).

2.3.2 Deuteration of the Glassing Matrix: Do's and Don'ts

Using deuterated glassing solvents could be beneficial in some cases leading to large increases in hyperpolarized ^{13}C NMR signal. In other instances, however, this method may negatively impact the ^{13}C DNP–NMR signal enhancement (Lumata, Merritt, & Kovacs, 2013). For example:

- *Do's: ^{13}C DNP samples doped with large ESR linewidth free radicals.* Use deuterated glassing solvents for ^{13}C samples doped with large ESR linewidth free radicals galvinoxyl, DPPH, and TEMPO. This sample preparation method was shown to double or triple the DNP-enhanced ^{13}C NMR signal in the frozen sample state. The idea is that deuterons in the glassing matrix are lesser heat loads than protons due to their smaller magnetic moments; thus, the former would lead to more efficient ^{13}C hyperpolarization (Kurdzesau et al., 2008; Lumata, Merritt, & Kovacs, 2013).
- *Don'ts: ^{13}C DNP samples doped with narrow-ESR linewidth free radicals.* Do not use deuterated glassing solvents for ^{13}C samples doped with narrow ESR linewidth free radicals such as trityl OX063 and BDPA. Doing so would lead to decreases in ^{13}C DNP levels by as much as 50%. Protons do not effectively couple to the electron dipolar system of trityl and BDPA, rendering them effectively absent from the DNP process. With glassing matrix deuteration, the narrow ESR linewidth free radical electrons have to polarize not only ^{13}C spins but also the deuterons, leading to less efficient ^{13}C DNP (Goertz, 2004; Lumata, Merritt, & Kovacs, 2013).

2.3.3 ^{13}C Enrichment of the Glassing Matrix

This technique is employed to expedite the ^{13}C hyperpolarization process, while the sample is inside the polarizer at cryogenic temperatures. For a typical ^{13}C DNP sample such as sodium [1-^{13}C]pyruvate in 1:1 vol/vol DMSO:water doped with trityl OX063, the microwave irradiation time needed to achieve the maximum DNP-enhanced ^{13}C polarization is about 2–3 h. By using ^{13}C-enriched DMSO in the glassing matrix, the hyperpolarization time can be two to three times faster, thus saving time and cost of cryogens. This phenomenon can be explained by a faster nuclear spin diffusion process as ^{13}C internuclear distance is decreased with the increasing ^{13}C concentration in the system. This is analogous to a domino effect in terms of nuclear polarization transfer (Lumata, Kovacs, Malloy, Sherry, & Merritt, 2011).

2.3.4 Frozen Pellets

This sample preparation practice is especially useful for preparing large DNP sample volumes where amorphous formation of the frozen sample is sometimes compromised. Normally, small aliquots (10–100 μL) of ^{13}C DNP samples are pipetted into a sample cup (made of polyetheretherketone or Teflon) with approximately 200 μL volume capacity. To ensure glass formation, these small sample aliquots are flash-frozen by dipping the sample cup into liquid nitrogen (LN_2). Difficulties arise with large samples (e.g., 200 μL) because of the time necessary to freeze the entire volume. This may allow a fraction of the sample to crystallize. Small droplets of the sample may be prepared by pipetting out small volumes (~10 μL) of ^{13}C sample solution into a styrofoam container with LN_2. These small volumes are flash-frozen immediately, then transferred using plastic tweezers into the sample cup, which is also kept in LN_2. Once filled with frozen droplets, the sample cup is immediately inserted into the polarizer at cryogenic temperature.

2.4 Operational Steps of ^{13}C Hyperpolarizer

2.4.1 HyperSense Commercial Polarizer

The HyperSense polarizer (Oxford Instruments, UK) is a highly automated commercial polarizer that operates at 3.35 T and at 1–1.6 K with a W-band (94 GHz) microwave source. The following are the experimental steps that we typically follow to polarize ^{13}C DNP samples:

- Prepare cooldown and filling of liquid Helium (LHe) into the cryostat sample space of the polarizer. This can be done by clicking "cooldown" on the software displace on the PC. This step will turn on the rotary vane pump connected to the polarizer as well as adjust the needle valve opening for LHe. Once the LHe level in the polarizer reaches 65% and the temperature is close to 1.4 K, the insert sample button on the PC is activated.
- Pipette out the desired ^{13}C DNP sample volume into the cup (typically 10–100 μL). Flash-freeze the sample cup into LN_2. Attach the sample cup snugly at the bottom of the insertion stick. When ready, press "insert sample" on the computer.
- The HyperSense polarizer will then automatically block the pump connection to the polarizer and overpressure the sample space of the cryostat close to 1 Atmosphere. The upper entry of cryostat will then open for sample insertion. Open the upper polarizer door for sample insertion.
- Quickly insert the sample stick into the bottom of the cryostat. Dislodge the sample cup inside the cryostat immediately by simultaneously holding the outer tube and pulling the inner tube of the sample insertion wand.

- Close the polarizer door immediately and click "finish" on the computer. Allow the polarizer to get back to the base temperature for a couple of minutes.
- Click "polarize" to start the ^{13}C hyperpolarization process. This will turn on the microwave source to irradiate the sample at 3.35 T and 1.4 K. Make sure that the optimum microwave irradiation frequency for ^{13}C with the relevant free radical is established before the experiment. It is also recommended to monitor the ^{13}C solid-state polarization buildup curve of the ^{13}C DNP sample for documentation purposes and to ensure that the sample is being properly hyperpolarized.
- Wait until the ^{13}C solid-state polarization buildup curve reaches a plateau or maximum value. The ^{13}C buildup curves are typically exponential with a waiting time of 1–3 h.
- Once the ^{13}C DNP sample reaches the maximum solid-state polarization, click the dissolution button on the computer.
- Open the polarizer door. Inject 4 mL of water or other approved solvent into the top of the automated dissolution wand. Close the upper valve of the dissolution wand.
- Align the bottom of the dissolution wand with the center opening of the cryostat. Close the polarizer door.
- Click "Start dissolution" for the process to start. The water or solvent in the small cylinder container of the dissolution wand will begin to heat and pressurize. This process takes about 3.5 min before the dissolution process begins.
- Once the temperature reaches 200 °C and pressure is at 10 bars, the microwave source is turned off, the pump is blocked from the polarizer, and the sample space of the cryostat is overpressured with He gas.
- The dissolution wand is automatically inserted in the cryostat, acquiring the sample cup and injecting superheated water or solvent into the sample via high pressure He gas.
- A diluted liquid solution containing hyperpolarized ^{13}C biomolecules comes out of the polarizer via a PTFE or Teflon tubing. In 8–10 s, approximately 4 mL of hyperpolarized liquid will be collected into a beaker. This hyperpolarized liquid is then ready for immediate administration into cells, perfused organs or *in vivo* into living subjects for dynamic assessment of metabolism.

Other notes:

- The frequency of microwave irradiation varies among different radicals and is dependent on the magnetic field. The optimal frequency of irradiation must be determined for each radical experimentally.
- The polarization times are specific to the sample being hyperpolarized. The glassing matrix and presence or absence of Gd^{3+} also influence the polarization time. For new analytes, several experiments optimizing the sample conditions such as glassing matrix and radical concentration should be performed to achieve the highest possible solid-state enhancement in the shortest possible time.

2.4.2 Homebuilt DNP Hyperpolarizer

The general operational steps for polarizing ^{13}C samples in our homebuilt 129 GHz DNP hyperpolarizer (Lumata et al., 2015) are very similar to that of the automated HyperSense commercial polarizer, although ours is operated manually. The sample cup in the homebuilt polarizer is optimized to handle up to 600 μL of ^{13}C DNP sample volume. This is advantageous for hyperpolarized ^{13}C NMR or MRI studies of animals larger than small rodents, due to requirement of higher concentration and/or volume of hyperpolarized liquids.

3. DYNAMIC ASSESSMENT OF METABOLISM IN CELLS

While a major advantage of hyperpolarized ^{13}C magnetic resonance is its ability to enable noninvasive monitoring of metabolism in intact tissues, there are several reasons to perform hyperpolarization experiments in cultured cell models as well. First, hyperpolarized substrates allow kinetic assessment of discrete metabolic reactions on a time scale that would be impractical using other methods. Second, cancer cells or other cell lines may provide convenient and affordable systems to assess candidate hyperpolarized substrates prior to deploying these molecules into more cumbersome biological models, including perfused organs and live animals.

The methods described below outline approaches to observe metabolism of hyperpolarized molecules in fresh suspensions of adherent cancer cells. We have demonstrated that some of the metabolic rates observed using hyperpolarized [1-^{13}C]pyruvate were maintained during short periods of suspension (Yang, Harrison, et al., 2014). It should be noted that bioreactor systems have also been used to observe metabolism of encapsulated adherent cells, providing an alternative to the simple suspension method outlined here (Keshari et al., 2010).

3.1 Preparation of Cultures for Hyperpolarization Experiments

3.1.1 Scale and Quality of Cells Used

- As metabolic rates vary substantially among established cell lines, pilot experiments may be required to optimize conditions for each cell line. The procedure described here has been applied successfully to multiple rapidly proliferating, highly glycolytic cancer cell lines.
- In a typical *in vitro* analysis using hyperpolarized [1-^{13}C]pyruvate, we use 40–80 million cells harvested during exponential growth. This has provided good reproducibility of metabolic rates among individual experiments.

3.1.2 Details of Cell Culture and Harvesting

- Specific plating protocols should be optimized for each cell line, but plating 5×10^6 to 8×10^6 cells into each of several 150-mm dish works well for most adherent cancer cell lines. These cultures can be expected to reach 80–90% confluency—the target for cell harvest—approximately 36–48 h after plating.
- Harvest $75–150 \times 10^6$ cells from three to five 150-mm dishes by using trypsin–EDTA (0.05% for most cell lines) to disengage from the dish. Inactivate the trypsin by diluting in serum-containing culture medium, centrifuge to pellet the cells, then briefly rinse once in 10 mL prewarmed phosphate buffered saline (PBS). Resuspend the culture at a concentration of $50–100 \times 10^6$ cells/mL in fresh medium lacking glucose and pyruvate, but containing 10% dialyzed fetal calf serum, 4 m*M* glutamine, and standard concentrations of other amino acids.
- Time the preparation of the cell suspension to minimize the time in suspension prior to the hyperpolarization experiment, typically no more than 15 min. In the interim, keep the cells in a conical tube in a 37 °C water bath with frequent mixing to protect against depletion of oxygen from the medium. Trypan blue staining can be performed on a small aliquot of cells right before the hyperpolarization experiment to ensure high viability.
- In the moments preceding dissolution of the hyperpolarized material, transfer the cell suspension into a 10-mL syringe connected with a long teflon tube which goes through a free cap which will later be fitted to a 5-mm NMR tube (see Fig. 2).

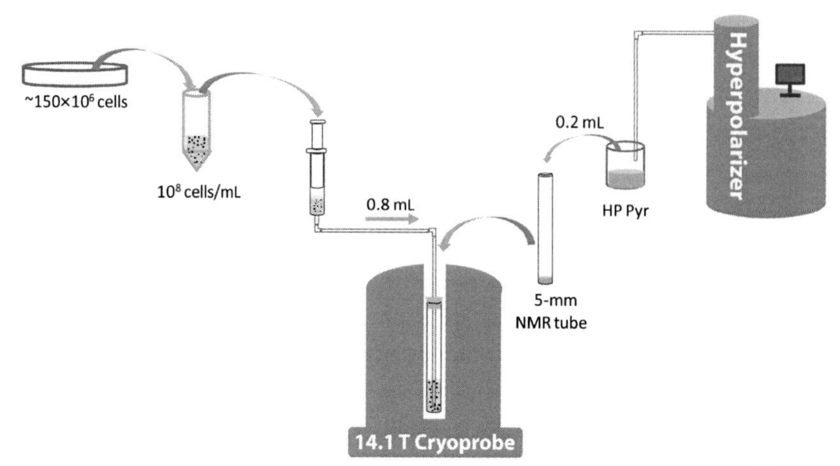

Figure 2 Schematic of a highly efficient system to perform hyperpolarized ^{13}C NMR experiments in cultured cells. The cells must be rapidly harvested for the experiment without changing their metabolism. Mixing of the cells with the hyperpolarized agent is obviously essential. The preferred method is placing a small volume of the HP solution into the bottom of the tube and subsequently injecting a large volume of cells into it to cause turbulent mixing. Injecting a small volume of HP solution into a large volume of cells does not accomplish this goal. To record the initial kinetics, mixing of the cells and the imaging agent should take place inside the magnet with the experiment already queued. (See the color plate.)

3.2 Administration of Hyperpolarized ^{13}C Substrate and Data Acquisition

- Dissolve hyperpolarized samples using an automated process like the one described above. For cell-based experiments, we use 4 mL of 15.3 mM sodium bicarbonate at 190 °C. Approximately 4 mL of hyperpolarized liquid (pH ~ 7) should be evacuated from the hyperpolarization chamber and collected in a beaker.
- Using a handheld p1000 pipettor, immediately transfer 0.2 mL of the hyperpolarized solution into the bottom of a 5-mm NMR tube.
- Affix the cap fitted with the Teflon tube connected to the syringe containing the cell suspension, as described above.
- Center the tube into a Varian 10-mm broadband probe tuned to ^{13}C in a 14.1-T magnet.
- Initiate acquisition of ^{13}C NMR spectra immediately after the NMR tube settles into the magnet.
- As the acquisition is initiated, transfer 0.8 mL of the cell suspension into the NMR tube using the syringe fitted with tubing. This produces a final

volume of 1 mL, with 6 m*M* pyruvate and a cell density of 0.4–0.8 × 10^8 cells/mL. Note that the rationale for this method is to enable observation of the initial period of metabolism as the cells first encounter the hyperpolarized material, as the initial rate of ^{13}C transfer is highly informative.

- Maximizing reproducibility of transfer may require some modifications, particularly for cell lines that become viscous when suspended at the concentrations used here. Following the cell suspension with 0.2–0.5 mL of air bubbling may improve the homogenization of the mixture. Alternatively, the cell suspension can be mixed thoroughly and precisely with the hyperpolarized material in the NMR tube outside the magnet, although this requires some sacrifice of acquiring NMR spectra over the initial phase of ^{13}C transfer.
- At the end of the ^{13}C NMR acquisition (usually no more than 3–4 min after mixing the cells with hyperpolarized material), pellet the cells by centrifugation, save the culture medium, and flash-freeze the pellet in LN_2. This enables complementary analysis using mass spectrometry or other methods to validate or aid in the interpretation of data acquired in the hyperpolarization experiment.

3.3 Techniques to Maximize Detection of Products

Our group has successfully combined hyperpolarization with two other techniques aimed to enhance ^{13}C NMR signals. These are described briefly below.

3.3.1 Selective Pulses

Shaped pulses are especially helpful in cases where the NMR signals emanating from products of interest are small compared to the ^{13}C signal of the parent compound. For instance, in *in vitro* hyperpolarized ^{13}C NMR experiments on cancer cell suspensions, the LDH-catalyzed product [1-^{13}C]lactate signal is dwarfed by the hyperpolarized ^{13}C signal from the parent compound, [1-^{13}C]pyruvate. Signal from hyperpolarized bicarbonate arising from decarboxylation of [1-^{13}C]pyruvate is typically much smaller than [1-^{13}C]lactate. To enhance the ^{13}C lactate signal, we applied a Gaussian-shaped selective radiofrequency (RF) excitation pulse at the frequency where [1-^{13}C]lactate is expected. For Varian or Agilent NMR spectrometers, we create the selective shaped pulses using the VNMRJ PBox software (Harrison et al., 2012). This shaped pulse will only excite [1-^{13}C]lactate spins with no or minimal RF excitation effect on the

[1-^{13}C]pyruvate spins. In cases where there are two or more tiny NMR signals of metabolic products of interest, double- or multiple Gaussian selective pulses can be generated to simultaneously interrogate the kinetics of the metabolic products in real time. An important consideration in using shaped pulses is that the NMR signal of the metabolic product should be located at a distinct distance from NMR signal of the parent compound, otherwise there will be substantial RF excitation of the ^{13}C spins from the parent compound. Moreover, the use of selective pulses prolongs the availability of hyperpolarized ^{13}C magnetization of the parent compound since the latter is not or only minimally excited by the selective shaped RF pulse.

3.3.2 Cryogenic Probes

In addition to hyperpolarization, the invention of cryogenically cooled probes (cryoprobes) for NMR is considered to be another important advancement in terms of sensitivity enhancement. This NMR technology is based on the fact that cooling the NMR coil and its tuning and matching components substantially reduces the random thermal motions of electrons termed as the Johnson–Nyquist noise. In addition, the preamplifier, filters, and transceiver electronics are also cooled to reduce the noise originating from the electronics (Kovacs, Moskau, & Spraul, 2005). The use of an NMR cryoprobe results in an immediate improvement of signal-to-noise ratio (SNR) by a factor of 3–5 relative to the SNR obtained in a conventional NMR probe. Unlike dissolution DNP, the improvement in NMR sensitivity produced by an NMR cryoprobe is persistent, provided that the cryogenic conditions of NMR electronics are maintained. We use a 10 mm ^{1}H-^{13}C Bruker CryoProbe system (Bruker Biospin, Billerica, MA) installed in a 600 MHz NMR magnet. This Cryoprobe system is located adjacent to the commercial HyperSense polarizer and a homebuilt DNP polarizer. When hyperpolarization is combined with the cryoprobe, the high ^{13}C SNR that we normally obtain for ^{13}C DNP is further improved by a factor of 4. To our knowledge, combining these two NMR technologies provides the highest ^{13}C NMR sensitivity currently achievable. Consistent with this idea, we have been able to use a hyperpolarization/cryoprobe combination to observe time-resolved signals previously achievable only using selective pulses (Yang, Ko, et al., 2014). Furthermore, the cryoprobe allows straightforward studies of the perfused mouse heart, an organ of such small size that only the combination of hyperpolarization and the cryoprobe has produced published results (Purmal et al., 2014).

4. DYNAMIC ASSESSMENT OF METABOLISM IN PERFUSED ORGANS

Perfused organs, particularly the heart and liver, have long been used as models to analyze intermediary metabolism, particularly with mass spectrometry and conventional ^{13}C NMR spectroscopy. We have applied hyperpolarization approaches to the perfused heart and liver from mice to observe dynamic metabolism of ^{13}C-labeled probes. Basic experimental setup for these experiments is described below.

4.1 Hyperpolarized ^{13}C NMR of Perfused Heart

4.1.1 Preparation of the Perfused Heart

Animals are anesthetized and a transabdominal incision is then performed to expose the thoracic cavity. The heart is removed from the chest cavity by excising the pulmonary artery, vena cavae, and aorta. Animal death occurs within seconds by exsanguination. The blood vessels are cut as close to the heart as possible without causing any damage to the heart. The excised heart is transferred to ice-cold saline to rinse off the blood and arrest the beating heart. All subsequent procedures prior to the start of perfusion are carried out on ice-cold saline.

Fat deposits and other contaminant tissue on the heart are removed surgically. After sufficient cleaning, a plastic cannula is inserted into the aorta. To provide robust cannulation and avoid any leaks, a surgical suture is used to tie the tissue around the cannula. Care must be taken to avoid any rupture of the tissue during the insertion of the cannula and subsequent application of the suture.

Special concerns

- After excising the heart, it is important to transfer the heart to ice-cold saline rapidly to prevent ischemic injury.
- Insertion of the cannula into the aorta must be performed carefully so that no air bubbles are introduced into the heart. This can typically be accomplished by leaving the cannula submerged in the ice-cold saline as the heart is being cleaned.
- Leaving the heart in ice-cold saline more than a few minutes often results in poor function. The time taken between the excision of the heart and start of perfusion should be minimized as much as practically possible.

4.1.2 Preparation of the Perfusion Rig

A straightforward setup to perfuse the hearts in constant pressure Langendorff mode is to utilize a vertical column providing the appropriate hydrostatic pressure (see Fig. 3). Although portable perfusion systems exist (with electronic pressure and flow rate control), the glass columns are more directly amenable to studying perfused hearts inside an NMR magnet. The perfusate of interest is maintained at a fixed height (in order to provide constant hydrostatic pressure) in the column using peristaltic pumps.

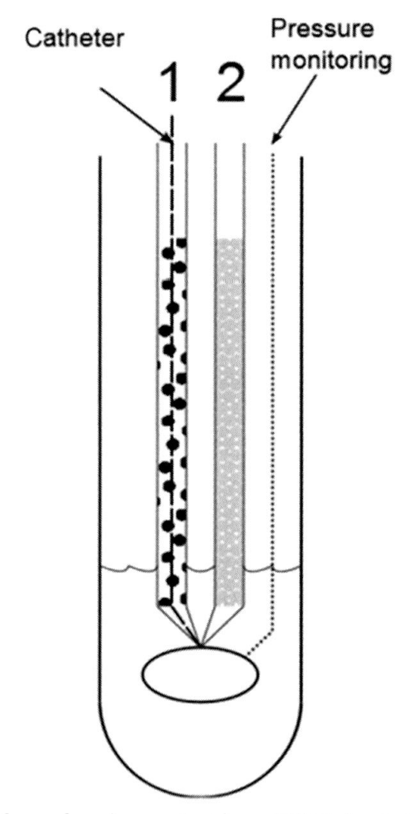

Figure 3 Schematic of a perfused organ inside an NMR tube. In this case, the perfusion rig contains two separate cannula that supply perfusate for either the heart or the liver. The presence of two cannula (1 or 2, or both) allows switching of perfusates to maintain a constant concentration of substrate when the hyperpolarized media is injected through a catheter placed inside the cannula. In the case of the heart, the developed pressure can be measured using tubing placed inside the aorta. The column height of the perfusate is set to provide the correct hydrostatic pressure for a heart. In the case of the liver, flow of the perfusate is set at 8 mL/min using a peristaltic pump distal from the MR magnet.

Continuous supply of oxygen is maintained by aerating the perfusate constantly with a mixture of 95% oxygen/5% carbon dioxide, while temperature is regulated using a heated water jacket. A simple balloon catheter is typically used as a valve to turn the flow of perfusate on or off in this setup.

Special concerns

- Depending on the perfusate, additional equipment may be needed. For example, a thin film oxygenator effectively oxygenates perfusates containing long-chain fatty acids and bovine serum albumin. Bubbling gas directly into such a perfusate will introduce excessive foaming in the perfusion column.
- With the dissolution of CO_2, pH of the perfusate is likely to be altered slightly. Hence, pH should be adjusted prior to attaching the heart to the perfusion rig.

4.1.3 Heart Perfusion and Setting Up the NMR Spectrometer

Once a bubble-free connection is established, perfusate flow is established by deflating the balloon catheter. The heart typically starts beating within a few seconds (typically less than 30). Heart function is monitored in either a continuous mode or at discrete intervals as the instrumentation permits. Heart function can be typically monitored by recording the heart rate and oxygen consumption.

For NMR experiments, an attachment comprised of an NMR tube is connected to the bottom of the perfusion rig and is lowered into the NMR magnet for acquisition of spectra (Fig. 3). By utilizing a sodium-free flush, the effluent from the beating heart is continuously removed. The flush is accomplished by inserting independent tubing to the bottom of the NMR tube. A sodium-free solution that is otherwise osmotically balanced is pumped into the region below the heart at a high rate to force the removal of the perfusate exiting the heart. In this setup, shimming can be accomplished by optimizing the linewidth of ^{23}Na signal from the heart itself. ^{23}Na linewidth of less than 12 Hz (corresponding to ^{13}C linewidth of $\sim$10 Hz) can be routinely obtained with mouse hearts at a field strength of 14.1 T.

Special concerns

- It is critical to ensure that the position of the heart in the NMR tube corresponds to the center of the RF coil of the NMR probe.
- Shimming the magnet prior to the experiment is optimal when carried out as a two-step procedure. In the first step, shimming of the magnet is accomplished using the strong ^{23}Na signal from both the perfusate and

the heart. Once a satisfactory peak shape and width is achieved, the sodium-free flush is utilized, and further shimming is carried out directly using the sodium signal from the heart.

- Shimming on the ^{23}Na signal is often sensitive to abnormalities in the heart function. For example, irregular heart beat can be correlated with fluctuations in the random and rapid variations in the signal which can be correlated with the changes in heart rate (if measured continuously).

For the dissolution, the hyperpolarized sample is usually dissolved using 4 mL of heated phosphate buffered saline and transferred into a beaker where it mixes with a predetermined volume of perfusate. This sample is injected gently, close to the heart, using a catheter that is inserted into the cannula immediately above the heart. Pyruvate metabolism is observed by measuring a series of time-resolved ^{13}C spectra with small flip angle RF pulses. Alternatively, variable flip angle schemes may be utilized if necessary (Xing, Reed, Pauly, Kerr, & Larson, 2013).

Special concerns

- Mixing with perfusate ensures that the osmolarity of the hyperpolarized sample is similar to that of the perfusate and that the sample is adequately oxygenated.
- The injection of the sample should be smooth and constant. Rapid injection can generate turbulence and uneven mixing of the injected substrate with the solution present in the perfusion rig proper.

4.2 Hyperpolarized ^{13}C NMR of Perfused Liver

4.2.1 Hepatectomy

- Heparinize the mouse with about 0.07 mL heparin injected intraperitoneally and anesthetize with 0.09 mL ketamine/xylocaine, also administered intraperitoneally.
- Weigh the mouse and transfer to a prep station. There will be blood and runoff perfusate. It helps to have a mini work table lined with plastic cling wrap.
- Tape the feet of the mouse to the cling wrap while allowing the head and shoulders to fall over the edge of the work table. This allows the liver to rest upward on the tissue lining the thoracic cavity, while the cannulation is performed.
- Pinch the skin above the abdomen with forceps. Lift up and make a wide incision along the midline of the abdomen, taking care not to puncture any organs.

- Fold the skin up over the sternum to expose the abdominal cavity. Slide the fat and intestines with your middle finger to the right and outside of the body cavity, exposing the portal vein.
- Create a path underneath the portal vein close to the liver with a pair of forceps. Feed two 3″ pieces of 4-0 silk suture under the vein and begin a "left-over-right" knot. Taking care not to close the knot, tie them down close to the vein and lay the ends of the piece closer to the head up between the arms.
- Cannulate the portal vein by inserting a 22G × 1″ Terumo Surflo IV Catheter. Use your left hand to move the fatty tissue and create tension on the vessel if needed. Take care not to advance the needle too far for risk of puncturing the vein. An assistant can pull on the two ends of suture closest to the head of the mouse to also create more tension for ease of catheter insertion.
- Once inside the vessel, hold the needle still and use your index finger to advance the catheter off of the needle further into the vein, about ¼″. Do not push the catheter as far as it will go inside the vessel as this may also puncture or create a stoppage of flow and induce ischemia.
- While holding the needle and catheter in place, have an assistant to tie the suture closest to the head of the animal down as tight as possible ensuring that the vein does not slip off the end of the catheter. Ideally, there will be about 5 mm between the end of the catheter tip and first suture knot.
- Let the needle lay in place, while you tie down the second suture. Double knot both sutures and cut off any excess suture material.
- Carefully remove the needle by holding the catheter hub in place and pulling the needle back. Blood should travel back into the hub. Very little to no blood should seep into the body cavity at this point.
- Displace any air in the hub with perfusate and insert your perfusion line while holding the hub firmly. Start your perfusion of the liver by starting the pump connected to the perfusate.
- Cut open the vena cava. The mouse should expire within seconds from exsanguination, and the liver should flush all of the blood out and turn from a red/maroon color to an off-yellow/mustard or tannish color.
- Livers are very delicate. Minimal time should elapse from cannulation to starting the perfusion pump.
- Dissect out the liver by removing first the kidneys followed by intestines and the stomach/pancreas.

- Carefully cut underneath the liver near the spine to separate from connective tissue.
- Switch to the head of the mouse and have an assistant hold the perfusate line in place.
- Cradle the head between your left thumb resting on the chin of the mouse and your index finger curled underneath the shoulders. Lift gently and roll the liver back down over itself toward its natural position.
- Start cutting away the connective tissues, beginning with the thoracic cavity tissue. The liver will slide freely on the plastic cling wrap once more tissue gets cut away.
- As you cut more connective tissues away from the liver, use your left hand to slide the carcass out and away from the liver. Take care in keeping the liver and perfusion line as still as possible. Quick movements with the liver will damage it and induce ischemia.
- Transfer the liver to experimental setup. Use a small cup for transport and take care not to introduce air while moving the catheter hub from setup to setup.

Once the liver has been successfully attached to the perfusion rig, the experimental procedure mirrors the heart protocol. The perfused mouse liver is of course larger than the heart, and as such, a minimum NMR tube size of ~18 mm must be used to accommodate the mouse liver in the NMR magnet. Due to its larger size, the use of a sodium-free flush can be more problematic, as excessively high flow rates can cause the liver to move, making shimming extremely difficult. However, with the proper flow rate for the flush, the two-step ^{23}Na shimming procedure used for the heart will produce similar results for the liver. In addition to oxygen consumption measurements, metabolic viability of the liver is also easily assessed by inspection. If the liver is turning white, it is likely not being perfused correctly and should be discarded.

5. CHALLENGES AND FUTURE DIRECTIONS

5.1 Hyperpolarized Substrates in Addition to [1-^{13}C]Pyruvate

Multiple qualities define the utility of new hyperpolarized substrates (Comment & Merritt, 2014). A primary condition is the solubility of the compound in the matrix which carries the radical used for the DNP process. If a solvent mixture is necessary for production of the combined substrate and radical, it must be thoroughly miscible, as partitioning of the sample into separate phases precludes the process of nuclear spin diffusion necessary for DNP to be effective throughout the sample (Lumata, Kovacs, et al., 2011). Connected to the sample preparation is a phenomenological observation that absolute enhancement of the sample must be high to allow kinetic data to be collected. A good rule of thumb is that enhancements of ~10,000 will provide an initial signal amplitude suitable for acquisition of time course data. Another primary limitation is the nuclear T_1 of the substrate and its metabolites, as discussed above. Interestingly, if metabolic flux is highly upregulated, as for glucose metabolism in cancer, even agents with short T_1s can yield important metabolic data and potentially biological insights; see the following discussion of hyperpolarized glucose as an imaging agent in cancer. It is most beneficial if the compound in question is either transported or rapidly diffuses into the same space where metabolism occurs. Finally, a multiplicity of metabolic fates will enhance the information yield of the experiment. For example, pyruvate can produce up to four different compounds in single enzyme-catalyzed reactions, enabling simultaneous analysis of multiple pathways, including some that are redox dependent and some that are not.

Another important consideration is the pool sizes of the metabolite derived from the hyperpolarized-imaging agent. Pyruvate again has unique qualifications, as its circulating concentration is fairly low but the size of the alanine and lactate pools are ~3 and ~10 times as large. The signal amplitude in hyperpolarization experiments is intimately connected to the pool sizes. This phenomenon is exacerbated in the case of exchange. When exogenous lactate was added to cells studied with hyperpolarized pyruvate, a nearly linear increase in hyperpolarized lactate signal intensity was observed (Day et al., 2007). The kinetics of isotope transport from the small intracellular pyruvate pool into the larger lactate pool is limited only by the amount of ^{13}C label in the lactate pool that is subject to the reverse reaction. In cases where hyperpolarized [1-^{13}C]lactate or [1-^{13}C]alanine has been used, the resulting hyperpolarized [1-^{13}C]pyruvate signal has been relatively small, confirming that fast enzyme kinetics is not sufficient for the production of a large hyperpolarized signal (Chen et al., 2008; Jensen et al., 2009). A large absolute number of HP nuclear spins associated with the detected metabolite is also necessary.

The preceding discussion explains the paucity of data collected with hyperpolarized lactate or alanine, though examples do exist (Bastiaansen, Yoshihara, Takado, Gruetter, & Comment, 2014). From the standpoint of a metabolic perturbation, lactate and alanine should both have fewer consequences than a bolus injection of the highly oxidized substrate pyruvate. However, due to the pool size effect, hyperpolarized [1-^{13}C]lactate produces only a small amount of [1-^{13}C]pyruvate and consequently a less intense hyperpolarized [^{13}C]bicarbonate signal (Chen et al., 2008). Alanine is confronted with even more difficult circumstances, as its transport does not occur through the high-capacity monocarboxylate transporters that enable lactate and pyruvate import but is electrogenic in nature. The attendant lactate and pyruvate signals generated by injection of hyperpolarized alanine are therefore quite small.

5.1.1 Dihydroxyacetone

Recently, [2-^{13}C]dihydroxyacetone (DHA) has been demonstrated for hyperpolarized imaging of the liver (Moreno et al., 2014). DHA has many qualities similar to pyruvate that suggest it may be successfully used as an HP imaging agent. First, it polarizes to an outstanding level in a water–glycerol matrix, exceeding even the standard sample of trityl radical solubilized directly into pyruvic acid. Second, it has a long nuclear T_1, c.a. 32 s at 9.4 T, which facilitates its delivery and subsequent metabolism while a large signal remains. Third, DHA produces a variety of metabolic products after its initial phosphorylation to dihydroxyacetone phosphate (DHAP). Unfortunately, DHAP has a chemical shift that is unresolvable from the parent compound. DHA metabolism has not yet been observed in organs other than the liver, indicating that an as yet unidentified transporter may be responsible for its uptake. Alternatively, other tissues may lack the ability to phosphorylate DHA to DHAP, which would prevent its subsequent metabolism.

The initial report on DHA as a contrast agent produced new insights into glycolysis, glycogenolysis, and gluconeogenesis in the liver. The traditional model of regulation in the Embden–Meyerhof pathway identifies the three most important regulatory nodes as phosphoenolpyruvate carboxykinase/pyruvate kinase, fructose 1,6-bisphosphatase/6-phosphofructo-1-kinase, and glucose-6-phosphatase/glucokinase. It is difficult to observe kinetics at each separate point of control in an intact system. However, due to the chemical selectivity of MR, almost all the upstream (toward glucose) and downstream (toward pyruvate) metabolites can be detected with a 2 s time resolution using hyperpolarized [2-^{13}C]DHA. In glycogenolytic versus gluconeogenic conditions, only pyruvate kinase produced an observable delay in the production of its product pyruvate and subsequent metabolites, while fructose-1,6-bisphosphatase and glucose-6-phosphatase did not. Prior to the introduction of hyperpolarization methods, insights like this were nearly impossible to generate, as the only option for generating kinetic measurements with this degree of time resolution would necessitate the serial, destructive acquisition of samples at each time point.

5.1.2 Glucose

Glucose itself is also a suitable hyperpolarized imaging agent, but only when deuteration of the ^{13}C-enriched carbon sites has been used to increase the T_1. The first hint of the power of hyperpolarized [U-^{13}C, U-^{2}H]glucose for studying metabolism was in the work of Meier et al. studying yeast metabolism (Meier, Karlsson, Jensen, Lerche, & Duus, 2011). Nearly, the entire glycolytic pathway from glucose to ethanol formation was observed. In addition, $^{13}CO_2$ and [^{13}C]bicarbonate were detected and taken as a marker of pyruvate dehydrogenase (PDH) flux. This implies that more than 12 separate enzymatic fluxes can be simultaneously observed. The same substrate was subsequently used to study tumor metabolism *in vivo* where lactate was the primary product observed (Rodrigues et al., 2014). In summary, the authors believe that hyperpolarization technology offers a tremendous opportunity to study enzyme kinetics in intact systems, increasing the likelihood that the insights generated could inform on metabolic control and flux in a variety of tissues and model systems.

5.2 Specific Challenges Associated with Using HP Substrates to Quantify Fluxes *In Vivo*

One of the potentially impactful deliverables of hyperpolarization research is the development of methods to extract quantitative flux data from *in vivo* experiments, ultimately in human patients. Here, we discuss some of the specific challenges associated with using hyperpolarization to definitively quantify metabolic fluxes. In the case of cell culture or isolated systems, this process is dramatically simplified because delivery of the hyperpolarized substrate and its initial concentration can be rigorously controlled. Knowledge of the substrate concentration is obviously a necessary precondition for absolute flux measurements, but extraction of simple rate constants might also provide metrics sensitive to metabolic changes. Estimates of rate constants or actual fluxes *in vivo* have been obtained by many labs using disparate methods. The first method introduced below has obvious counterparts in the biochemical literature. The other studies rely upon magnetic resonance phenomena to facilitate estimates of flux and exchange.

Michaelis–Menten kinetics is typically used to describe single substrate enzyme-catalyzed reactions. Seminal work by Zierhut et al. measured dose–response curves for hyperpolarized [1-^{13}C]pyruvate in mice and rats in normal tissue and in a model of prostate cancer (Zierhut et al., 2010). Using doses ranging from 50 to 725 μmol/kg, V_{max} and K_m values were measured using the appearance of hyperpolarized [1-^{13}C]alanine and [1-^{13}C]lactate. Tumors were shown to have larger *k* values (initial rates for forward flux) for lactate production as predicted by the Warburg description of aerobic glycolysis in cancer (Warburg, 1956a). Modeling of the kinetic curves was accomplished using equations that describe only flux from pyruvate into the lactate and alanine pools. While this is an important assumption, other experiments subsequently showed that due to the finite lifetime of the hyperpolarized substrate and its metabolites, the kinetic curves are largely insensitive to exchange, when simple, constant repeat time, and small flip angle experimental protocols are used for data collection (Harrison et al., 2012). The straightforward experimental design of this study produced estimates of values that are readily understood by the larger research community, and as such, this paper served an important place in communicating how HP substrates could give insights into *in vivo* kinetics.

Shortly after this study, Xu et al. reported an alternative method for producing estimates of K_m and V_{max} that used incrementally higher excitation angles for the MR experiment (Xu et al., 2011). Hyperpolarized nuclei are not in thermal equilibrium with surrounding spins in the system being studied. As such, the hyperpolarization state is in a constant state of decay back to the thermal equilibrium value. Also, the detection pulses used for observation of the magnetic resonance signal perturb the spin state in such a way that the hyperpolarization is not recoverable. This latter phenomenon allows specific insights to be generated by starting with low flip angles and incrementing them to a maximum value of 90° at each time point. Using this protocol, K_m and V_{max} can be estimated from a single injection of the hyperpolarized substrate. Intuitively, this experimental design can be explained by analogy to standard dose–response experiments. Since a 90° pulse consumes all the polarization in the region of interest, the subsequent delay before the protocol is repeated allows new hyperpolarized material to flow in. Since the imaging experiment is queued upon injection, each time point reflects the initial delivery and kinetics of metabolite formation as a function of [pyruvate] in the observed tissue. Due to the specifics of the modified Michaelis–Menten equations used for this study, estimates of K_m are subject to effects derived from reverse flux from either alanine or lactate back to the exogenous pyruvate pool. However, quantification of the exchange is still problematic. Paired with the fast chemical shift imaging protocol used for detection of the metabolites, this progressive excitation method could have tremendous utility in cases where the researcher anticipates large differences in K_m and V_{max}, such as in malignant tissues.

In an effort to gain better insights into the exchange phenomena between pyruvate, lactate, and alanine as measured by hyperpolarized pyruvate, alternative MR protocols have been proposed. Larson et al. demonstrated that a stimulated echo acquisition mode (STEAM) excitation sequence could be tuned to make quantitative estimates of both forward and reverse flux *in vivo* while at the same time producing localized spectra (Swisher et al., 2014). The underlying phenomenon that allows exchange to be monitored is the frequency difference between pyruvate and its metabolites in the MR spectrum. This difference in frequency will manifest as a phase change in the spectra, when the delays of the STEAM sequence are chosen appropriately. While the method is incredibly powerful, deconvolution of the effects of exchange between multiple pools is not straightforward.

Experimentally, a more challenging protocol that is based upon more traditional methods of measuring flux by MR is the inversion transfer method proposed by Kettunen et al. (2010). Selective inversion of either pyruvate or lactate signals produces modulation of the exchanging metabolite that can be modeled according to the modified Bloch equations. This method relies upon mathematical fitting of the exchange curves generated from the time-dependent spectra. The magnetization inversion preparation phase of the sequence can be placed in front of any imaging sequence desired and is therefore at least as flexible as the method proposed by Xu. The primary drawback of these methods is insuring the inversion pulse is properly setup and calibrated. A potentially more robust method of extracting the same kinetic information about pyruvate–lactate exchange was proposed by the same group (Kennedy, Kettunen, Hu, & Brindle, 2012). In this case, [U-^{2}H]lactate is the hyperpolarized substrate. Due to the exchange of protons that is inherent when lactate transits lactate dehydrogenase to for pyruvate, a simple spin echo method can be used to modulate the directly detected ^{13}C MR signals. The modulation again can be modeled using the Bloch equations, yielding accurate estimates of the rate constants describing the pyruvate–lactate equilibrium.

A fundamental challenge for the field of HP applications has been establishing methods for independently verifying the kinetic data observed. Steady-state isotope tracer methods for measuring metabolic flux are extremely well established, and with proper experimental design, they can be used to not only confirm HP results but also, when integrated with the HP data, produce a powerful tool for assessing global energy metabolism. The combination of HP and steady-state tracer methods was first used to rigorously prove that HP [^{13}C]bicarbonate produced by the heart arose exclusively from PDH flux (Merritt et al., 2007). By monitoring the ^{13}C enrichment of glutamate extracted from the perfused heart, it was established that the addition of a competing fatty acid to the perfusate effectively quenched pyruvate oxidation and hence the production of HP [^{13}C]bicarbonate by PDH flux. Further development of this approach led to the demonstration that HP [^{13}C]bicarbonate production could be quantitatively modeled in cell culture (Yang, Harrison, et al., 2014). This single absolute flux measurement could then be used to normalize a complete set of relative flux measurements obtained by tracer NMR methods. The result was a global, quantitative picture of Krebs cycle turnover. The juxtaposition of oxidative versus anaplerotic handling of pyruvate illustrated just how powerful this new method could be for establishing metabolic phenotypes in cancer cells. Future work marrying hyperpolarization with other quantitative methods should significantly enhance the interpretation of *in vivo* HP data as well as produce new insights into intermediary metabolism.

6. CONCLUSION

Hyperpolarization has vastly expanded the applicability of MR-based methods for assessing metabolism due to the $\sim 10^4$ gain in sensitivity. Combined with the traditionally understood strengths of MR, i.e., the chemical selectivity and its applicability to living, functioning systems, powerful new insights into the kinetics of metabolic processes are being routinely generated. We believe that as the technology becomes more widely disseminated, it will find an increasing number of applications in basic science as well as in medical research.

ACKNOWLEDGMENTS

The authors would like to thank these funding agencies for financial support of this work: Department of Defense PCRP Grant no. W81XWH-14-1-0048 (L.L.); N.I.H. Grant CA157996-01 (R.J.D.); Robert A. Welch Foundation Grant nos. I-1733 (R.J.D.) and AT-1877 (L.L.); Cancer Prevention and Research Institute of Texas Grant nos. RP140021-P3, NIH P41 EB015908, R37 HL34557, and R21 EB016197 (M.E.M.).

REFERENCES

Abragam, A., & Goldman, M. (1978). Principles of dynamic nuclear polarisation. *Reports on Progress in Physics*, *41*, 395.

Andronesi, O. C., Kim, G. S., Gerstner, E., Batchelor, T., Tzika, A. A., Fantin, V. R., et al. (2012). Detection of 2-hydroxyglutarate in IDH-mutated glioma patients by in vivo spectral-editing and 2D correlation magnetic resonance spectroscopy. *Science Translational Medicine*, *4*, 116ra4.

Ardenkjaer-Larsen, J. H., Fridlund, B., Gram, A., Hansson, G., Hansson, L., Lerche, M. H., et al. (2003). Increase in signal-to-noise ratio of >10,000 times in liquid-state NMR. *Proceedings of the National Academy of Sciences of the United States of America*, *100*, 10158–10163.

Bastiaansen, J. M., Yoshihara, H. I., Takado, Y., Gruetter, R., & Comment, A. (2014). Hyperpolarized ^{13}C lactate as a substrate for in vivo metabolic studies in skeletal muscle. *Metabolomics*, *10*, 986–994.

Bohndiek, S. E., Kettunen, M. I., Hu, D.-e., Kennedy, B. W. C., Boren, J., Gallagher, F. A., et al. (2011). Hyperpolarized [1-^{13}C]-ascorbic and dehydroascorbic acid: Vitamin C as a probe for imaging redox status in vivo. *Journal of the American Chemical Society*, *133*, 11795–11801.

Busch, R., Neese, R. A., Awada, M., Hayes, G. M., & Hellerstein, M. K. (2007). Measurement of cell proliferation by heavy water labeling. *Nature Protocols*, *2*, 3045–3057.

Chen, A. P., Kurhanewicz, J., Bok, R., Xu, D., Joun, D., Zhang, V., et al. (2008). Feasibility of using hyperpolarized [1-^{13}C] lactate as a substrate for in vivo metabolic ^{13}C MRSI studies. *Magnetic Resonance Imaging*, *26*, 721–726.

Choi, C., Ganji, S. K., DeBerardinis, R. J., Hatanpaa, K. J., Rakheja, D., Kovacs, Z., et al. (2012). 2-hydroxyglutarate detection by magnetic resonance spectroscopy in IDH-mutated patients with gliomas. *Nature Medicine*, *18*, 624–629.

Choudhary, C., Weinert, B. T., Nishida, Y., Verdin, E., & Mann, M. (2014). The growing landscape of lysine acetylation links metabolism and cell signalling. *Nature Reviews Molecular Cell Biology*, *15*, 536–550.

Comment, A., & Merritt, M. E. (2014). Hyperpolarized magnetic resonance as a sensitive detector of metabolic function. *Biochemistry*, *53*, 7333–7357.

Day, S. E., Kettunen, M. I., Gallagher, F. A., De-En, H., Lerche, M., Wolber, J., et al. (2007). Detecting tumor response to treatment using hyperpolarized ^{13}C magnetic resonance imaging and spectroscopy. *Nature Medicine*, *13*, 1382–1387.

DeBerardinis, R. J., & Thompson, C. B. (2012). Cellular metabolism and disease: What do metabolic outliers teach us? *Cell*, *148*, 1132–1144.

Elkhaled, A., Jalbert, L. E., Phillips, J. J., Yoshihara, H. A., Parvataneni, R., Srinivasan, R., et al. (2012). Magnetic resonance of 2-hydroxyglutarate in IDH1-mutated low-grade gliomas. *Science Translational Medicine*, *4*, 116ra5.

Gallagher, F. A., Kettunen, M. I., Day, S. E., Hu, D. E., Ardenkjaer-Larsen, J. H., Zandt, R., et al. (2008). Magnetic resonance imaging of pH in vivo using hyperpolarized ^{13}C-labelled bicarbonate. *Nature*, *453*, 940–943.

Gallagher, F. A., Kettunen, M. I., Day, S. E., Lerche, M., & Brindle, K. M. (2008). ^{13}C MR spectroscopy measurements of glutaminase activity in human hepatocellular carcinoma cells using hyperpolarized ^{13}C-labeled glutamine. *Magnetic Resonance in Medicine*, *60*, 253–257.

Gallagher, F. A., Kettunen, M. I., Hu, D. E., Jensen, P. R., Zandt, R. I., Karlsson, M., et al. (2009). Production of hyperpolarized [1,4-$^{13}C_2$]malate from [1,4-$^{13}C_2$]fumarate is a marker of cell necrosis and treatment response in tumors. *Proceedings of the National Academy of Sciences of the United States of America*, *106*, 19801–19806.

Gallamini, A., Zwarthoed, C., & Borra, A. (2014). Positron emission tomography (PET) in oncology. *Cancers*, *6*, 1821–1889.

Gerriets, V. A., Kishton, R. J., Nichols, A. G., Macintyre, A. N., Inoue, M., Ilkayeva, O., et al. (2015). Metabolic programming and PDHK1 control CD4+ T cell subsets and inflammation. *The Journal of Clinical Investigation*, *125*, 194–207.

Goertz, S. T. (2004). The dynamic nuclear polarization process. *Nuclear Instruments and Methods in Physics Research Section A: Accelerators, Spectrometers, Detectors and Associated Equipment*, *526*, 28–42.

Golman, K., Zandt, R. I., Lerche, M., Pehrson, R., & Ardenkjaer-Larsen, J. H. (2006). Metabolic imaging by hyperpolarized ^{13}C magnetic resonance imaging for in vivo tumor diagnosis. *Cancer Research*, *66*, 10855–10860.

Hanahan, D., & Weinberg, R. A. (2011). Hallmarks of cancer: The next generation. *Cell*, *144*, 646–674.

Harrison, C., Yang, C., Jindal, A., DeBerardinis, R. J., Hooshyar, M. A., Merritt, M., et al. (2012). Comparison of kinetic models for analysis of pyruvate-to-lactate exchange by hyperpolarized ^{13}C NMR. *NMR in Biomedicine*, *25*, 1286–1294.

Hu, S., Balakrishnan, A., Bok, R. A., Anderton, B., Larson, P. E., Nelson, S. J., et al. (2011). ^{13}C-pyruvate imaging reveals alterations in glycolysis that precede c-Myc-induced tumor formation and regression. *Cell Metabolism*, *14*, 131–142.

Jensen, P. R., Karlsson, M., Meier, S., Duus, J. Ø., & Lerche, M. H. (2009). Hyperpolarized amino acids for in vivo assays of transaminase activity. *Chemistry: A European Journal*, *15*, 10010–10012.

Kaelin, W. G., Jr., & McKnight, S. L. (2013). Influence of metabolism on epigenetics and disease. *Cell*, *153*, 56–69.

Kennedy, B. W. C., Kettunen, M. I., Hu, D.-E., & Brindle, K. M. (2012). Probing lactate dehydrogenase activity in tumors by measuring hydrogen/deuterium exchange in hyperpolarized l-[1-^{13}C, U-^{2}H]lactate. *Journal of the American Chemical Society*, *134*, 4969–4977.

Keshari, K. R., Kurhanewicz, J., Jeffries, R. E., Wilson, D. M., Dewar, B. J., Van Criekinge, M., et al. (2010). Hyperpolarized ^{13}C spectroscopy and an NMR-compatible bioreactor system for the investigation of real-time cellular metabolism. *Magnetic Resonance in Medicine*, *63*, 322–329.

Keshari, K. R., & Wilson, D. M. (2014). Chemistry and biochemistry of ^{13}C hyperpolarized magnetic resonance using dynamic nuclear polarization. *Chemical Society Reviews*, *43*, 1627–1659.

Kettunen, M. I., Hu, D. E., Witney, T. H., McLaughlin, R., Gallagher, F. A., Bohndiek, S. E., et al. (2010). Magnetization transfer measurements of exchange between hyperpolarized [1-^{13}C]pyruvate and [1-^{13}C]lactate in a murine lymphoma. *Magnetic Resonance in Medicine*, *63*, 872–880.

Kovacs, H., Moskau, D., & Spraul, M. (2005). Cryogenically cooled probes—A leap in NMR technology. *Progress in Nuclear Magnetic Resonance Spectroscopy*, *46*, 131–155.

Kurdzesau, F., van den Brandt, B., Comment, A., Hautle, P., Jannin, S., van den Klink, J. J., et al. (2008). Dynamic nuclear polarization of small labelled molecules in frozen water–alcohol solutions. *Journal of Physics D: Applied Physics*, *41*, 155506.

Kurhanewicz, J., Vigneron, D. B., Brindle, K., Chekmenev, E. Y., Comment, A., Cunningham, C. H., et al. (2011). Analysis of cancer metabolism by imaging hyperpolarized nuclei: Prospects for translation to clinical research. *Neoplasia*, *13*, 81–97.

Lumata, L., Jindal, A. K., Merritt, M. E., Malloy, C. R., Sherry, A. D., & Kovacs, Z. (2011). DNP by thermal mixing under optimized conditions yields >60,000-fold enhancement of 89Y NMR signal. *Journal of the American Chemical Society*, *133*, 8673–8680.

Lumata, L., Kovacs, Z., Malloy, C., Sherry, A. D., & Merritt, M. (2011). The effect of ^{13}C enrichment in the glassing matrix on dynamic nuclear polarization of [1-^{13}C]pyruvate. *Physics in Medicine and Biology*, *56*, N85.

Lumata, L., Kovacs, Z., Sherry, A. D., Malloy, C., Hill, S., van Tol, J., et al. (2013). Electron spin resonance studies of trityl OX063 at a concentration optimal for DNP. *Physical Chemistry Chemical Physics*, *15*, 9800–9807.

Lumata, L. L., Martin, R., Jindal, A. K., Kovacs, Z., Conradi, M. S., & Merritt, M. E. (2015). Development and performance of a 129-GHz dynamic nuclear polarizer in an ultra-wide bore superconducting magnet. *Magnetic Resonance Materials in Physics, Biology and Medicine*, *28*, 195–205.

Lumata, L., Merritt, M. E., Hashami, Z., Ratnakar, S. J., & Kovacs, Z. (2012). Production and NMR characterization of hyperpolarized 107,109Ag complexes. *Angewandte Chemie, International Edition*, *51*, 525–527.

Lumata, L., Merritt, M., Khemtong, C., Ratnakar, S. J., van Tol, J., Yu, L., et al. (2012). The efficiency of DPPH as a polarising agent for DNP-NMR spectroscopy. *RSC Advances*, *2*, 12812–12817.

Lumata, L., Merritt, M. E., & Kovacs, Z. (2013). Influence of deuteration in the glassing matrix on ^{13}C dynamic nuclear polarization. *Physical Chemistry Chemical Physics*, *15*, 7032–7035.

Lumata, L., Merritt, M., Malloy, C., Sherry, A. D., & Kovacs, Z. (2012a). Fast dissolution dynamic nuclear polarization NMR of ^{13}C-enriched 89Y-DOTA complex: Experimental and theoretical considerations. *Applied Magnetic Resonance*, *43*, 69–79.

Lumata, L., Merritt, M. E., Malloy, C. R., Sherry, A. D., & Kovacs, Z. (2012b). Impact of Gd^{3+} on DNP of [1-^{13}C]pyruvate doped with trityl OX063, BDPA, or 4-oxo-TEMPO. *The Journal of Physical Chemistry A*, *116*, 5129–5138.

Lumata, L. L., Merritt, M. E., Malloy, C. R., Sherry, A. D., van Tol, J., Song, L., et al. (2013). Dissolution DNP-NMR spectroscopy using galvinoxyl as a polarizing agent. *Journal of Magnetic Resonance*, *227*, 14–19.

Lumata, L., Ratnakar, S. J., Jindal, A., Merritt, M., Comment, A., Malloy, C., et al. (2011). BDPA: An efficient polarizing agent for fast dissolution dynamic nuclear polarization NMR spectroscopy. *Chemistry: A European Journal*, *17*, 10825–10827.

Maher, E. A., Marin-Valencia, I., Bachoo, R. M., Mashimo, T., Raisanen, J., Hatanpaa, K. J., et al. (2012). Metabolism of [U-^{13}C]glucose in human brain tumors in vivo. *NMR in Biomedicine*, *25*, 1234–1244.

Marin-Valencia, I., Yang, C., Mashimo, T., Cho, S., Baek, H., Yang, X. L., et al. (2012). Analysis of tumor metabolism reveals mitochondrial glucose oxidation in genetically diverse human glioblastomas in the mouse brain in vivo. *Cell Metabolism*, *15*, 827–837.

Meier, S., Karlsson, M., Jensen, P. R., Lerche, M. H., & Duus, J. Ø. (2011). Metabolic pathway visualization in living yeast by DNP-NMR. *Molecular BioSystems*, *7*, 2834–2836.

Merritt, M. E., Harrison, C., Sherry, A. D., Malloy, C. R., & Burgess, S. C. (2011). Flux through hepatic pyruvate carboxylase and phosphoenolpyruvate carboxykinase detected by hyperpolarized ^{13}C magnetic resonance. *Proceedings of the National Academy of Sciences of the United States of America*, *108*, 19084–19089.

Merritt, M. E., Harrison, C., Storey, C., Jeffrey, F. M., Sherry, A. D., & Malloy, C. R. (2007). Hyperpolarized ^{13}C allows a direct measure of flux through a single enzyme-catalyzed step by NMR. *Proceedings of the National Academy of Sciences of the United States of America*, *104*, 19773–19777.

Metallo, C. M., & Vander Heiden, M. G. (2013). Understanding metabolic regulation and its influence on cell physiology. *Molecular Cell*, *49*, 388–398.

Moreno, K. X., Satapati, S., DeBerardinis, R. J., Burgess, S. C., Malloy, C. R., & Merritt, M. E. (2014). Real-time detection of hepatic gluconeogenic and glycogenolytic states using hyperpolarized [2-^{13}C]dihydroxyacetone. *The Journal of Biological Chemistry*, *289*, 35859–35867.

Nelson, S. J., Kurhanewicz, J., Vigneron, D. B., Larson, P. E., Harzstark, A. L., Ferrone, M., et al. (2013). Metabolic imaging of patients with prostate cancer using hyperpolarized [1-^{13}C]pyruvate. *Science Translational Medicine*, *5*, 198ra108.

Oz, G., Alger, J. R., Barker, P. B., Bartha, R., Bizzi, A., Boesch, C., et al. (2014). Clinical proton MR spectroscopy in central nervous system disorders. *Radiology*, *270*, 658–679.

Plas, D. R., & Thompson, C. B. (2005). Akt-dependent transformation: There is more to growth than just surviving. *Oncogene*, *24*, 7435–7442.

Pope, W. B., Prins, R. M., Albert Thomas, M., Nagarajan, R., Yen, K. E., Bittinger, M. A., et al. (2012). Non-invasive detection of 2-hydroxyglutarate and other metabolites in IDH1 mutant glioma patients using magnetic resonance spectroscopy. *Journal of Neuro-Oncology*, *107*, 197–205.

Purmal, C., Kucejova, B., Sherry, A. D., Burgess, S. C., Malloy, C. R., & Merritt, M. E. (2014). Propionate stimulates pyruvate oxidation in the presence of acetate. *American Journal of Physiology Heart and Circulatory Physiology*, *307*, H1134–H1141.

Rodrigues, T. B., Serrao, E. M., Kennedy, B. W., Hu, D. E., Kettunen, M. I., & Brindle, K. M. (2014). Magnetic resonance imaging of tumor glycolysis using hyperpolarized ^{13}C-labeled glucose. *Nature Medicine*, *20*, 93–97.

Sunny, N. E., Parks, E. J., Browning, J. D., & Burgess, S. C. (2011). Excessive hepatic mitochondrial TCA cycle and gluconeogenesis in humans with nonalcoholic fatty liver disease. *Cell Metabolism*, *14*, 804–810.

Swisher, C. L., Larson, P. E., Kruttwig, K., Kerr, A. B., Hu, S., Bok, R. A., et al. (2014). Quantitative measurement of cancer metabolism using stimulated echo hyperpolarized carbon-13 MRS. *Magnetic Resonance in Medicine*, *71*, 1–11.

Warburg, O. (1956a). On respiratory impairment in cancer cells. *Science*, *124*, 269–270.

Warburg, O. (1956b). On the origin of cancer cells. *Science*, *123*, 309–314.

Ward, P. S., & Thompson, C. B. (2012). Metabolic reprogramming: A cancer hallmark even warburg did not anticipate. *Cancer Cell*, *21*, 297–308.

Xing, Y., Reed, G. D., Pauly, J. M., Kerr, A. B., & Larson, P. E. (2013). Optimal variable flip angle schemes for dynamic acquisition of exchanging hyperpolarized substrates. *Journal of Magnetic Resonance*, *234*, 75–81.

Xu, T., Mayer, D., Gu, M., Yen, Y. F., Josan, S., Tropp, J., et al. (2011). Quantification of in vivo metabolic kinetics of hyperpolarized pyruvate in rat kidneys using dynamic ^{13}C MRSI. *NMR in Biomedicine*, *24*, 997–1005.

Yang, C., Harrison, C., Jin, E. S., Chuang, D. T., Sherry, A. D., Malloy, C. R., et al. (2014). Simultaneous steady-state and dynamic ^{13}C NMR can differentiate alternative routes of pyruvate metabolism in living cancer cells. *The Journal of Biological Chemistry*, *289*, 6212–6224.

Yang, C., Ko, B., Hensley, C. T., Jiang, L., Wasti, A. T., Kim, J., et al. (2014). Glutamine oxidation maintains the TCA cycle and cell survival during impaired mitochondrial pyruvate transport. *Molecular Cell*, *56*, 414–424.

Ying, H., Kimmelman, A. C., Lyssiotis, C. A., Hua, S., Chu, G. C., Fletcher-Sananikone, E., et al. (2012). Oncogenic Kras maintains pancreatic tumors through regulation of anabolic glucose metabolism. *Cell*, *149*, 656–670.

Yuneva, M. O., Fan, T. W., Allen, T. D., Higashi, R. M., Ferraris, D. V., Tsukamoto, T., et al. (2012). The metabolic profile of tumors depends on both the responsible genetic lesion and tissue type. *Cell Metabolism*, *15*, 157–170.

Zierhut, M. L., Yen, Y. F., Chen, A. P., Bok, R., Albers, M. J., Zhang, V., et al. (2010). Kinetic modeling of hyperpolarized ^{13}C1-pyruvate metabolism in normal rats and TRAMP mice. *Journal of Magnetic Resonance*, *202*, 85–92.

CHAPTER THREE

Metabolic Tracing Using Stable Isotope-Labeled Substrates and Mass Spectrometry in the Perfused Mouse Heart

Matthieu Ruiz[*,†], Roselle Gélinas[†,‡], Fanny Vaillant[§,¶], Benjamin Lauzier[||], Christine Des Rosiers[*,†,‡,1]

[*]Department of Nutrition, Université de Montréal, Montreal, Quebec, Canada
[†]Montreal Heart Institute, Université de Montréal, Montreal, Quebec, Canada
[‡]Department of Medicine, Université de Montréal, Montreal, Quebec, Canada
[§]IHU Institut de Rythmologie et Modélisation Cardiaque, Fondation Bordeaux, Université de Bordeaux, Bordeaux, France
[¶]Inserm U1045 Centre de Recherche Cardio-Thoracique de Bordeaux, Université de Bordeaux, Bordeaux, France
[||]Institut du thorax, Université de Nantes, Nantes, France
[1]Corresponding author: e-mail address: christine.des.rosiers@umontreal.ca

Contents

Methods in Enzymology, Volume 561
ISSN 0076-6879
http://dx.doi.org/10.1016/bs.mie.2015.06.026

Abstract

There has been a resurgence of interest for the field of cardiac metabolism catalyzed by evidence demonstrating a role of metabolic dysregulation in the pathogenesis of heart disease as well as the increased need for new therapeutic targets for patients with these diseases. In this regard, measuring substrate fluxes is critical in providing insight into the dynamics of cellular metabolism and in delineating the regulation of metabolite production and utilization. This chapter provides a comprehensive description of concepts, guidelines, and tips to assess metabolic fluxes relevant to energy substrate metabolism using ^{13}C-labeled substrates and ^{13}C-isotopomer analysis by gas chromatography–mass spectrometry (GC–MS), and the *ex vivo* working heart as study model. The focus will be on the mouse and on flux parameters, which are commonly assessed in the field, namely, those relevant to substrate selection for energy metabolism, specifically the relative contribution of carbohydrate (glucose, lactate, and pyruvate) and fatty acid oxidation to acetyl-CoA formation for citrate synthesis, glycolysis, as well as anaplerosis. We provide detailed procedures for the heart isolation and perfusion in the working mode as well as for sample processing for metabolite extraction and analysis by GC–MS and subsequent data processing for calculation of metabolic flux parameters. Finally, we address practical considerations and discuss additional applications and future challenges.

1. INTRODUCTION

While the role of dysregulated myocardial energy substrate metabolism in disease pathogenesis is now well recognized (for reviews see, Neubauer, 2007; Bugger & Abel, 2014), much remains to be understood about the underlying mechanisms under various (patho)physiologically relevant conditions in order to better translate this knowledge into therapeutic strategies. In this regard, research on cardiac metabolism has been facilitated by the development of techniques for perfusion of the beating heart *ex vivo*, which enables the study of intrinsic heart function in the absence of external neurohormonal influences, without changes in peripheral resistance and at fixed pre- and afterload pressures. This study model, first introduced in 1895, involved retrograde perfusion of the mammalian heart with a physiological buffer solution, which is referred to as the Langendorff mode (Langendorff, 1895; for review see, Taegtmeyer, 1995). Despite the fact that *ex vivo* Langendorff-perfused heart remains commonly used in the field of cardiovascular research, the *ex vivo* working heart model, initially developed by Otto Frank et al. for the frog (Frank, 1959) and subsequently by Neely, Libermeister, Battersby, and Morgan (1967) for the rat, is more relevant to the *in vivo* situation in terms of workload, which is a major determinant of energy substrate metabolism.

The *ex vivo* working heart has been adapted for investigations of cardiac metabolism using radioactive-labeled substrates in the rat (Lopashuck & Tsang, 1987; Taegtmeyer, Hems, & Krebs, 1980) and mouse (Belke, Larsen, Lopaschuk, & Severson, 1999). For metabolic investigations in any biological systems, including the heart, the use of stable isotopes and carbon 13 (^{13}C) analysis by nuclear magnetic resonance (NMR) (Jeffrey, Roach, Storey, Sherry, & Malloy, 2002) and gas chromatography–mass spectrometry (GC–MS) offers, however, several advantages. In this regard, we have previously published comprehensive and critical reviews on the use of ^{13}C-isotopomer analysis by GC–MS and/or NMR as applied to investigations of cardiac metabolism (Des Rosiers & Chatham, 2005; Des Rosiers, Labarthe, Lloyd, & Chatham, 2011; Des Rosiers, Lloyd, Comte, & Chatham, 2004). These articles also highlight the advantages linked to the measurements of metabolic fluxes versus static measurements of metabolites, protein, or RNA to probe the dynamics of cellular metabolism and to better deciphering the complex regulation of metabolite production and utilization.

The principle and theory behind our metabolic flux measurements using ^{13}C-labeled substrates as applied to the investigation of energy substrate metabolism in the *ex vivo* perfused working heart are building on previous work in the perfused liver (Brunengraber, Kelleher, & Des Rosiers, 1997; Des Rosiers & Fernandez, 1995; Des Rosiers, Fernandez, David, & Brunengraber, 1994; Des Rosiers et al., 1995). First described in 1997 for Langendorff-perfused rat hearts (Comte, Vincent, Bouchard, & Des Rosiers, 1997; Comte, Vincent, Bouchard, Jetté, et al., 1997), this approach was subsequently adapted to the working rat (Vincent, Khairallah, Bouchard, & Des Rosiers, 2003) and mouse (Khairallah et al., 2004) heart. In these published studies and review articles (Des Rosiers et al., 2004, 2011; Des Rosiers & Chatam, 2005), methods are described for the assessment of the following flux parameters relevant, respectively, to mitochondrial and cytosolic energy metabolism, namely: (i) flux ratios reflecting the relative contributions of fatty acid (FA) oxidation, pyruvate decarboxylation (PDC), and carboxylation to citrate formation and (ii) rates of lactate and pyruvate efflux, as a proxy for glycolysis. None of these publications, however, provides a consolidated and detailed description of all methods needed to enable one to conduct such metabolic flux studies in the *ex vivo* perfused working mouse heart, and which are currently routinely used. This chapter aims at filling this gap by providing an updated and complete description of the theory behind these metabolic flux measurements as well as of protocols for extraction and analysis of metabolites from heart tissues and perfusates by GC–MS. Finally, we consider assumptions behind these flux measurements as well as how this approach can be transposed to assess other metabolic systems.

2. STUDY MODEL: MOUSE HEART PERFUSION *EX VIVO* IN THE WORKING MODE WITH SEMI-RECIRCULATING BUFFER

In this section, we will describe the procedures for surgical isolation of the heart from mice as well as the setup for its *ex vivo* perfusion in the working mode using semi-recirculating buffer. This setup was developed and has been routinely used in our laboratory over the past 10 years, specifically for experiments with stable isotope-labeled substrates. Since *ex vivo* perfusion in the working mode requires prior perfusion in the Langendorff mode, general and specific considerations about perfusion conditions and buffers will be covered for both modes. The details of these procedures may, however, vary from one laboratory to another with respect to anesthesia, the setup and apparatus for *ex vivo* heart perfusion as well as the buffer composition. It is noteworthy that this study model is technically demanding; it may require an unexperienced person several months before mastering all steps. The success rate for an experienced person using mice having body weight >20 g is ~75%.

2.1 Experimental Setup

Figure 1 depicts the experimental setup for *ex vivo* mouse heart perfusion in the semi-recirculating mode and the legend lists all its components, which are identified by numbers (**in bold**). Although optional, this setup may be enclosed in a homemade wood cabinet with Plexiglas doors to facilitate temperature control, while protecting the entire system from dust. Note that the perfusion pressure in the Langendorff perfusion mode is set by the height of the water column; hence, at least 1 m must be left between ceiling and the perfusion cabinet. The setup is a modification of that previously described for perfusion of the working heart in the recirculating mode (Belke et al., 1999), except for the followings. The coronary effluent perfusate, which contains metabolites released by the heart, is not recirculated but continuously collected for various assays; only the aortic flow is recirculated into the buffer reservoir (**2** in Fig. 1). Consequently, the total volume needed for a

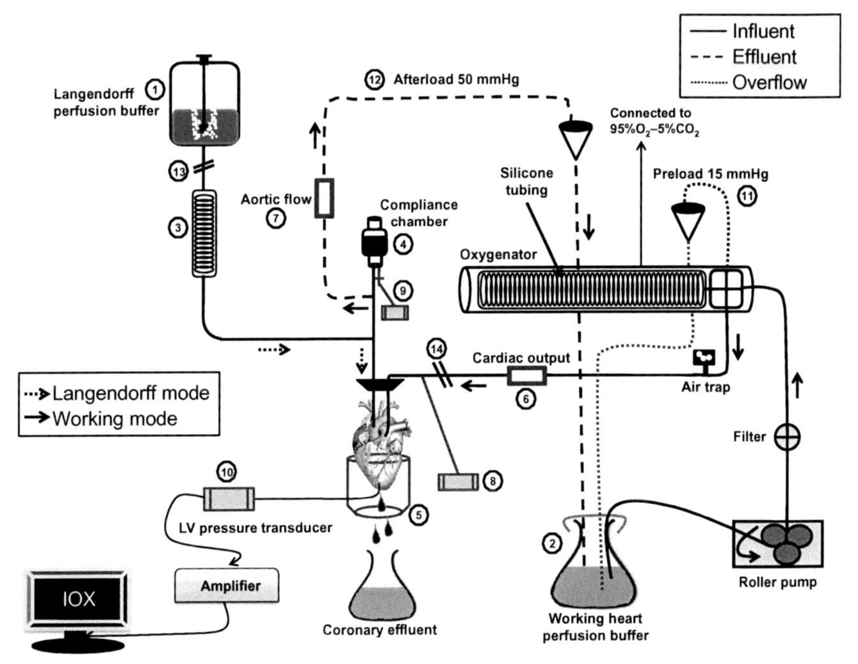

Figure 1 Schematic overview of the semi-recirculating working heart. See Section 2.1 for details. The number refers to the following items: **1**: Langendorff buffer reservoir, **2**: working mode perfusion buffer reservoir, **3**: helical glass coil, **4**: compliance chamber, **5**: opened heart chamber, **6**: electromagnetic flow probes (atrial inflow), **7**: electromagnetic flow probes (aortic outflow), **8** and **9**: pressure transducers (preload and afterload pressures), **10**: pressure transducer (intraventricular pressures), **11**: preload line (15 mmHg), **12**: afterload line (50 mmHg), **13**: three-way valve (aortic line), and **14**: three-way valve (atrial line). Abbreviation: LV = Left Ventricle. (See the color plate.)

given perfusion experiment, about 200 ml, is greater than for a similar heart perfusion in the recirculating mode.

The setup includes the following components, indicated by numbers (**in bold**) referring to Fig. 1:

(1) *Water jacketed glassware*: Two buffer reservoirs (**1**: >1 l; and **2**: ~200 ml), a helical glass coil (**3**), a compliance chamber (**4**: 1.5 cm^3 of air; inserted directly into the afterload line; to set arterial compliance); an opened heart chamber (**5**), and an oxygenator (cf. below for details), which are all interconnected using gas impermeable tubing (PVC tubing, ID 3.2 mm, OD 6.4 mm, wall 1.6 mm; TYGON S-50-HL). The outer jacket of all glasswares is connected using silicone tubing (ID 6.35 mm, OD 9.53 mm, wall 1.59 mm; SILASTIC Laboratory) to a recirculating heated water bath. The temperature of the bath is set to enable the perfused heart to reach a physiological temperature of 37.5 °C (assessed using a thermocouple), which is critical for heart's function and capacity to withstand stress (Templeton, Wildenthal, Willerson, & Reardon, 1974).

(2) *A gassing system to maintain perfusion buffers at pH 7.4 and a* $pO_2 > 500$ *mmHg*: A gas cylinder (95% O_2–5% CO_2) is connected through gas impermeable silicone tubing to: (A) a fritted glass bubbler inserted into the albumin-free perfusion buffer reservoir (**1**) and (B) a water jacketed glass oxygenator, with ~7 m of oxygen permeable silicone tubing (ID 1.58 mm, OD 2.41 mm) enrolled tightly in the gassed chamber (Gamcisk, Forder, Millis, & McGovern, 1996), for the albumin-containing buffer (**2**). N.B. The buffer needs to be continuously pumped through the oxygenator, while excess buffer is returned to the main reservoir through the oxygenator overflow outlet.

(3) *Several probes for continuous monitoring of physiological parameters*: (i) Two electromagnetic flow probes are inserted in the atrial inflow (**6**) and aortic outflow (**7**) lines, respectively. (ii) Three pressure transducers (**8**, **9**, and **10**) are connected: (a) to the atrial (**8**: via PVC tubing, ID 3.2 mm, OD 6.4 mm, wall 1.6 mm; TYGON S-50-HL) and aortic (**9**) line to monitor preload and afterload pressures, which are typically set at physiological level, namely, 15 and 50 mmHg (cf. **11** and **12**, respectively) and (b) to a PE-50 intraventricular cannula via a PE-60 polyethylene tubing (ID 0.76 mm, OD 1.22 mm, see below for details) to monitor left ventricular function (**10**).

N.B. Practical considerations and limitations of our setup include: (1) The aforementioned probes are connected to the following modules: (i) For flows: Transonic system with Transonics PXN Inline flow probes, (ii) For pre- and afterload pressures: Micromed blood pressure analyzer, and (iii) intraventricular pressure: EMKA technologies. There are, however, other companies proposing equivalent recording instruments and softwares (e.g., ADInstruments). (2) To assess left ventricular function, we use a fluid-filled catheter (PE-50 polyethylene, Fig. 2) as our intraventricular cannula instead of the Millar catheter system since it is more affordable. We are, however, limited to a low resonant frequency, which could lead to slight underestimation of some functional values and especially concerning the first derivate of maximum left ventricular systolic pressure. (3) Our working mouse heart model enables assessment of left ventricular function only. A biventricular isolated heart model has recently been described and will assess both left and

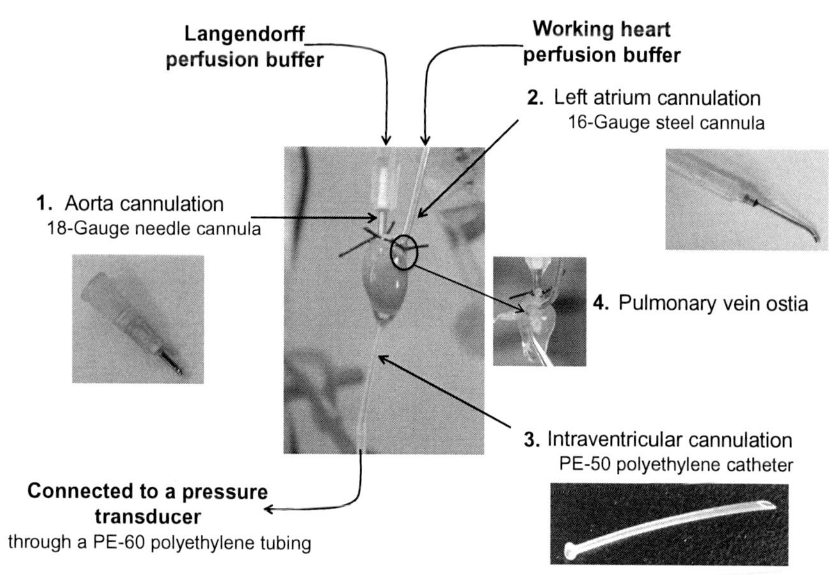

Figure 2 Close-up view of the cannulated isolated heart. See Section 2.3 for details.

right ventricular functions (Asfour, Wengrowski, Jaimes, Swift, & Kay, 2012); to the best of our knowledge, this model has, however, not yet been used for metabolic flux studies.

(4) *General care of the perfusion system*: At the end of each day, the entire perfusion system should be extensively washed with >2 l of hot water prior to rinsing with 2 additional liters of distilled water. All silicone tubing must also be changed regularly to avoid bacterial contamination. The glass oxygenator and pressure transducers are cleaned using 1 N HCl or 100% ethanol, respectively, prior to rinsing with distilled water.

2.2 Perfusion Buffers

2.2.1 General Considerations

All perfusion buffers are prepared on the day of the experiment. Since the concentrations of ions, particularly free calcium, and substrates (carbohydrates (CHOs): glucose, lactate, pyruvate, FAs, etc.) will impact on heart contractility and metabolism, care should be taken in preparing these solutions. Prior to any given sets of experiments, decisions need also to be made about the buffer substrate composition, which depends on the study's objective. For the initial perfusion in the Langendorff mode, we used a modified Krebs–Henseleit buffer, the composition of which is based on previous publications (for review see, Liao, Podesser, & Lim, 2012). For perfusion of the working heart, the buffer composition is chosen to mimic physiological conditions on the basis of the following considerations.

(1) *Free Ca^{2+} concentration*: This is set at 1.5 m*M* and is routinely assessed for a given set of perfusions experiments (e.g., 1.6 ± 0.3 m*M*; Gelinas et al., 2008). This is important since some compounds of interest may have calcium-chelating properties; this was the case for hydroxycitrate and 1,2,3-benzene tricarboxylic acid, which are inhibitors of ATP-citrate lyase (Poirier et al., 2002), and of the tricarboxylic acid transporter (Vincent, Comte, Poirier, & Des Rosiers, 2000), respectively. Although we initially used higher free Ca^{2+} concentration (2.5 m*M*; Khairallah et al., 2004), this was found not to be suitable for a transgenic mouse model that displays increased susceptibility to necrosis such as the *mdx* mice (Khairallah et al., 2007).

(2) *Substrates, hormones, and cofactors*: We have assessed plasma concentrations for some substrates in various strains of mice both in the fed and fasted state (Gelinas et al., 2008, 2011; Khairallah et al., 2004). Reference values can also be found in the "phenome database" (Jackson Laboratories, phenome.jax.org). In addition to glucose, lactate and pyruvate are added at a physiological concentration and ratio (~10) to ensure adequate supply of CHOs (Bunger, Mallet, & Hartman, 1989; Chatham, Des Rosiers, & Forder, 2001) and minimize cytosolic redox state perturbations (Corkey & Shirihai, 2012). The concentration of glucose and FAs is typically between 7–15 and 0.4–1.2 m*M*, respectively, to mimic a given (patho)physiological state, such as fed, fasting, or diabetes. Among FAs, the long chain FAs palmitate or oleate, which are the most abundant ones in plasma, are used in the form of an albumin complex (in a final molar ratio 4:1 and 6:1, respectively) (Gelinas et al., 2011; Khairallah et al., 2004). While a mixture of FAs would be more representative of the *in vivo* situation, data interpretation may, however, be more complicated. Of note, the addition of FAs has been found to affect cardiac function even in the healthy normoxic working heart (Vincent, Bouchard, Khairallah, & Des Rosiers, 2004). We also routinely add insulin and carnitine, which facilitates glucose and FA transport, respectively. Moreover, the addition of carnitine compensates for its potential loss during heart isolation (Rodgers, Christe, Tremblay, Babson, & Daniels, 2001). The addition of other substrates should also be considered such as ketone bodies (Cotter, Schugar, & Crawford, 2013) and amino acids (Marazzi, Rosanio, Caminiti, Dioguardi, & Mercuro, 2008), particularly glutamine, the concentration of which is depleted upon *ex vivo* perfusion (Lauzier et al., 2013). Nevertheless, it should be kept in mind that despite providing what are generally accepted in term of physiological levels of workload and nutrients, the *ex vivo* perfusion environment fails to reproduce the complexity of the *in vivo* situation as evidenced by differences in myocardial citric acid cycle (CAC) intermediate levels following *ex vivo* perfusion versus *in situ* levels (Gelinas et al., 2008; Khairallah et al., 2007).

2.2.2 Buffer for Perfusion in the Langendorff Mode

A modified Krebs–Henseleit buffer is used, which contains: 110 m*M* NaCl, 4.7 m*M* KCl, 2.1 m*M* $CaCl_2$, 0.24 m*M* KH_2PO_4, 0.48 m*M* K_2HPO_4, 0.48 m*M* Na_2HPO_4, 1.2 m*M* $MgSO_4$, 25 m*M* $NaHCO_3$, 11 m*M* glucose, and 0.1 m*M* EDTA supplemented with 0.8 n*M* insulin, 0.2 m*M* pyruvate, 1.5 m*M* lactate, and 50 μ*M* carnitine (Gelinas et al., 2008). Salts and glucose are added in the powder form. Stock solutions are prepared in water prior to any sets of experiments for pyruvate (1 *M*), lactate (1 *M*), carnitine (50 m*M*), and insulin (100 μ*M*). The concentrations of lactate and pyruvate solutions are validated by standard enzymatic assay. All these solutions, for which the pH is adjusted to 7.4, are stored at −20 °C in small aliquots to avoid frequent freeze-thawing cycles for several months (to conduct a given set of experiments), except for insulin, which is unstable, and for which we recommend <2 weeks. When preparing the buffer solution, the first three salts are added in water prior to gassing with CO_2 for a few minutes, until complete dissolution and before adding the remaining salts and glucose, to avoid calcium precipitation. Thereafter, the buffer solution is filtered (cellulose nitrate membrane filters, 5 μm) using a vacuum pump in order to remove impurities, which could obstruct the coronary vessels. The buffer is then poured into the appropriate reservoir and gassed with a mixture of 95% O_2–5% CO_2 to achieve a pH of 7.4. For the Langendorff perfusion step, 1 l of buffer is prepared to ensure adequate perfusion pressure and poured into reservoir (**1** in Fig. 1).

2.2.3 Buffer for Perfusion in the Working Mode

For the working mode perfusion step, a minimum of 200 ml of buffer is prepared for a 30-min perfusion period. The composition of the perfusion buffer for the working heart is essentially the same as for the Langendorff-perfused heart, except for the following additions, which are made after the filtration step: (i) FA complexed with dialyzed albumin (3% final concentration for albumin) and (ii) ^{13}C-labeled substrate(s) for metabolic flux analysis, which for any given perfusion is replacing its corresponding unlabeled substrate(s). We also routinely add epinephrine at a physiological concentration (5 n*M*; from a 50 μ*M* stock solution prepared on the day of perfusion, protected from light, oxygen, and kept at +4 °C), since it was found to stabilize heart rate without the need for pacing, given that the isolated heart is denervated.

2.2.3.1 Dialysis of Albumin (±20% in Final Volume of 1.5 l)

300 g albumin (BSA, fraction V, FA free, Intergen) is dissolved at 4 °C in 1.2 l modified Krebs–Henseleit bicarbonate buffer prepared as described above, but without EDTA and substrates/cofactors/hormone, under gentle agitation in a beaker. After complete dissolution (~24 h), the volume is adjusted to 1.5 l and 30 ml of a FA–albumin complex solution, for instance 20 m*M* oleate or 12 m*M* palmitate (cf. Section 2.2.3.2 for preparation of FA–albumin complex solution) is added prior to pouring of solution in dialysis membranes (molecular weight cutoff 6000–8000, 100 ml of albumin in 20 cm long membrane). The dialysis is achieved at 4 °C against 25 l of the same buffer for ~28 h with agitation and constant bubbling of CO_2. The albumin solution is gassed with CO_2 to remove oxygen and avoid oxidation prior to storage in 200–500 ml aliquots at −20 °C. Prior to any perfusion experiment, the frozen albumin solution is slowly thawed at 4 °C overnight; it should not be heated >60 °C to avoid denaturation. The concentration of albumin in the final solution is assessed using a standard protein assay and that of FA using GC–MS or a commercial kit.

2.2.3.2 FA–Albumin Complex

The following protocol describes how to complex palmitate (12 m*M* final concentration) or oleate (20 m*M* final concentration) with dialyzed FA-free albumin (±20%, prepared as described in Section 2.2.3.1 but without the addition of FA). A 60 m*M* palmitate solution (or 100 m*M* for oleate) is prepared by dissolving the FA powder in 0.5 N NaOH at 70 °C for 10 min by mixing gently and then completed with warm water (75:15, v/v; NaOH/H_2O) with gentle mixing (do not vortex). After dissolution, the FA solution is slowly cool down to 54 °C, mixed with the dialyzed FA-free albumin solution kept at 37 °C (FA/albumin mole ratio: 4:1 for palmitate or 6:1 for oleate), and finally kept at ambient temperature under agitation prior to storage in 40 ml aliquots at −20 °C. The requested number of aliquots for a given perfusion is thawed overnight at 4 °C. The FA concentration of the solution is assessed by GC–MS.

2.2.3.3 ^{13}C-Labeled Substrates

For metabolic flux studies, the ^{13}C-labeled substrates used should be of the highest chemical and isotopic purity in order to minimize error propagation. Of note, the reported enrichment of commercially available ^{13}C-labeled substrates is typically 99%, but this value reflects atom percent enrichment (APE), which is not equal to molar percent enrichment (MPE), particularly for uniformly ^{13}C-labeled substrates. For instance, if the APE of [U-$^{13}C_6$]glucose is 99%, this is not equal an MPE M6 of 99% but may correspond to an MPE M6 of 94% and MPE M5 of 6%. It is thus good practice to assess the MPE of the solution of ^{13}C-labeled substrates prior to conducting any given sets of perfusion experiments. For the purpose of this chapter, we will focus on the following uniformly ^{13}C-labeled substrates to assess their contribution to various metabolic pathways relevant to energy metabolism, namely: (i) [U-$^{13}C_{16}$]labeled palmitate or [U-$^{13}C_{18}$]labeled oleate to assess the contribution of FA β-oxidation to acetyl-CoA formation relative to citrate synthesis, (ii) [U-$^{13}C_3$]labeled lactate plus pyruvate to assess pyruvate partitioning between mitochondrial oxidation to acetyl-CoA and anaplerotic carboxylation to oxaloacetate (OAA) relative to citrate synthesis, and (iii) [U-$^{13}C_6$]glucose to assess the formation of ^{13}C-labeled lactate plus pyruvate from glycolysis. Stock solutions of all ^{13}C-labeled lactate, pyruvate, and FAs are prepared at concentrations similar to their corresponding unlabeled substrates and stored in aliquots at −20 °C to avoid repetitive freeze-thawing cycles. An aliquot of the stock solution is thawed on the day of the experiment, except for [U-$^{13}C_6$]glucose which is added to the buffer as a powder. These ^{13}C-labeled substrates are added to the perfusion buffer at different initial MPE, the choice of which represents a compromise between costs of perfusion versus precision in MPE measurements of downstream metabolites of interest. Typically, [U-$^{13}C_3$]labeled lactate and pyruvate are supplied at 100%, [U-$^{13}C_6$]glucose at 50% and [U-$^{13}C_{16}$]palmitate or [U-$^{13}C_{18}$]oleate at 25% MPE, and the appropriate amount of the corresponding unlabeled substrate is added to achieve the desired final concentration in the perfusion reservoir (**2** in Fig. 1).

2.3 Heart Isolation and *Ex Vivo* Perfusion

2.3.1 Preparation for a Heart Perfusion Experiment

On the day of the perfusion experiment:

- Switch on the water heating bath;
- Prepare 0.5 l of saline solution (0.9% NaCl), which is kept in plastic wash bottle at 4 °C and will be used during surgery to chill the heart to minimize contraction and hence ATP depletion.

- Prepare the cardioplegic solution, a commercial lactated Ringer's solution (273 mOsm/l, pH 6.7, high potassium 34 mEq/l), which is kept at 4 °C in a 50-ml syringe positioned onto an infusion pump set to deliver at a flow rate of ~2 ml/min.
- Prepare the Langendorff and working heart buffers and pour them into the appropriate reservoirs (**1** and **2**, respectively; Fig. 1) and start pumping of the working heart perfusion buffer through the system.
- Start gassing the buffers with a mixture of 95% O_2–5% CO_2 (8000 cc/min) and 100% CO_2 (±20 cc/min) to obtain a $_pO_2 > 500$ mmHg and maintain a pH of 7.4 (~5 and 30 min for reservoir **1** and **2**, respectively). The compliance chamber is filled with 1.5 cm^3 of air (**4** in Fig. 1).
- Fill and calibrate the pressure transducers and flow probes.

2.3.2 Surgery for Heart Isolation and Perfusion in the Langendorff Mode

- Prepare a clean surface with surgical scissors, cannula for the aorta (a 18-gauge steel cannula for which the needle is cut diagonally and buffed to remove the sharp edge, **1** in Fig. 2) and the left atrium (a 16-gauge bended cannula; **2** in Fig. 2); a 10 to 15-mm polyethylene catheter (PE-50) with a fluted end (**3** in Fig. 2); the pump with the 50-ml syringe containing the cold cardioplegic solution.
- Mouse (body weight >20 g) is anesthetized with sodium thiopental (100 mg/kg ip) or a mixture (1µl/g ip) or a mixture of ketamine (100 mg/ml) and xylazine (20 mg/ml)), heparinized (5000 U/kg sc; to avoid thrombus formation in the aorta and coronary vessels) 15 min prior to the surgery, and fixed in a supine position.
- Promptly cut through the abdominal skin and wall using scissors. Once the abdomen is opened, gently pull the liver aside and make an incision on both sides of the thorax (antero-lateral thoracotomy). Immediately chill the exposed heart by flushing of the cold saline solution to stop contraction; flushing is maintained thereafter for the remaining of the procedure.
- Rapidly cut through the diaphragm and set the sternum aside. Clear the heart from the pericardium and using forceps carefully remove the thymus and fat around the aorta.
- Using a clip, pull a suture wire (black braided silk nonabsorbable surgical suture, 2-0) underneath the aorta using fine surgical forceps and prepare for ligation. Gently pull the aorta between the innominate and the

subclavian arteries and practice a hemisection with a 2.5 mm cutting edge scissors. Gently place the cannula into the aorta and make a ligature on the cannula. Finally, cut and remove heart and lungs as a whole by cutting across the aorta, vena cava, pulmonary veins, and trachea.

- Slowly insert into the aorta an 18-gauge cannula (**1** in Fig. 2), which is connected using polyethylene tubing (75 cm length) to the 50-ml syringe containing cold (4 °C) cardioplegic solution. (N.B. Be careful not to force the cannula beyond the aortic valve.) When the cannula is correctly positioned into the aorta, the coronary arteries become translucent. The cardioplegic solution is retroperfused into the large vessels and the blood will be flushed. The cannula is fixed onto the aorta using a double knot (Fig. 2).
- The heart is then excised by slowly cutting underneath tissues, but keeping part of the lungs to preserve the integrity of the left atrium. While maintaining cannula in the aorta–heart axis, quickly place the cannula with the heart onto the setup, on the three-way valve which is connected to the compliance chamber and the Langendorff line. Prior to positioning of the heart, allow buffer from reservoir **1** to flow through the system to avoid formation of air bubbles, which could clog the coronary arteries. The heart should start beating within a few seconds upon perfusion with buffer at 37 °C.
- While the heart is being perfused in the retrograde or Langendorff mode (Fig. 1, green arrow) via the aorta (Fig. 2), carefully and quickly remove remaining lung and fat tissues from the heart, while preserving the integrity of the left atrium, in order to have access to the pulmonary veins.

2.3.3 Switching from Langendorff to the Working Mode

- N.B. Preparing the Langendorff-perfused heart for switching to the working mode requires about 5 min.
- Step 1: Insertion of the intraventricular cannula for monitoring of left ventricular function (Fig. 1): A polyethylene catheter (PE-50; 10–15 mm long; fluted end) is slowly and carefully inserted through the pulmonary veins ostia into the left ventricle, via the mitral valve, and anchored at the apex of the heart through the fluted end. The end of the catheter is connected to the calibrated fluid-filled pressure transducer (**10** in Fig. 1 and **3** in Fig. 2) via a PE-60 polyethylene catheter. N.B. Be careful not to force catheter through the septum to avoid damaging the His bundle, which may elicit arrhythmias.

- Step 2: Insertion of the left atrial cannula (**2** in Fig. 2): A 16-gauge steel needle is inserted into the left atrium passing through the pulmonary vein ostia (**4** in Fig. 2) and a surgical thread (black braided silk nonabsorbable surgical suture, 2-0) with a loose knot is positioned around this steel cannula. Connective tissue is slowly pulled around the cannula, while the assistant tightens the knot to secure the cannula in place. Before securing with a double knot, it is important to check for leaks since this will result in overestimation of coronary flow, while decreasing the preload. This is achieved by visual inspection and by asserting that the cardiac output (CO) is within acceptable values, namely, ~10 ml/min. If there is evidence of a leak, the left atrial cannula needs to be repositioned and secured with another knot.
- Step 3: Switching from the Langendorff to the working mode (Fig. 1, red arrow): Simultaneously close the three-way valve from the aortic line receiving the Langendorff buffer (**13** in Fig. 1), while opening that from the left atrium receiving the working buffer (**14** in Fig. 1) via the preload reservoir. The heart should resume contractile function within 1–2 min. Both preload and afterload pressures are checked and adjusted, respectively, to 15 and 50 mmHg. N.B. These pressures are continuously monitored throughout the perfusion and may need some fine readjustment, which is achieved by repositioning of the height of the preload and afterload lines.
- N.B. Commonly encountered sources of abnormal cardiac function are (i) Pinching of the pulmonary artery, which results in right atrium and ventricle swelling; this can be fixed by promptly performing a small incision in the pulmonary artery. (ii) Inadequate heart temperature or pinching of the right atrium (sinus node), which results in irregular or slow heart beat; this can be fixed by rapid repositioning or replacing the thread around the cannula.

2.4 Evaluating Contractile Function and Integrity of *Ex Vivo* Working Mouse Heart

Physiological parameters, which can be continuously monitored and recorded using computer software, as well as additional parameters that can be calculated from them, can be divided in three categories:

(1) *Cardiac flows*: CO and aortic flow are continuously monitored using calibrated electromagnetic flow probes (**6** and **7** in Fig. 1) and coronary flow (CF) is calculated from:

$$\mathrm{CF(ml/min)} = \mathrm{CO(ml/min)} - \mathrm{AF(ml/min)} \quad (1)$$

(2) *Cardiac function*: The intraventricular catheter (**3** in Fig. 2) enables the acquisition of left ventricular pressures as well as parameters relevant to the cardiac cycle. In our setup, the catheter is connected to a pressure transducer that transmits signals to an amplification module (EMKA technologies) and the associated IOX software (2.4.5 version, EMKA technologies) is used for data acquisition, analysis, and storage. Recorded parameters include heart rate (calculated by the software from the number of heart cycles per minute), systolic and diastolic pressures, indices of systolic function and contractility, namely, maximum left ventricular systolic pressure ($LVSP_{max}$), maximum value for the first derivative of $LVSP_{max}$ ($+dP/dt_{max}$), isovolumic contraction time and systolic ejection time and indices of diastolic function and relaxation, namely, minimal left ventricular pressure (LVP_{min}), left ventricular end diastolic pressure (LVeDP), Tau 1/2, minimum value for the first derivative of $LVDP_{min}$ ($-dP/dt_{min}$), isovolumic relaxation time and diastolic filling time (listed in Table 1).

The following functional parameters, which are indices of cardiac work, can also be calculated: (i) left ventricular developed pressure (LVDevP), (ii) rate pressure product (RPP), (iii) cardiac power (CP), and (iv) stroke volume (SV = blood volume pumped by the left ventricle per beat) using the following equations (listed in Table 1):

$$\mathrm{LVDevP(mmHg)} = \mathrm{LVSPmax(mmHg)} - \mathrm{LVeDP(mmHg)} \quad (2)$$

$$\mathrm{RPP(mmHg\ beats/min)} = \mathrm{LVDevP(mmHg)} \times \mathrm{HR(beats/min)} \quad (3)$$

$$\mathrm{CP(mW)} = \mathrm{CO}\left(\mathrm{m^3/s}\right) \times \mathrm{LVDevP(Pa)} \times 10^3 \quad (4)$$

whereas factor of 133.32 Pa/mmHg is used to convert LVDP from mmHg to Pa and the following equation to convert CO in m^3/s: CO (m^3/s) = CO (ml/min)/60/1,000,000.

$$\mathrm{SV(ml/beats)} = \mathrm{CO(ml/min)}/\mathrm{HR(beats/min)} \quad (5)$$

(3) *Other physiological parameters*: (i) Cardiac membrane integrity is assessed from lactate dehydrogenase (LDH) release in aliquots of coronary effluent perfusates, which are stored on ice until processed on the same day for standard enzymatic assay. (ii) pO_2, pCO_2, free Ca^{2+}, and pH are monitored in the atrial influent and coronary effluent perfusates using a pH, blood gas, and electrolyte analyzer (ABL 80 series, Radiometer).

Table 1 Physiological Parameters Measured in *Ex Vivo* Working Hearts from Control Healthy Mice with a Mixed C57Bl6/129 Background Perfused with 11 m*M* Glucose, 1.5 m*M* Lactate, 0.2 m*M* Pyruvate, and 0.7 m*M* Palmitate Bound to 3% Dialyzed Albumin

Physiological Parameters	Mean ± SEM
HR (beats/min)	344 ± 14
Systolic function	
$LVSP_{max}$ (mmHg)	107 ± 3
$+dP/dt_{max}$ (mmHg/s)	5509 ± 139
Isovolumic contraction time (ms)	35.9 ± 1.7
Systolic ejection time (ms)	42.4 ± 0.8
Diastolic function	
LVP_{min} (mmHg)	−6.14 ± 2.01
LVeDP (mmHg)	6.71 ± 0.64
Tau 1/2 (ms)	2.57 ± 0.20
$-dP/dt_{min}$ (mmHg/s)	−4058 ± 123
Isovolumic relaxation time (ms)	13.7 ± 0.6
Diastolic filling time (ms)	85.7 ± 8.54
LVDevP (mmHg) (Eq. 2)	113 ± 5
Cardiac flow	
Cardiac output (ml/min)	11.3 ± 0.5
Aortic flow (ml/min)	8.99 ± 0.50
Coronary flow (ml/min) (Eq. 1)	2.33 ± 0.40
Stroke volume (ml/beat) (Eq. 5)	0.003 ± 0.001
Cardiac work	
Cardiac power (mW) (Eq. 4)	2.81 ± 0.15
MVO_2 (μmol/min) (Eq. 6)	0.78 ± 0.14
Cardiac efficiency (mW min/μmol) (Eq. 7)	5.05 ± 1.66
RPP (mmHg/beats/min) (Eq. 3)	38,898 ± 2061
Other	
LDH (mU/min)	6.14 ± 1.27

Values are expressed as means ± SEM of seven hearts for the 20 to 25-min perfusion period. HR, heart rate; $LVSP_{max}$, maximum left ventricular systolic pressure; LVP_{min}, minimal left ventricular pressure; LVeDP, left ventricular end diastolic pressure; LVDP, left ventricular developed pressure; RPP, rate pressure product; MVO_2, myocardial oxygen consumption; LDH, lactate dehydrogenase.

Myocardial oxygen consumption (MVO_2) and cardiac efficiency (CE) are estimated from:

$$MVO_2(\mu mol/min) = [O_2\,influent(mM) - O_2\,effluent(mM)] \times CF(ml/min) \quad (6)$$

where [O_2(m*M*)] is calculated using pO_2 values assessed using the gas analyzer and the following equation: (pO_2*1.06)/760,where 1.06 m*M* is the concentration of dissolved O_2 at 100% saturation and 760 is an approximate value for atmospheric pressure in mmHg. N.B. A fiber-optic oxygen probe can also be placed in the pulmonary trunk for online recording of pO_2 in the coronary effluate (How et al., 2005, 2006). These references also provide insightful considerations about measurements of cardiac efficiency.

$$CE(mW\ \mu mol/min) = CP(mW) \times MVO_2(\mu mol/min) \quad (7)$$

Table 1 (last column) reports representative values for all physiological parameters, which were assessed in *ex vivo* working normoxic hearts isolated from 20-week-old fed, healthy control mice with a mixed genetic background; C57Bl6/129 perfused at physiological preload and afterload with a mix of CHOs as described above and with 0.7 m*M* palmitate as the FA substrate.

N.B. (1) Based on our experience, we recommend conducting between 10 and 15 perfusion experiments for a given set of conditions to obtain reliable functional parameters, while a minimum of five perfusions is required to obtain reliable values for flux parameters for a given ^{13}C-labeled substrate. (2) Electrical pacing of the heart can also be used to limit spontaneous variations in heart rate and to study the impact of increasing heart rate or the heart's susceptibility to atrial or ventricular arrhythmia (Baker, London, Choi, Koren, & Salama, 2000; Choisy, Arberry, Hancox, & James, 2007). For this, two electrodes are positioned on the right atrium (to keep sinus node activation) and the reference electrode on the aorta. It is noteworthy that electrical pacing can be achieved at a higher, albeit not at a lower frequency than the heart's spontaneous frequency. To pace the hearts at a lower frequency than that achieved spontaneously, the sino-atrial node should be removed to prevent its spontaneous electrical activity. (3) It is noteworthy that cardiac function varies with biological sex (Wittnich, Tan, Wallen, & Belanger, 2013), circadian clock (Young, 2006), as well as with mouse strains (Barnabei, Palpant, & Metzger, 2010; Vaillant et al., 2014). For transgenic mouse studies, it is therefore crucial to use mouse littermates as controls.

3. METABOLIC FLUX MEASUREMENTS IN THE PERFUSED MOUSE HEART USING ^{13}C-LABELED SUBSTRATES AND Gas Chromatography-Mass Spectrometry

This chapter provides a consolidated and updated description of our previously published methods for the assessment of the following flux parameters relevant, respectively, to mitochondrial and cytosolic energy metabolism, namely: (1) flux ratios reflecting the relative contributions of FA oxidation, PDC, and carboxylation to citrate formation and (2) rates of lactate and pyruvate efflux, as proxy for glycolysis. This includes the theory behind these metabolic flux measurements (Sections 3.1 and 3.2) as well of methods for extraction and analysis of metabolites from heart tissues and perfusates by GC–MS (Section 3.3).

3.1 Theory

Metabolic flux studies using ^{13}C-labeled substrates require some prior knowledge of biochemical reactions involved in the pathway as well as in the tissue of interest. For a detailed description of metabolic pathways relevant to energy metabolism and their regulation in the heart, please refer to recent review articles (Doenst, Nguyen, & Abel, 2013; Kolwicz, Purohit, & Tian, 2014). We will illustrate the theory behind flux analysis related to CAC metabolism by focusing sequentially on the following [^{13}C]labeled substrates and metabolic flux parameters: (1) [U-$^{13}C_{16}$]palmitate to assess the relative contribution of exogenous long-chain FA oxidation to citrate formation via β-oxidation, (2) [U-$^{13}C_3$]labeled lactate plus pyruvate (L + P) to assess the relative contribution of CHOs to citrate formation through decarboxylation (oxidation) and carboxylation (anaplerosis), and (3) [U-$^{13}C_6$]glucose for estimation of rates of lactate and pyruvate efflux rates as a proxy of glycolysis. The metabolism of these [U-^{13}C]labeled substrates will be described, emphasizing the nature of the various ^{13}C-labeled metabolites formed using the following terminology (see Brunengraber et al., 1997 for details): "positional and mass isotopomers" to refer to molecules that differ, respectively, in the position and number of heavy atoms, specifically carbon 13 atoms in this chapter. MS allows measurements of the mass distribution of molecules or molecular fragments, of different mass isotopomers, since the latter result in different molecular weights. For example, the different mass isotopomers of pyruvate are either unlabeled or labeled ^{13}C atoms on 1, 2, or

3 carbons; these will be referred to as (M), (M + 1), (M + 2), and (M + 3) pyruvate or—as used in this chapter—as M, M1, M2, and M3 pyruvate.

3.1.1 Nature of ^{13}C-Labeled Isotopomers Formed

Figures 3 and 4 illustrate the ^{13}C-labeling of citrate and other CAC intermediates resulting from the metabolism of [U-$^{13}C_{16}$]palmitate and [U-$^{13}C_3$]L + P to acetyl-CoA and/or OAA at the first turn of the CAC, respectively. For simplicity, we assume that only one substrate, [U-$^{13}C_{16}$] palmitate, or group of substrates [U-$^{13}C_3$](L + P), is labeled with ^{13}C in each condition and that glucose is unlabeled.

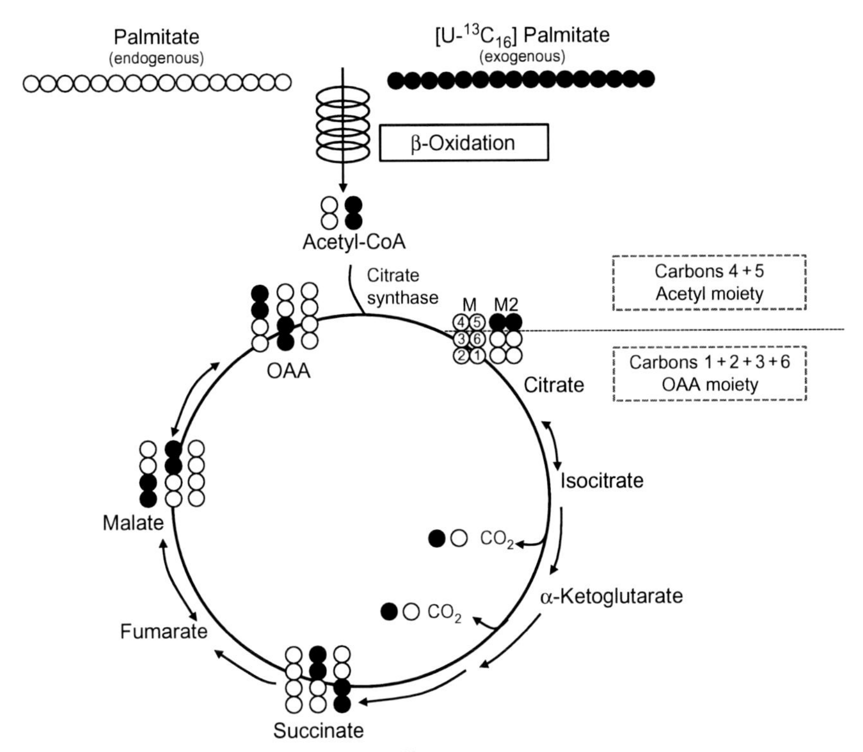

Figure 3 Schematic demonstrating the ^{13}C-labeling of citrate and other citric acid cycle (CAC) intermediates with [U-$^{13}C_{16}$]palmitate as substrate, at the end of the first turn of the CAC. Endogenous unlabeled and exogenous ^{13}C-labeled palmitate are β-oxidized and enter the CAC as unlabeled (M) and [1,2-$^{13}C_2$]labeled (M2) acetyl-CoA, respectively, which through condensation with unlabeled oxaloacetate (OAA) forms unlabeled and [4,5-^{13}C2]citrate (M2). Filled circles, ^{13}C; open circles, ^{12}C. See Section 3.1.1 for details.

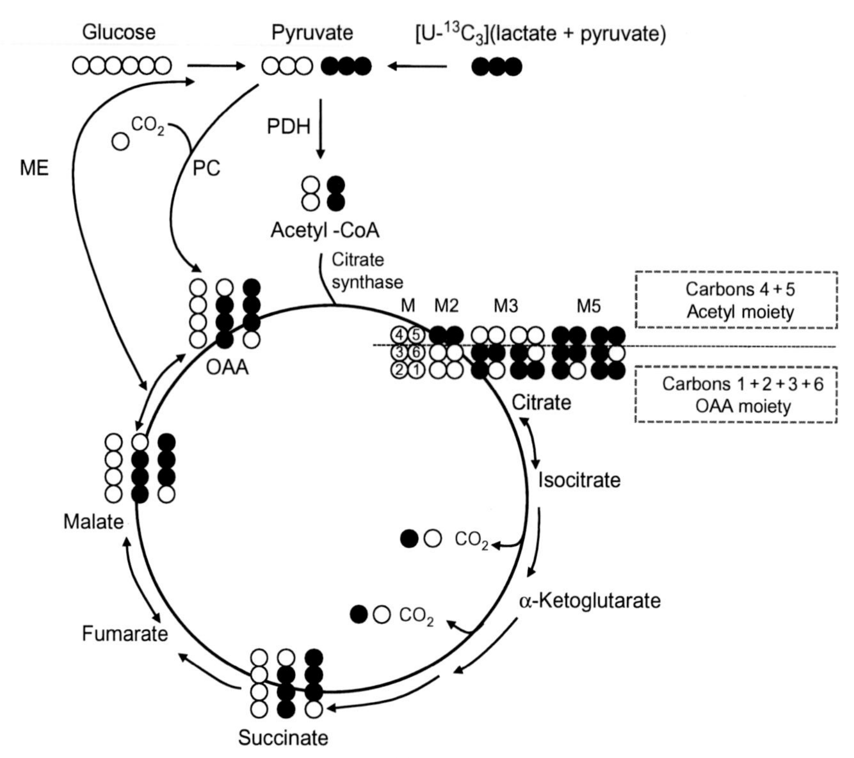

Figure 4 Schematic demonstrating the ^{13}C-labeling of citrate and other CAC intermediates with [U-$^{13}C_3$]lactate + pyruvate (L + P) as substrates, at the end of the first turn of the CAC cycle. Unlabeled and ^{13}C-labeled pyruvate enters the CAC via (1) decarboxylation to unlabeled (M) and 1,2-$^{13}C_2$]acetyl-CoA (M2), respectively or (2) carboxylation via the anaplerotic pathway to unlabeled and [1,2,3-^{13}C]- or [2,3,4-$^{13}_{2}$]OAA (M3), respectively. The condensation of acetyl-CoA (M and M2) with OAA (M and M3) leads to the formation of various isotopomers of citrate labeled in its acetyl (C-4 + 5) and/or OAA (carbon-1 + 2 + 3 + 6). Filled circles, ^{13}C; open circles, ^{12}C; PDH, pyruvate dehydrogenase; PC, pyruvate carboxylase; ME, malic enzyme. See Section 3.1.1 for details.

[U-$^{13}C_{16}$]palmitate is β-oxidized to [1,2-$^{13}C_2$] acetyl-CoA, a M2 isotopomer. After condensation with unlabeled OAA catalyzed by citrate synthase will form citrate molecules labeled on carbons 4 and 5 or [4,5-$^{13}C_2$]citrate (M2), which arise, respectively, from even and odd carbon numbers of the palmitate molecules at each turn of the β-oxidation cycle (Fig. 3). Similarly, [U-$^{13}C_3$]L + P, a M3 isotopomer, is decarboxylated to [1,2-$^{13}C_2$]acetyl-CoA by pyruvate dehydrogenase and labels carbons 4 and 5 of citrate, corresponding to carbons 2 and 3 of pyruvate, respectively.

In addition, [U-$^{13}C_3$]L + P is carboxylated to [1,2,3-$^{13}C_3$]OAA (M3) by pyruvate carboxylase or [1,2,3-$^{13}C_3$]malate by NADP-malic enzyme. Because of the reversibility of the fumarase and malate dehydrogenase reactions, [2,3,4-$^{13}C_3$]OAA (M3) will also be formed; the proportion of [1,2,3-$^{13}C_3$]- versus [2,3,4-$^{13}C_3$]-labeled OAA molecules depends on the degree of isotope randomization resulting from these two reactions, which we assume to be fully equilibrated. These two positional isotopomers will label citrate on carbons 6 + 3 + 2 and 3 + 2 + 1, respectively, resulting in the formation of M3 citrate isotopomers provided that there is negligible condensation between labeled acetyl-CoA and OAA. Upon further metabolism in the CAC, these citrate isotopomers will become M2 and M1 isotopomers of OAA. Condensation between labeled acetyl-CoA and OAA will result in a mixture of different citrate mass and positional isotopomers. For example, condensation between M2 acetyl-CoA and M3, M2 and M1 OAA results in M5, M4, and M3 citrate isotopomers, respectively, which upon further metabolism in the CAC form a mixture of M4, M3, M3, and M1 positional isotopomers of OAA. Theoretically, up to 64 possible isotopomers of citrate can be formed, labeled in their acetyl (carbons 4 + 5: C-4 + 5) and OAA (C-1 + 2 + 3 + 6) moieties; this is calculated from the equation: 2^n, where n is the number of carbons in a given molecule. In practice, however, this number of citrate isotopomers formed is lower. For example, in hearts perfused with [U-^{13}C]labeled substrates leading to the formation of M + 2 acetyl-CoA, there is negligible formation of citrate isotopomers labeled with only one ^{13}C atom on carbon 4 or 5 reducing the number of possible citrate isotopomers. (N.B. In practice, a low level of M1 acetyl-CoA can be detected and is attributed to the 98–99% APE of most commercially available ^{13}C-labeled substrates.) Furthermore, the formation of citrate isotopomers M4–M6 depends on the probability of condensation of labeled acetyl-CoA and OAA, estimated by multiplying their respective fractional enrichments (cf. Section 3.2.1 below for definition), which in hearts perfused with [U-^{13}C]labeled LCFAs as sole-labeled substrate, is low because OAA ^{13}C-enrichment is low.

3.2 Calculation of Substrate Fluxes

Despite the apparent complexity of the metabolism of ^{13}C-labeled citrate, its specific isotopomer distribution is determined in a predictable manner by the relative contributions of different unlabeled or ^{13}C-labeled substrates entering the CAC via (1) acetyl-CoA, via FA β-oxidation or PDC and

(2) OAA, via pyruvate carboxylation (PC), as well as the influx of unlabeled carbons entering the CAC via anaplerotic pathways other than PC. These various, relative flux parameters must be assessed under conditions of metabolic and isotopic steady state; this means that tissue levels of CAC intermediates and related metabolites are constant and that their ^{13}C-enrichment is maximal and stable for a given condition. This requires perfusion of the heart for 25–30 min with ^{13}C-labeled substrates entering at the level of OAA, although 15 min is sufficient for labeling of acetyl-CoA or pyruvate and lactate (Comte et al., 1997). The equations that were developed to calculate substrate flux ratios relevant to using GC–MS are based on ^{13}C-isotopomer balance, which were initially described for perfused livers (Des Rosiers, David, Garneau, & Brunengraber, 1991; Des Rosiers & Fernandez, 1995; Des Rosiers et al., 1994, 1995) and subsequently adapted for the heart (Comte et al., 1997). All equations are described in terms of fractional flux contribution (FC) parameters and metabolite isotopomer distribution.

3.2.1 Notation

The following notation and definitions are used to calculate substrate fluxes using the various equations:

(i) *Metabolites*: AC, acetyl-CoA; CIT, citrate; OAA, oxaloacetate; PAL, palmitate; PYRi and PYRe, intracellular and extracellular pyruvate; AC^{CIT}, the acetyl moiety of citrate, equivalent to carbons 4 and 5 of citrate (C-4+5); OAA^{CIT}, the OAA moiety of citrate equivalent to C-1+2+3+6.

(ii) *Isotopomer specifications*: PYR_{Mi}, mass isotopomer of a given metabolite, pyruvate in this example, labeled with i atoms of ^{13}C; MF, mole fraction in a given mass isotopomer (Mi) of a metabolite calculated as: MF (Mi) $= A_{Mi}/(\Sigma A_{Mi} + A_M)$, where A represents the peak area of each fragmentogram, determined by computer integration and corrected for naturally occurring heavy isotopes (cf. Section 3.3.3 below for details); MPE in a given mass isotopomer of a metabolites, equivalent to MF $\times$ 100.

(iii) *Flux rates*: $FC_{PYRi \rightarrow OAA}$: FC of one metabolite to total flux of the other, in this example intracellular pyruvate to OAA; for a given reaction, the sum of the different FCs equals 1.

3.2.2 Equations

This section reports the step-by-step calculation of the various flux ratios relevant to (1) the contributions of either exogenous FAs or CHOs to citrate formation using, respectively, [U-$^{13}C_6$]palmitate or [U-$^{13}C_3$]L+P as-labeled-substrate and (2) rates of lactate and pyruvate and efflux (from glycolysis) rates evaluated with [U-$^{13}C_6$]glucose. N.B. Calculation of M2 AC^{CIT} (Eq. 27) is given in Section 3.3.3 below.

3.2.2.1 Relative Contribution of FAs and CHOs to Citrate Formation Through Acetyl-CoA

These relative fluxes are inferred from the ^{13}C-labeling of the acetyl moiety of citrate (AC_{CIT}=C-4+5 of citrate). Using [U-$^{13}C_{16}$]palmitate, the balance of M2 isotopomers of acetyl-CoA and adjacent metabolites yields Eq. (8):

$$(FC_{PAL\rightarrow AC}) = (M2\,AC)/(M16\,PAL) \tag{8}$$

where $FC_{PAL\rightarrow AC}$ is the fractional contribution of exogenous palmitate to acetyl-CoA formation, and M2 AC and M16 PAL are the MF in M2 of intracellular acetyl-CoA and in M16 of the incoming palmitate tracer, respectively. For a given set of reactions, the sum of all FC terms equals 1, then $(1 - FC_{PAL\rightarrow AC})$ represents the acetyl-CoA molecules coming from all other sources, namely, decarboxylation of pyruvate formed from various sources and β-oxidation of FAs released from endogenous triglyceride hydrolysis. Since M2 acetyl-CoA is the only source of M2 citrate labeled in its acetyl moiety (AC^{CIT}), the balance of M2 isotopomers of acetyl moiety of citrate and adjacent metabolites yields Eq. (9):

$$(FC_{AC\rightarrow CIT}) = \left(M2\,AC^{CIT}\right)/(M2\,AC) \tag{9}$$

where $FC_{AC\rightarrow CIT}$ is the fractional contribution of acetyl-CoA to citrate via citrate synthase, and M2 AC^{CIT} is the MF in M2 of the acetyl moiety of citrate. Multiplying Eq. (8) by Eq. (9) yields the FC of exogenous palmitate to citrate via β-oxidation ($FC_{PAL\rightarrow AC(CIT)}$), expressed relative to flux through citrate synthesis (CS), which is set equal to 1. This flux ratio is referred to as PAL/CS:

$$\begin{aligned} PAL/CS &= \left(FC_{PAL\rightarrow AC(CIT)}\right) = (FC_{PAL\rightarrow AC}) \times (FC_{AC\rightarrow CIT}) \\ &= M2\,AC^{CIT}/M16\,PAL \end{aligned} \tag{10}$$

N.B. Eq. (10) assumes that palmitate is completely β-oxidized to eight acetyl-CoA molecules in the mitochondria. This is likely to be valid for palmitate or oleate under most conditions of aerobic perfusions, but FA with longer chain length may undergo a few cycles of peroxisomal β-oxidation resulting in preferential labeling of cytosolic acetyl-CoA (Reszko et al., 2004). N.B. This calculation abrogates the need to evaluate direct acetyl-CoA labeling, which can be challenging with LC–MS instrumentation.

Similar reasoning yields the relative contribution of extracellular pyruvate to citrate via PDC using $[U\text{-}^{13}C_3]L+P$ through the following Eqs. (11)–(13):

$$\left(FC_{PYRe \to AC(CIT)}\right) = \left(FC_{PYRe \to AC}\right) \times \left(FC_{AC \to CIT}\right) = M2\,AC^{CIT}/M3\,PYRe \quad (11)$$

where $FC_{M3PYRe \to AC(CIT)}$, $FC_{PYRe \to AC}$, and $FC_{AC \to CIT}$ are, respectively, the fractional contribution of exogenous pyruvate to the acetyl moiety of citrate, exogenous pyruvate to acetyl-CoA, and acetyl-CoA to citrate; M2 AC^{CIT} and M3 PYRe are the MF in M2 of the acetyl moiety of citrate and M3 of the incoming pyruvate tracer. The fractional contribution of intracellular pyruvate (PYRi) to the acetyl moiety of citrate $FC_{PYRi \to AC(CIT)}$, referred to as PDC/CS, is obtained from Eq. (12):

$$PDC/CS = \left(FC_{PYRi \to AC(CIT)}\right) = \left(FC_{PYRi \to AC}\right) \times \left(FC_{AC \to CIT}\right) = M2\,AC^{CIT}/M3\,PYRi \quad (12)$$

where M3 PYRi is the MF of intracellular pyruvate. Note that $FC_{PYRi \to AC}$ represents the decarboxylation of pyruvate derived from both exogenous and endogenous sources of pyruvate. The fractional contribution of extracellular pyruvate to tissue pyruvate ($FC_{PYRe \to PYRi}$) is obtained from Eq. (13):

$$\left(FC_{PYRe \to PYRi}\right) = (M3\,PYRi)/(M3\,PYRe) \quad (13)$$

The contribution of unlabeled pyruvate arising from other sources, such as exogenous unlabeled glucose and glycogen through glycolysis, is obtained by $(1 - FC_{PYRe \to PYRi})$. Note that in this example, hearts are perfused with uniformly labeled lactate plus pyruvate; hence, the contribution from unlabeled lactate is considered negligible. The sum of Eqs. (10) and (12) reflects the relative combined contributions of palmitate oxidation and PDC to the acetyl moiety of citrate relative to citrate synthase; the contribution of other sources (OS/CS; $FC_{OS \rightarrow AC(CIT)}$), such as endogenous FAs and/or leucine, is given by Eq. (14):

$$\left(FC_{OS \rightarrow AC(CIT)}\right) = 1 - \left[\left(FC_{PYRi \rightarrow AC(CIT)}\right) + \left(FC_{PAL \rightarrow AC(CIT)}\right)\right] \quad (14)$$

3.2.2.2 Relative Contribution of CHOs to Citrate Formation Through Anaplerotic OAA

The relative contribution of pyruvate to citrate formation through anaplerotic carboxylation (PC) is inferred from the ^{13}C-labeling of the OAA moiety of citrate ($OAA^{CIT} = C\text{-}1+2+3+6$ of citrate) and referred to as PC/CS. Note that this flux ratio reflects carboxylation of pyruvate metabolism via both pyruvate carboxylase and NADP-linked malic enzyme. To the best of our knowledge, none of the currently available isotopic methods distinguishes between these two pathways.

First, for simplicity, we assume that the carboxylation of [U-$^{13}C_3$]pyruvate is the only pathway forming M3 OAA. Then, we consider the possible formation of M3 OAA through the metabolism in the CAC of some ^{13}C-labeled citrate isotopomers. Following a reasoning similar to the calculation of the flux ratio PDC/CS, the flux ratio PC/CS is calculated by multiplying the fractional contribution (i) of pyruvate to OAA formation ($FC_{PYRi \rightarrow OAA}$) by carboxylation by that (ii) of OAA to citrate via citrate synthase ($FC_{OAA \rightarrow CIT}$):

$$\begin{aligned} PC/CS &= \left(FC_{PYR \rightarrow OAA}\right) \times \left(FC_{OAA \rightarrow CIT}\right) \\ &= M3\,OAA^{CIT}/M3\,PYRi \end{aligned} \quad (15)$$

where $FC_{PYRi \rightarrow OAA(CIT)}$ equals (M3 OAA)/(M3 PYRi), $FC_{OAA \rightarrow CIT}$ equals (M3 OAA^{CIT})/(M3 OAA), M3 OAA^{CIT} and M3 PYRi are the MF in M3 of tissue OAA moiety of citrate and pyruvate, respectively.

Equation (15) is, however, valid only when pyruvate is the sole source of M3 OAA. In practice, this condition rarely prevails when hearts are perfused with [$^{13}C_3$]labeled lactate plus pyruvate since a significant proportion of M3 OAA molecules are formed through the metabolism in the CAC of some ^{13}C-labeled citrate isotopomers formed from the condensation of ^{13}C-labeled acetyl-CoA and OAA. Hence, the measured MF M3 of the OAA moiety of citrate needs to be corrected; this is estimated using Eq. (16):

$$M3\,OAA^{CIT^*} = \text{measured}\,M3\,OAA^{CIT} - \left(M3\,OAA^{PR}/DF\right) \quad (16)$$

where OAA^{CIT*} is the corrected MF M3 value of the OAA moiety of citrate, M3 OAA^{PR} is the MF M3 in citrate isotopomer precursors of M3 OAA, and DF is the ^{13}C dilution of citrate isotopomer precursors of M3 OAA in the CAC due to the entry of unlabeled molecules.

The MF M3 of OAA^{PR} is estimated from the measured mass isotopomer distribution (MID) of the acetyl and OAA moiety of citrate. This calculation is based on the probability of condensation between labeled acetyl-CoA and labeled OAA molecules and on prior knowledge of the metabolism of ^{13}C-labeled citrate isotopomers in the CAC. For example, there are four possible citrate isotopomers formed by condensation of M2 acetyl-CoA isotopomers (labeling C-4 + 5 of citrate) and M1 OAA isotopomers (labeling carbon 1, 2, 3, or 6 of the OAA moiety of citrate): [1, 4, 5-^{13}C]citrate (A), [2, 4, 5-^{13}C]citrate (B), [3, 4, 5-^{13}C] citrate (C), and [4, 5, 6-^{13}C] citrate (D). For simplicity, it is assumed that citrate isotopomers A–D are similarly enriched due to complete randomization of ^{13}C label in the CAC at the level of succinate and fumarate. Upon CAC metabolism, only B and C citrate isotopomers will form M3 OAA, whereas isotopomers A and D forms M2 OAA due to decarboxylation of 6 and 1 of citrate at the NAD^+-linked isocitrate dehydrogenase and α-ketoglutarate dehydrogenase reactions, respectively. Then, the ^{13}C-enrichment of citrate isotopomers B and C is estimated from: (M2 AC^{CIT}) × (M1 OAA^{CIT}) × 0.5, where M2 AC^{CIT} and M1 OAA^{CIT} are determined experimentally and the number 0.5 takes into account that only half of the citrate isotopomers formed through the condensation of M2 acetyl-CoA and M1 OAA are metabolized in the CAC to M3 OAA. This constitutes the first term of Eq. (17) below. A similar reasoning was applied for each term of this equation:

$$\begin{aligned} \mathrm{M3\,OAA^{PR}} = {} & [0.5 \times (\mathrm{M2\,AC^{CIT}} \times \mathrm{M1\,OAA^{CIT}})] \\ & + [2/3 \times (\mathrm{M2\,AC^{CIT}} \times \mathrm{M2\,OAA^{CIT}})] \\ & + [0.5 \times (\mathrm{M2\,AC^{CIT}} \times \mathrm{M3\,OAA^{CIT}})] \end{aligned} \tag{17}$$

DF is estimated from the total ^{13}C-enrichment in tissue citrate (ΣMi CIT) and succinate (ΣMi SUC) using Eq. (18):

$$\mathrm{DF} = \left[\Sigma\,\mathrm{Mi\,CIT} - \left(f \times \mathrm{M1\,OAA^{CIT}}\right)\right] / [\Sigma\,\mathrm{Mi\,SUC}] \tag{18}$$

where the factor f takes into account that a fraction of all citrate molecules with a ^{13}C in any one carbon of its OAA moiety are converted to unlabeled succinate. The magnitude of f depends on the nature of the ^{13}C-labeled substrate. For examples, f equals 0.5 for ^{13}C-labeled substrates forming

[4,5-$^{13}C_2$] citrate such as [U-$^{13}C_3$]labeled lactate plus pyruvate. The minimal value of DF is 1, which corresponds to no ^{13}C dilution. In practice, we found that in the heart, the DF value is often close to 1 and has little impact on the value of the estimated M3 OAA^{PR}.

3.2.2.3 Lactate and Pyruvate Efflux Rate as a Proxy of Glycolysis

This flux rate is assessed in hearts perfused with [U-$^{13}C_6$]glucose. The efflux rate of [U-$^{13}C_3$]labeled L + P (M3), which reflects glycolysis from exogenous glucose, is estimated from (1) the difference between their "arterial" influent and "venous" effluent perfusate concentrations determined by GC–MS and standard enzymatic assays and (2) the coronary flow rate. For GC–MS assay, perfusate samples are treated with $NaBD_4$ to reduce unlabeled (M) and [U-$^{13}C_3$]labeled (M3) pyruvate into corresponding [^2H]lactate (M1) and [^2H, U-$^{13}C_3$]lactate (M4), respectively. The four different mass isotopomers of lactate can be distinguished and their respective ^{13}C-enrichment simultaneously assessed by GC–MS: [^{12}C]lactate → [^{12}C] lactate (M), [U-$^{13}C_3$]lactate → [U-$^{13}C_3$] lactate (M3), [^{12}C]pyruvate → [^{12}C]lactate deuterated (M1), and [U-$^{13}C_3$]pyruvate → [U-$^{13}C_3$]lactate deuterated (M4). The enzyme assay enables assessment of total lactate concentration (which includes both M and M3 lactate isotopomers) in untreated perfusate samples. N.B. To assay lactate concentration, we found that a separate enzyme assay led to less error in the calculated flux rate than a single GC–MS assay of samples spiked with an internal standard of [$^2H_6$] lactate (M6).

The concentration of M3 lactate ([M3 LAC]) in perfusate samples is calculated from GC–MS determined corrected peak area of M and M3 lactate (LAC^{CPA}) and enzymatically determined lactate concentration (LAC^{ENZ}) using Eq. (19):

$$[\mathrm{M3\,LAC}]\,\mu\mathrm{mol/ml} = \left[\mathrm{LAC^{ENZ}}\right]\mu\mathrm{mol/ml} \times \left[\mathrm{M3\,LAC^{CPA}}/\left(\mathrm{M3\,LAC^{CPA}} + \mathrm{M\,LAC^{CPA}}\right)\right] \tag{19}$$

Similarly, the concentration of M3 pyruvate [M3 PYR] is calculated using Eq. (20) from GC–MS determined corrected peak area of M1 (M1 LAC^{CPA}, from unlabeled pyruvate reduced to ^2H-lactate) and M4 lactate (M4 LAC^{CPA}, from [U-$^{13}C_3$]pyruvate reduced to [^2H,$^{13}C_3$]lactate) (LAC^{CPA}).

$$[\mathrm{M3\,PYR}]\mu\mathrm{mol/ml} = \left[\mathrm{LAC^{ENZ}}\right]\mu\mathrm{mol/ml} \times \left[\mathrm{M4\,LAC^{CPA}} + \mathrm{M1\,LAC^{CPA}}\right] \tag{20}$$

Equations (19) and (20) are used to calculate the concentration of M3 lactate and pyruvate in "arterial" influent perfusates (Inf) and in the "venous" effluent perfusates (Eff) collected at a given perfusion time. Efflux rates of lactate plus pyruvate (μmol/min) are obtained by multiplying these concentrations by the coronary flow (ml/min).

$$\text{Lactate efflux}\,(\mu\mathrm{mol/min}) = \mathrm{CF}\,(\mathrm{ml/min}) \times \left\{\left([\mathrm{M3\,LAC}]_{\mathrm{inf}}\right)(\mu\mathrm{mol/min})\right\} \tag{21}$$

$$\text{Pyruvate efflux}\,(\mu\mathrm{mol/min}) = \mathrm{CF}\,(\mathrm{ml/min}) \times \left\{\left([\mathrm{M3\,PYR}]_{\mathrm{inf}}\right)(\mu\mathrm{mol/min})\right\} \tag{22}$$

3.3 Methods

This section describes the analytical procedures for extraction and analysis of unlabeled and ^{13}C-labeled CAC intermediates and related metabolites by GC–MS for the determination of the various flux parameters. A global overview of the sample processing workflow is depicted in Fig. 5.

3.3.1 Chemicals, Materials, and Solutions

Chemicals and materials: NaCl; anhydrous sodium sulfate (Na_2SO_4); anhydrous ethyl ether and ethyl acetate LC–MS grade solution; *N*-methyl-*N*-(*t*-butyldimethylsilyl)-trifluoacetamide (MTBSTFA); citrate lyase (0.4 units/mg); glass tubes (16 × 125 mm); GC–MS vials.

Solutions: 1 *M* hydroxylamine hydrochloride; 100 m*M* methoxylamine hydrochloride; 1 *M* sodium borodeuteride ($NABD_4$) and 1 *M* sodium borohydride ($NABH_4$, prepared fresh before use); 1 *M* NaOH and HCl; 12.8 N HCl; saturated and 16% sulfosalicylic acid (SSA); 100 and 500 m*M* triethanolamine (TEA).

3.3.2 Sample Processing

Foreword: The smaller size of the mouse heart compared to that of rats (±0.2 vs. 1 g) or other species poses some methodological challenges for the precision of ^{13}C-enrichment data of some metabolites. We noticed that with lower sample size, there is a higher background noise signal for unlabeled lactate and succinate, and to a lesser extent for citrate and malate. If not accounted for, the endogenous contamination of succinate will result in

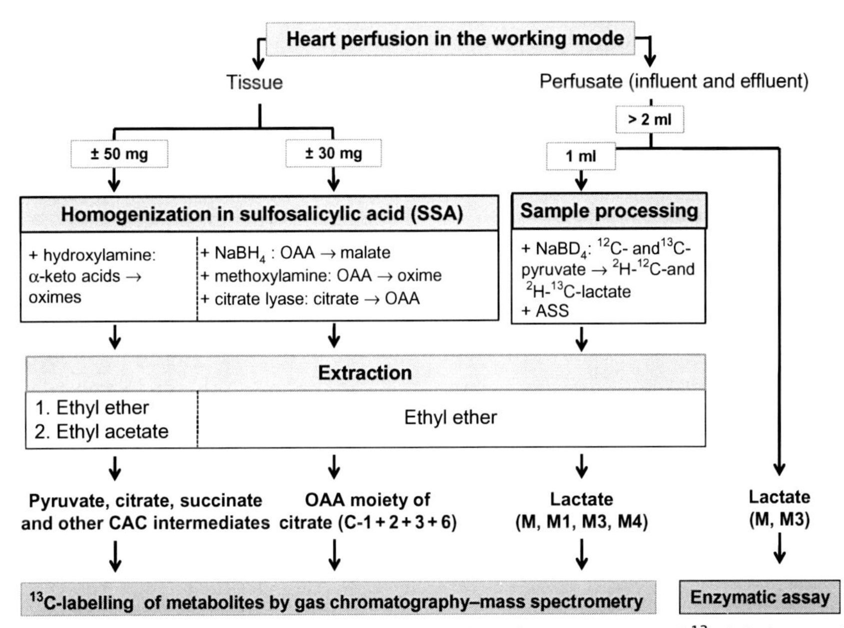

Figure 5 Overview of sample processing workflow for assessment of ^{13}C-labeling and concentration of metabolites used for calculation of flux parameters. See Section 3.3 for details.

overestimation of the dilution factor (DF) (Eq. 18) and, hence, of the PC/CS flux ratio (Eq. 15); while that for citrate will result in underestimation of the MF M2 of the calculated acetyl moiety of citrate. Hence, we recommend monitoring the level of background contamination by processing blanks consisting of: (i) water as negative control and (ii) water spiked with internal standard of labeled CAC intermediates, namely, 25 nmol lactate, 10 nmol succinate, and 8 nmol citrate. (N.B. Whenever there is substantial contamination of tissue with unlabeled succinate, one may substitute the ^{13}C-enrichment of tissue succinate for that of α-ketoglutarate (ΣMi α-KG) to estimate the DF (Eq. 19), albeit this underestimates the ^{13}C dilution since it neglects the entry of unlabeled carbons at the level of succinyl-CoA and succinate.) Finally, we also routinely process a sample of pulverized heart, which had not been perfused with labeled substrates, to verify the validity of our correction matrix for natural abundance.

3.3.2.1 Relative Flux Parameters Relevant to Citrate Formation

The relative contribution of FAs and CHOs to citrate formation through acetyl-CoA is calculated from the measured MID of the following tissue metabolites: (i) citrate and its OAA moiety (OAA_{CIT}), from which we extrapolate the acetyl moiety of citrate (AC_{CIT}) and (ii) pyruvate. Calculation of the relative contribution of pyruvate to citrate formation through OAA (anaplerosis) requires the additional measurement of the MID of succinate. The inclusion of ^{13}C-enrichment data on other CAC intermediates, namely, α-ketoglutarate, fumarate, and malate, provides also valuable insight for data interpretation.

Subsequent to each perfusion period, hearts are freeze-clamped with metal tongs chilled in liquid nitrogen (N_2) and stored at −80 °C until further analysis. The frozen heart is pulverized at the temperature of liquid N_2 to a fine powder. GC–MS analysis of the MID of the aforementioned metabolites require two separate sample extraction steps.

Extraction 1: MID of all metabolites, except the OAA moiety of citrate: For tissue homogenization, we used a mechanical homogenizer (Omni Bead Ruptor Homogenizer). The pulverized heart powder (50 mg) is added into a 2-ml screwable tube, cooled at the temperature of liquid N_2, containing ceramic beads (2.8 mm) in 300 μl of cold 16% SAA (to precipitate proteins) solution and 300 μl 1 *M* hydroxylamine (pH 7.6; to convert α-keto acids, namely, pyruvate and α-ketoglutarate into their more stable oxime derivatives; Laplante, Comte, & Des Rosiers, 1995). Homogenates kept on ice are subsequently centrifuged at 4000 × *g* for 20 min at 4 °C. The supernatants are transferred into glass tubes (16 × 125 mm), adjusted to pH 7–8 with dropwise addition of 1 *M* NaOH while vortexing, and sonicated for 1 min in a sonication bath to ensure complete conversion of α-keto acids to their oxime derivatives. Supernatants are acidified to pH to ~1 with few drops 12.8 N HCl (to convert organic salts to their free acid form) and saturated with NaCl and then extracted consecutively three times with 3 ml ethyl ether (this extracts pyruvate and succinate) followed by three times with 3 ml of ethyl acetate (this extracts citrate and other CAC intermediates). Each time, the solution is vortexed for 3 min in a Pulse Vortex Glas-Col and centrifugated for 20 s at 1000 × *g* and the organic (upper) phase is transferred into a glass tube (16 × 125 mm). The combined organic extracts are dried by adding a scoop of Na_2SO_4 and transferred into a clean glass tube. (N.B. This step may need to be repeated two or three times in order to remove all traces of water, which interfere with the derivatization reaction. Glass tubes may also be rinsed with 0.5 ml anhydrous ethyl ether solution and 0.5 ml ethyl acetate to enhance extraction recovery.) The combined organic extracts are evaporated under a stream of nitrogen gas in a water bath at 30 °C until ~0.5 ml and transferred into GC–MS vials to dryness. The dried residue is reacted with 100 μl MTBSTFA for 4 h at 90 °C for optimal conversion of all metabolites into their corresponding *t*-butyldimethyl (TBDMS) ether derivatives.

Extraction 2: MID of the OAA moiety of citrate (OAA_{CIT}): The pulverized heart powder (30 mg) is homogenized as described above except for the addition of 200 μl of SSA 16% and 200 μl 100 m*M* TEA (pH 7.6) prior to centrifugation at 4000 × *g* for 20 min at 4 °C. The supernatants are transferred into glass tubes (16 × 125 mm) and neutralized to pH 7–8 by dropwise addition, with constant vortexing, of a mixture of 5 N KOH/1.25 N TEA before the addition of a molar excess of sodium borohydride ($NaBH_4$, 200 μl of a 1 *M* solution) to reduce endogenous OAA to malate. After 30 min at room temperature, the solution is acidified with 125 μl saturated ASS to destroy residual $NaBH_4$. (N.B. Small bubbles should be observed due to hydrogen gas formation.) The pH is readjusted to 7.6 ± 0.1, and sample are treated with a molar excess of methoxylamine (175 μl of a 100 m*M* solution: pH 7.6; to convert OAA to its oxime derivative) before the addition of the following mixture for citrate cleavage to OAA (by sample): 2 ml 500 m*M* TEA buffer, 13.5 μl 500 m*M* $MgCl_2$, 0.2 ml 250 m*M* EDTA, and 1 mg of citrate lyase (0.4 U). Samples are incubated 30 min at 37 °C under gentle agitation and sonicated for 2 min (for complete conversion of OAA to its oxime derivatives: mtoxOAA). After addition of 400 μl saturated SSA and centrifugation at 6000 × *g* for 15 min at room temperature, samples are acidified to pH 1 with three drops 12.8 N HCl to pH 1, saturated with NaCl, and extracted three times with 10 ml anhydrous ethyl ether. The combined organic phases are dried with Na_2SO_4 and evaporated to dryness as described above before addition of 100 μl MTBSTFA and incubation for 4 h at 90 °C.

3.3.2.2 Lactate and Pyruvate Efflux Rates as Proxy of Glycolysis

During heart perfusion with [U-$^{13}C_6$]glucose, influent (arterial) and effluent (venous) perfusates (2 ml minimum) are collected on ice at 20–25 and 25–30 min, of which 1 ml is immediately treated with 50 μl 1 *M* $NaBD_4$ for 30 min on ice. The two 1 ml aliquots are stored at −20 °C until further analysis. Perfusate samples (1 ml) are treated with 375 μl saturated SSA and centrifuged 15 min at 4000 × *g*. Supernatants are transferred into glass tubes (16 × 125 mm), adjusted to pH ~1 with 12.8 *M* HCl, saturated with NaCl, and extracted three times with 4 ml anhydrous ethyl ether. The combined organic phases are dried with anhydrous Na_2SO_4 and evaporated dryness in

GC–MS vials as described above before addition of 100 μl MTBSTFA and incubation for 4 h at 90 °C.

GC–MS conditions: All metabolites are analyzed as their TBDMS derivatives by GC–MS. The following conditions are routinely used in our laboratory but are only indicative since they will be specific for a given instrument. Samples (1–2 μl) are injected onto a Agilent 6890 N gas chromatograph equipped with an HP-5 capillary column (5% phenyl methyl siloxane; 50 m ×; 0.2 mm i.d; 0.50 μm thickness) and coupled to a Agilent 5975 mass selective detector operating in the electron impact and selected ion monitoring mode under the following conditions: ion source temperature at 230 °C, emission current 34.6 μA, and electron energy 70 eV. GC conditions are helium is used as carrier gas at a flow rate of 23.3 ml/min, split ratio is 28:1, injection port is 280 °C, and transfer line is 320 °C. The following GC program provides good resolution of many TBDMS derivatives of CAC intermediates (citrate, succinate, α-ketoglutarate, malate, and fumarate) and related metabolites (pyruvate, lactate): The column temperature is initially set at 150 °C for 5 min, increased by 10 °C/min until 300 °C for 7 min and by 20 °C/min to 320 °C for 3 min. (N.B. The retention time (RT) and ion set to be monitored for each TBDMS derivative has to be determined with standards for a given instrument.) The *m/z* has also to be determined with a precision of 0.1 mass unit to achieve maximal peak area. Appropriate ion sets are monitored with a dwell time of 50–75 ms/ion. The following ion sets and RT were determined under our specific conditions and are given as indicative values only. (i) For relative substrate flux to citrate formation, values are pyruvate M–M3: *m/z* 274.1–277.1 (RT: 13.5 min), citrate M–M6: *m/z* 458.2–464.2 (RT: 25.1 min), succinate M–M4: *m/z* 289.1–293.1 (RT: 15.6 min), and OAA_{mtox} M–M4: *m/z* 332.3–336.3 (RT: 13.3 min). Values for other CAC intermediates are α-ketoglutarate M–M5: *m/z* 447.2–451.2 (RT: 21 min), malate M–M4: *m/z* 419.2–423.2 (RT: 19.4 min), and fumarate *m/z*: 287.1–287.1 (RT: 15.8 min). (ii) For lactate and pyruvate efflux rates, values are lactate M–M4: *m/z* 261.1–265.1 (RT: 12.2 min).

3.3.3 GC–MS Data Processing

Areas under each fragmentogram are determined by computer integration and corrected for naturally occurring heavy isotopes (^{13}C, ^{2}H, ^{15}N, ^{28}Si, ^{18}O) as described previously in detail (Brunengraber et al., 1997; Des Rosiers et al., 1988; Fernandez, Des Rosiers, Previs, David, & Brunengraber, 1996). Briefly, this involves solving the following matrix:

$$A_c{}^T = A_r{}^T[\mathrm{CM}]^{-1} \tag{23}$$

where $A_c{}^T$ and $A_r{}^T$ represent the vectors of the corrected and uncorrected areas, respectively, and CM is the correction matrix. The CM is composed of n rows and columns, where n represents the number of unlabeled plus ^{13}C-labeled mass isotopomers for a given metabolite. For example, for citrate, the matrix consists of seven rows and seven columns. Rows are ratios of ion intensities relative to base peak for incrementally ^{13}C-labeled citrate. These intensity ratios are calculated from measured intensity ratios for the corresponding unlabeled standard, e.g., citrate, using a computer program developed by Fernandez et al. (Des Rosiers et al., 2004; Fernandez et al., 1996). The latter article also provides an overview of the different approaches to correct GC–MS data for natural abundance in heavy isotopes.

The corrected areas are used to calculate total, absolute, and relative MPE of individual ^{13}C-labeled mass isotopomers (Mi) of a given metabolite as follows:

$$\text{Total MPE} = \%\,\Sigma A_{\mathrm{Mi}}/[A_{\mathrm{M}} + \Sigma A_{\mathrm{Mi}}] \tag{24}$$

$$\text{Absolute MPE(Mi)} = \%\,A_{\mathrm{Mi}}/[A_{\mathrm{M}} + \Sigma A_{\mathrm{Mi}}] \tag{25}$$

$$\text{Relative MPE(Mi)} = \%A_{\mathrm{Mi}}/\Sigma A_{\mathrm{Mi}} \tag{26}$$

where A_{Mi} and A_{M} represent the peak areas of ion chromatograms, corrected for natural abundance in heavy isotopes, corresponding to the unlabeled (M) and ^{13}C-labeled (Mi) mass isotopomers, respectively. N.B. Because Eq. (26) does not include in its denominator the term A_{M}, the relative MPE is not sensitive to simple dilution by unlabeled metabolites. MF in a given mass isotopomer (Mi) of a metabolite is equivalent to absolute MPE/100.

MF values are used to calculate the various flux ratios relevant to the contribution of substrates to the formation of citrate via acetyl-CoA and to extrapolate the MPE of the acetyl moiety of citrate ($\mathrm{AC^{CIT}}$) from the measured MID of tissue citrate and its OAA moiety. For hearts perfused with ^{13}C-labeled substrates forming M2 acetyl-CoA either [U-^{13}C]palmitate or [U-^{13}C]labeled lactate plus pyruvate, the MPE M2 of the acetyl moiety of citrate is calculated from:

$$\mathrm{MPE\,M2\,AC^{CIT}} = 0.5 \times \left[(\Sigma\,\mathrm{MF(Mi \times i)citrate}) - \left(\Sigma\,\mathrm{MF(Mi \times i\,)OAA^{CIT}}\right)\right] \tag{27}$$

where $\mathrm{Mi \times i}$ represents the total isotopic enrichment of citrate and $\mathrm{OAA^{CIT}}$, the factor 0.5 is to take into account that there are two ^{13}C atoms per molecule of M2 $\mathrm{AC^{CIT}}$.

4. PRACTICAL CONSIDERATIONS AND FUTURE CHALLENGES

Studies conducted over the past 15 years using ^{13}C-labeled substrates and isotopomer analysis by GC–MS in various study models and over a range of (patho)physiological conditions provide examples of applications of the metabolic flux analysis using equations described in this chapter (see for review Des Rosiers et al., 2011). Results obtained were consistent across species (rat, mouse, pig), models (*ex vivo* vs. *in situ*) and with the known regulation of cardiac energy substrate metabolism, while revealing unexpected findings emphasizing the complex network of regulatory mechanisms governing flux through the various measured pathways and anaplerotic PC. As it is the case for any study model, however, one has to make compromises between advantages and limitations and the latter needs to be kept in mind in data interpretation. This subject has recently been reviewed in details for metabolic flux experiments in cells (Buescher et al., 2015). Therefore, in this section, we are focusing on considerations, which are relevant to our approach and to the presented metabolic flux equations, which have not been addressed in the previous (section 3).

4.1 Metabolic Flux Parameters

The reliability of the various flux parameters calculated from the measured MID data of the various tissue (citrate, its OAA and acetyl moieties, pyruvate, and succinate) and perfusate (pyruvate and lactate) metabolites depends (i) on the validity of assumptions on which the aforementioned equations are developed and (ii) on the precision of the measured and calculated enrichment data.

4.1.1 Validity of Assumptions

One general underlying assumption of our approach, which is also inherent to all metabolic investigations using labeled tracers, is that isotopic effects on reaction rates and equilibria can be neglected; this appears justified for ^{13}C. It is also assumed that there is homogeneous ^{13}C-labeling of intracellular metabolites; in other words, metabolite channeling or inter- or intracellular compartmentation is not considered, but are among factors that influence the fate of metabolites. The whole heart includes cardiomyocytes (85% of its wet mass, but also a substantial noncardiomyocyte cell population, e.g., endothelial cells and fibroblasts) (deBlois, Orlov, & Hamet, 2001).

Compartmentation of glucose, pyruvate, and lactate metabolism has been previously reported in the heart, and this may explain the small differences in the flux ratio PC/PDC calculated from individual ^{13}C-labeled CHOs, [U-$^{13}C_6$]glucose versus [U-$^{13}C_3$]lactate versus [U-$^{13}C_3$]pyruvate that we found in the healthy mouse heart (Khairallah et al., 2004). There are also evidence for the existence of subpools of acetyl-CoA (Reszko et al., 2004) and succinate (Des Rosiers et al., 2004).

4.1.2 Precision of ^{13}C-enrichment Data

Precision in the ^{13}C-enrichment data of metabolites, which include computer determined raw data for all mass isotopomers by GC–MS as well as of correction factors for natural abundance in heavy isotopes have been previously addressed (Antoniewicz, Kelleher, & Stephanopoulos, 2007; Brunengraber et al., 1997; Des Rosiers et al., 1988; Fernandez et al., 1996). Given that small errors in MID propagate to large errors inestimated fluxes, based on sensitivity studies of realistic metabolic networks, Antionemicz et al. determined that measurement errors should be <0.5 mol% (Antoniewicz et al., 2007). In practice, low precision on ^{13}C-enrichment data of metabolites, particularly for MPE values below 5% at M1 and M2 ions, may result from the high natural background of the *TBDMS* derivatives of citrate and OAA precision and is further decreased when these low enrichments occur at a mass adjacent to one which is highly enriched. Additional considerations about the precision of ^{13}C-enrichment data specifically linked to the calculation of the flux parameters reported in this chapter include the followings: (1) The M3 enrichment of intracellular pyruvate (M3 PYRi) is probably slightly overestimated, since it is measured in whole tissue; hence, the flux ratios PDC/CS and PC/CS are likely slightly underestimated. (2) The precision of ^{13}C-enrichment values for the acetyl moiety of citrate could be improved by the development of a MS method for its direct measurement. It is noteworthy, however, that direct analysis of the moiety of citrate as (i) acetate released by cleavage of citrate with citrate lyase or (ii) as acetyl-CoA released by cleavage with ATP-citrate lyase as described previously was found not to be feasible because of contamination with unlabeled acetate or too low concentration of citrate, respectively (Comte et al., 1997). Alternatively, one may consider direct analysis of the ^{13}C-enrichment of whole-tissue acetyl-CoA by LC–MS although this may not always reflect that of the acetyl moiety of citrate because of reported ^{13}C-labeling heterogeneity between mitochondrial versus cytosolic acetyl-CoA.

4.2 Additional Considerations and Challenges

The focus of this chapter was on the measurements in the *ex vivo* working mouse heart model of flux parameters relevant, respectively, to mitochondrial and cytosolic energy metabolism. The presented flux measurements reflect, however, only part of the complex metabolic networks involved in cardiac energy substrate metabolism and which can be probed with ^{13}C-labeled substrates and ^{13}C-labeled mass isotopomer analysis by GC–MS. For example, while these measurements represent an estimation of the relative contribution of other sources to the formation of intracellular pyruvate (Eq. 14: OS/CS) or acetyl-CoA for citrate formation (M2 AC^{CIT}), these sources are not identified. In some instances, changes in the flux ratio OS/CS in a given condition can be hypothesis generating as it was the case for hearts from transgenic mice with a cardiomyocyte-specific expression of guanylate cyclase. An increase in this ratio led us to postulate that unlabeled FA released from endogenous triglyceride stores was contributing to acetyl-CoA formation and evidence was subsequently obtained to support this hypothesis (Khairallah et al., 2008).

Second, the proposed ^{13}C-labeled substrates, measurements, and equations also reflect only part of the complexity of the metabolic network at the pyruvate branch point. In addition to pyruvate formation through cytosolic glycolysis, its mitochondrial decarboxylation (oxidation) and carboxylation (anaplerosis), additional metabolic processes that can be probed include lactate and pyruvate uptake using [U-^{13}C]labeled lactate and/or pyruvate, but also lactate–pyruvate interconversion by LDH and/or alanine transamination (Chatham, 2002; Khairallah et al., 2004; Vincent et al., 2003).

Third, the contribution of pyruvate carboxylation to OAA forming citrate (i.e., the PC/CS flux ratio) is only one of many anaplerotic pathways that can potentially replenish the pool of intermediates of the CAC in the heart. Indeed, there are several other reactions/pathways which participate in the entry or removal of CAC intermediates in the heart. These include: (1) transamination between OAA and α-ketoglutarate and their corresponding amino acids, aspartate and glutamate, respectively and (2) formation of succinyl-CoA from propionyl-CoA precursors such as the branched chain amino acids, valine or isoleucine (for a recent review, see Des Rosiers et al., 2011). The contribution of these various pathways can be probed using specific ^{13}C-labeled substrates. Using an approach similar to that described in this chapter, we have found that relative contribution of [U-^{13}C]labeled glutamate or glutamine to the formation of ^{13}C-labeled α-ketoglutarate is at most 5% (Comte, Vincent, Bouchard, Benderdour, & Des Rosiers, 2002; Lauzier et al., 2013). Irrespective of the ^{13}C-labeled substrates used, given the complexity of the metabolic network regulating cardiac anaplerosis, flux data should be combined with additional measurements of the concentrations and ^{13}C-enrichment of all CAC intermediates and related metabolites in order to obtain a comprehensive view of metabolic regulation.

Finally, it is noteworthy that except for the lactate and pyruvate efflux rates, the calculated metabolic flux parameters reflect fractional contributions of FA versus CHO oxidation to citrate formation expressed relative to the rate of citrate synthase, which is set to 1. To convert these flux ratios into absolute fluxes, one needs to assess the CAC flux rate using another isotopic approach under nonsteady-state conditions. Alternatively, absolute CAC flux rate can be extrapolated from measured MVO_2 values and the stoichiometric relationships between oxygen consumption and citrate formation from CHOs and FAs (Khairallah et al., 2004; Vincent et al., 2004). Similarly, total rates of ATP production can be calculated from (1) the various absolute flux rates by using theoretical yields of ATP per mole of substrate oxidized and (2) rates of anaerobic glycolysis determined from the efflux rate of [U-$^{13}C_3$]lactate measured in hearts perfused with [U-$^{13}C_6$] glucose. It should be kept in mind, however, that these extrapolated absolute flux values are approximate and rely on the validity of the assumptions underlying the various equations used in their calculations.

4.3 Conclusion

The use of stable isotope labeling by GC/LCMS for investigations of cardiac metabolism enables the assessment of substrate flux through complex metabolic processes, beyond substrate fuel selection for energy metabolism. These processes are critical to the overall energy balance and metabolic integrity of the heart, and their role and regulation remain to be better understood.

ACKNOWLEDGMENTS

We would like to acknowledge people who have contributed to the implementation of these methods, namely, trainees, Drs. Blandine Comte, Geneviève Vincent, Maya Khairallah, Ramzi Khairallah and François Labarthe; research assistant, Bertrand Bouchard, as well as collaborators, Drs. Henri Brunengraber, Joanne Kelleher, and John Chatham. We also thank Dr. Christian Metallo for his insightful comments. Finally, we are indebted to the Canadian Institutes of Health Research for supporting financially this work.

REFERENCES

Antoniewicz, M. R., Kelleher, J. K., & Stephanopoulos, G. (2007). Accurate assessment of amino acid mass isotopomer distributions for metabolic flux analysis. *Analytical Chemistry*, *79*(19), 7554–7559.

Asfour, H., Wengrowski, A. M., Jaimes, R., Swift, L. M., & Kay, M. W. (2012). NADH fluorescence imaging of isolated biventricular working rabbit hearts. *Journal of Visualized Experiments*, *65*, 4115.

Baker, L. C., London, B., Choi, B. R., Koren, G., & Salama, G. (2000). Enhanced dispersion of repolarization and refractoriness in transgenic mouse hearts promotes reentrant ventricular tachycardia. *Circulation Research*, *86*(4), 396–407.

Barnabei, M. S., Palpant, N. J., & Metzger, J. M. (2010). Influence of genetic background on *ex vivo* and *in vivo* cardiac function in several commonly used inbred mouse strains. *Physiological Genomics*, *42A*, 103–113.

Belke, D. D., Larsen, T. S., Lopaschuk, G. D., & Severson, D. L. (1999). Glucose and fatty acid metabolism in the isolated working mouse heart. *The American Journal of Physiology*, *277*, R1210–R1217.

Brunengraber, H., Kelleher, J. K., & Des Rosiers, C. (1997). Applications of mass isotopomer analysis to nutrition research. *Annual Review of Nutrition*, *17*, 559–596.

Buescher, J. M., Antoniewicz, M. R., Boros, L. G., Burgess, S. C., Brunengraber, H., Clish, C. B., et al. (2015). A roadmap for interpreting ^{13}C metabolite labeling patterns from cells. *Current Opinion in Biotechnology*, *34*, 189–201.

Bugger, H., & Abel, E. D. (2014). Molecular mechanisms of diabetic cardiomyopathy. *Diabetologia*, *57*, 660–671.

Bunger, R., Mallet, R. T., & Hartman, D. A. (1989). Pyruvate-enhanced phosphorylation potential and inotropism in normoxic and postischemic isolated working heart. Near-complete prevention of reperfusion contractile failure. *European Journal of Biochemistry*, *180*, 221–233.

Chatham, J. C. (2002). Lactate—The forgotten fuel! *The Journal of Physiology*, *542*(Pt. 2), 333.

Chatham, J. C., Des Rosiers, C., & Forder, J. R. (2001). Evidence of separate pathways for lactate uptake and release by the perfused rat heart. *American Journal of Physiology. Endocrinology and Metabolism*, *281*, E794–E802.

Choisy, S. C., Arberry, L. A., Hancox, J. C., & James, A. F. (2007). Increased susceptibility to atrial tachyarrhythmia in spontaneously hypertensive rat hearts. *Hypertension*, *49*(3), 498–505.

Comte, B., Vincent, G., Bouchard, B., Benderdour, M., & Des Rosiers, C. (2002). Reverse flux through cardiac NADP(+)-isocitrate dehydrogenase under normoxia and ischemia. *American Journal of Physiology. Heart and Circulatory Physiology*, *283*(4), H1505–H1514.

Comte, B., Vincent, G., Bouchard, B., & Des Rosiers, C. (1997). Probing the origin of acetyl-CoA and oxaloacetate entering the citric acid cycle from the ^{13}C labeling of citrate released by perfused rat hearts. *The Journal of Biological Chemistry*, *272*, 26117–26124.

Comte, B., Vincent, G., Bouchard, B., Jetté, M., Cordeau, S., & Des Rosiers, C. (1997). A ^{13}C mass isotopomer study of anaplerotic pyruvate carboxylation in perfused rat hearts. *The Journal of Biological Chemistry*, *272*(42), 26125–26131.

Corkey, B. E., & Shirihai, O. (2012). Metabolic master regulators: Sharing information among multiple systems. *Trends in Endocrinology and Metabolism*, *23*, 594–601.

Cotter, D. G., Schugar, R. C., & Crawford, P. A. (2013). Ketone body metabolism and cardiovascular disease. *American Journal of Physiology. Heart and Circulatory Physiology*, *304*, H1060–H1076.

deBlois, D., Orlov, S. N., & Hamet, P. (2001). Apoptosis in cardiovascular remodeling—Effect of medication. *Cardiovascular Drugs and Therapy*, *15*(6), 539–545.

Des Rosiers, C., & Chatham, J. C. (2005). Myocardial phenotyping using isotopomer analysis of metabolic fluxes. *Biochemical Society Transactions*, *33*, 1413–1417.

Des Rosiers, C., David, F., Garneau, M., & Brunengraber, H. (1991). Nonhomogeneous labeling of liver mitochondrial acetyl-CoA. *The Journal of Biological Chemistry*, *226*(3), 1574–1578.

Des Rosiers, C., Di Donato, L., Comte, B., Laplante, A., Marcoux, C., David, F., et al. (1995). Isotopomer analysis of citric acid cycle and gluconeogenesis in rat liver. Reversibility of isocitrate dehydrogenase and involvement of ATP-citrate lyase in gluconeogenesis. *The Journal of Biological Chemistry*, *270*, 10027–10036.

Des Rosiers, C., & Fernandez, C. A. (1995). Modeling of citric acid cycle and gluconeogenesis based on mass distribution analysis of intermediates. *The Journal of Biological Chemistry*, *270*, 10037–10042.

Des Rosiers, C., Fernandez, C. A., David, F., & Brunengraber, H. (1994). Reversibility of the mitochondrial isocitrate dehydrogenase reaction in the perfused rat liver. Evidence from isotopomer analysis of citric acid cycle intermediates. *The Journal of Biological Chemistry*, *269*(44), 27179–27182.

Des Rosiers, C., Labarthe, F., Lloyd, S. G., & Chatham, J. C. (2011). Cardiac anaplerosis in health and disease: Food for thought. *Cardiovascular Research*, *90*, 210–219.

Des Rosiers, C., Lloyd, S., Comte, B., & Chatham, J. C. (2004). A critical perspective of the use of ^{13}C-isotopomer analysis by GCMS and NMR as applied to cardiac metabolism. *Metabolic Engineering*, *6*, 44–58.

Des Rosiers, C., Montgomery, J. A., Desrochers, S., Garneau, M., David, F., Mamer, O. A., et al. (1988). Interference of 3-hydroxyisobutyrate with measurements of ketone body concentration and isotopic enrichment by gas chromatography–mass spectrometry. *Analytical Biochemistry*, *173*(1), 96–105.

Doenst, T., Nguyen, T. D., & Abel, E. D. (2013). Cardiac metabolism in heart failure: Implications beyond ATP production. *Circulation Research*, *113*, 709–724.

Fernandez, C. A., Des Rosiers, C., Previs, S. F., David, F., & Brunengraber, H. (1996). Correction of 13C mass isotopomer distributions for natural stable isotope abundance. *Journal of Mass Spectrometry*, *31*, 255–262.

Frank, O. (1959). On the dynamics of cardiac muscle. *American Heart Journal*, *58*, 282–317.

Gamcisk, M. P., Forder, J. R., Millis, K. K., & McGovern, K. A. (1996). A versatile oxygenator and perfusion system for magnetic resonance studies. *Biotechnology and Bioengineering*, *49*(3), 348–354.

Gelinas, R., Labarthe, F., Bouchard, B., McDuff, J., Charron, G., Young, M. E., et al. (2008). Alterations in carbohydrate metabolism and its regulation in PPARalpha null mouse hearts. *American Journal of Physiology. Heart and Circulatory Physiology*, *294*, H1571–H1580.

Gelinas, R., Thompson-Legault, J., Bouchard, B., Daneault, C., Mansour, A., Gillis, M. A., et al. (2011). Prolonged QT interval and lipid alterations beyond beta-oxidation in very long-chain acyl-CoA dehydrogenase null mouse hearts. *American Journal of Physiology. Heart and Circulatory Physiology*, *301*, H813–H823.

How, O. J., Aasum, E., Kunnathu, S., Severson, D. L., Myhre, E. S., & Larsen, T. S. (2005). Influence of substrate supply on cardiac efficiency, as measured by pressure-volume analysis in ex vivo mouse hearts. *American Journal of Physiology Heart and Circulatory Physiology*, *288*(6), H2979–H2985.

How, O. J., Aasum, E., Severson, D. L., Chan, W. Y., Essop, M. F., & Larsen, T. S. (2006). Increased myocardial oxygen consumption reduces cardiac efficiency in diabetic mice. *Diabetes*, *55*(2), 466–473.

Jeffrey, J. M., Roach, J. S., Storey, C. J., Sherry, A. D., & Malloy, C. R. (2002). ^{13}C isotopomer analysis of glutamate by tandem mass spectrometry. *Analytical Biochemistry*, *300*(2), 192–205.

Khairallah, R. J., Khairallah, M., Gelinas, R., Bouchard, B., Young, M. E., Allen, B. G., et al. (2008). Cyclic GMP signaling in cardiomyocytes modulates fatty acid trafficking and prevents triglyceride accumulation. *Journal of Molecular and Cellular Cardiology, 45*, 230–239.

Khairallah, M., Khairallah, R., Young, M. E., Dyck, J. R., Petrof, B. J., & Des Rosiers, C. (2007). Metabolic and signaling alterations in dystrophin-deficient hearts precede overt cardiomyopathy. *Journal of Molecular and Cellular Cardiology, 43*, 119–129.

Khairallah, M., Labarthe, F., Bouchard, B., Danialou, G., Petrof, B. J., & Des Rosiers, C. (2004). Profiling substrate fluxes in the isolated working mouse heart using ^{13}C-labelled substrates: Focusing on the origin and fate of pyruvate and citrate carbons. *American Journal of Physiology. Heart and Circulatory Physiology, 286*, H1461–H1470.

Kolwicz, S. C., Jr., Purohit, S., & Tian, R. (2014). Cardiac metabolism and its interaction with contraction, growth, and survival of cardiomyocytes. *Circulation Research, 113*(5), 603–616.

Langendorff, O. (1895). Untersuchugen am überlebenden Säugethierherzen. *Pflügers Archiv, 61*, 291.

Laplante, A., Comte, B., & Des, Rosiers C. (1995). Assay of blood and tissue oxaloacetate and alpha-ketoglutarate by isotope dilution gas chromatography–mass spectrometry. *Analytical Biochemistry, 224*(2), 580–587.

Lauzier, B., Vaillant, F., Merlen, C., Gelinas, R., Bouchard, B., Rivard, M. E., et al. (2013). Metabolic effects of glutamine on the heart: Anaplerosis versus the hexosamine biosynthetic pathway. *Journal of Molecular and Cellular Cardiology, 55*, 92–100.

Liao, R., Podesser, B. K., & Lim, C. C. (2012). The continuing evolution of the Langendorff and ejecting murine heart: New advances in cardiac phenotyping. *American Journal of Physiology. Heart and Circulatory Physiology, 303*(2), H156–H167.

Lopashuck, G. D., & Tsang, H. (1987). Metabolism of palmitate in isolated working hearts from spontaneously diabetic "BB" wistar rats. *Circulation Research, 61*(6), 853–858.

Marazzi, G., Rosanio, S., Caminiti, G., Dioguardi, F. S., & Mercuro, G. (2008). The role of amino acids in the modulation of cardiac metabolism during ischemia and heart failure. *Current Pharmaceutical Design, 14*, 2592–2604.

Neely, J. R., Libermeister, H., Battersby, E. J., & Morgan, H. E. (1967). Effect of pressure development on oxygen consumption by isolated heart. *The American Journal of Physiology, 212*, 802–814.

Neubauer, S. (2007). The failing heart—An engine out of fuel. *The New England Journal of Medicine, 356*, 1140–1151.

Poirier, M., Vincent, G., Reszko, A. E., Bouchard, B., Kelleher, J. K., Brunengraber, H., et al. (2002). Probing the link between citrate and malonyl-CoA in perfused rat hearts. *American Journal of Physiology. Heart and Circulatory Physiology, 283*(4), H1379–H1386.

Reszko, A. E., Kasumov, T., David, F., Jobbins, K. A., Thomas, K. R., Brunengraber, H., et al. (2004). Peroxisomal fatty acid oxidation is a substantial source of the acetyl moiety of malonyl-CoA in rat heart. *The Journal of Biological Chemistry, 279*(19), 19574–19579.

Rodgers, R. L., Christe, M. E., Tremblay, G. C., Babson, J. R., & Daniels, T. (2001). Insulin-like effects of a physiological concentration of carnitine on cardiac metabolism. *Molecular and Cellular Biochemistry, 226*(1–2), 97–105.

Taegtmeyer, H. (1995). One hundred years ago: Oscar Langendorff and the birth of cardiac metabolism. *The Canadian Journal of Cardiology, 11*(11), 1030–1035.

Taegtmeyer, H., Hems, R., & Krebs, H. A. (1980). Utilization of energy-providing substrates in the isolated working rat heart. *The Biochemical Journal, 186*, 701–711.

Templeton, G. H., Wildenthal, K., Willerson, J. T., & Reardon, W. C. (1974). Influence of temperature on the mechanical properties of cardiac muscle. *Circulation Research, 34*, 624–634.

Vaillant, F., Lauzier, B., Poirier, I., Gelinas, R., Rivard, M. E., Robillard Frayne, I., et al. (2014). Mouse strain differences in metabolic fluxes and function of ex vivo working hearts. *American Journal of Physiology. Heart and Circulatory Physiology, 306*, H78–H87.

Vincent, G., Bouchard, B., Khairallah, M., & Des Rosiers, C. (2004). Differential modulation of citrate synthesis and release by fatty acids in perfused working rat hearts. *American Journal of Physiology. Heart and Circulatory Physiology*, *286*, H257–H266.

Vincent, G., Comte, B., Poirier, M., & Des Rosiers, C. (2000). Citrate release by perfused rat hearts: A window on mitochondrial cataplerosis. *American Journal of Physiology. Endocrinology and Metabolism*, *278*(5), E846–E856.

Vincent, G., Khairallah, M., Bouchard, B., & Des Rosiers, C. (2003). Metabolic phenotyping of the diseased rat heart using 13C-substrates and ex vivo perfusion in the working mode. *Molecular and Cellular Biochemistry*, *242*, 89–99.

Wittnich, C., Tan, L., Wallen, J., & Belanger, M. (2013). Sex differences in myocardial metabolism and cardiac function: An emerging concept. *Pflügers Archiv*, *465*, 719–729.

Young, M. E. (2006). The circadian clock within the heart: Potential influence on myocardial gene expression, metabolism, and function. *American Journal of Physiology. Heart and Circulatory Physiology*, *290*, H1–H16.

CHAPTER FOUR

Probing Metabolism in the Intact Retina Using Stable Isotope Tracers

Jianhai Du*,†, Jonathan D. Linton*,†, James B. Hurley*,†,1
*Departments of Biochemistry, University of Washington, Seattle, Washington, USA
†Department of Ophthalmology, University of Washington, Seattle, Washington, USA
[1]Corresponding author: e-mail address: jbhhh@u.washington.edu

Contents

Abstract

Vertebrate retinas have several characteristics that make them particularly interesting from a metabolic perspective. The retinas have a highly laminated structure, high energy demands, and they share several metabolic features with tumors, such as a strong Warburg effect and abundant pyruvate kinase M2 isoform expression. The energy demands of retinas are both qualitatively and quantitatively different in light and darkness and metabolic dysfunction could cause retinal degeneration. Stable isotope-based metabolic analysis with mass spectrometry is a powerful tool to trace the dynamic metabolic reactions and reveal novel metabolic pathways within cells and between cells in retina. Here, we describe methods to quantify retinal metabolism in intact retinas and discuss applications of these methods to the understanding of neuron–glia interaction, light and dark adaptation, and retinal degenerative diseases.

Methods in Enzymology, Volume 561
ISSN 0076-6879
http://dx.doi.org/10.1016/bs.mie.2015.04.002

1. INTRODUCTION

Vertebrate retinas are light-sensitive neural tissues with at least seven types of cells. Rods and cones are photoreceptors that detect light. Bipolar cells, horizontal cells, and amacrine cells are interneurons that process signals initiated by the photoreceptors. Ganglion cells relay the processed signals through the optic nerve to the brain (Dowling, 1970; Wassle & Boycott, 1991).

Photoreceptors, interneurons, and ganglion cells are confined to distinct layers in the retina. Müller glia radiate across all layers of the retina (Wassle & Boycott, 1991). Figure 1 illustrates structural features of the retina. Neurons and glia in the retina have a symbiotic relationship (Lindsay et al., 2014; Newman & Reichenbach, 1996). The laminated structure of the retina can be used to facilitate investigations of intracellular and intercellular metabolic pathways.

Nutrients enter the eye through blood vessels that enter the back of the eye along with the optic nerve. Blood flows into the choroid layer, which lines the inside surface of the sclera, the white part of the eye. Nutrients flow from the choroid layer, through the retinal pigment epithelial (RPE), through the interphotoreceptor matrix to the retina. In some retinas blood vessels on the inner surface of the retina or within the inner retina also provide a source of nutrients (Campbell & Humphries, 2012; Cunha-Vaz, Bernardes, & Lobo, 2011).

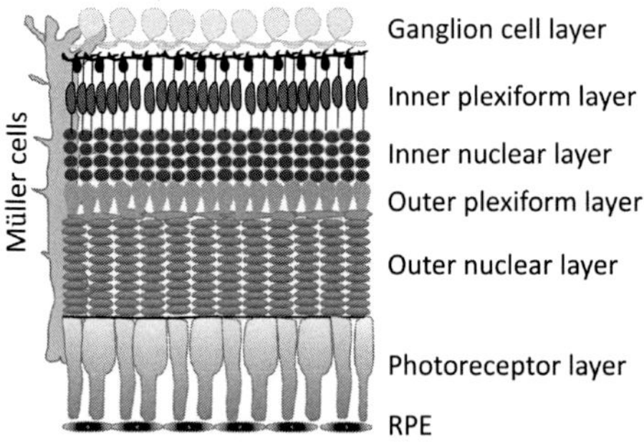

Figure 1 Laminar structure of vertebrate retina. The photoreceptor layer contains the outer segments and cell bodies of rods and cone cell bodies and the outer nuclear layer contains their nuclei; the inner nuclear layer contains nuclei of bipolar, amacrine, and horizontal and Müller cells; the inner plexiform layer contains dendrites and synapses of the inner retinal neurons and ganglion cells; the ganglion cell layer consists of ganglion cells; and Müller glia radiate across all layers of the retina. RPE is the retinal pigment epithelium. (See the color plate.)

Retinas have extraordinarily high energy demands. In darkness, active ion channels in the plasma membranes of photoreceptors drive ATP consumption by stimulating the activity ion pumps needed to maintain normal ion gradients. In light, metabolic energy is used to fuel phototransduction and to support synthesis of lipid and proteins (Ames, Li, Heher, & Kimble, 1992; Cornwall et al., 2003; Okawa, Sampath, Laughlin, & Fain, 2008).

Energy metabolism in the vertebrate retina is dominated by aerobic glycolysis, also called the Warburg effect. Between 80% and 96% of glucose consumed by an isolated retina is made into lactate even when O_2 is abundant (Ames et al., 1992; Du, Cleghorn, Contreras, Linton, et al., 2013; Wang, Kondo, & Bill, 1997; Winkler, 1981). Glucose is essential for retinal function. Removal of glucose, inhibition of glycolysis, or deprivation of oxygen rapidly diminishes the ability of the retina to respond to light (Ames et al., 1992; Kang Derwent & Linsenmeier, 2000). Deficiencies of key enzymes and transporters in glucose metabolism cause retinal degeneration (Bui, Kalloniatis, & Vingrys, 2004; Maurer, Schonthaler, Mueller, & Neuhauss, 2010; Ochrietor & Linser, 2004). Neurons in the retina require metabolic support from glial cells. Either disruption of Müller cell metabolism or genetic ablation of Müller cells can cause retinal degeneration (Jablonski & Iannaccone, 2000; Shen, Zhu, Lee, Chung, & Gillies, 2013). An important reason to investigate retinal metabolism is that its disruption is likely to be the basis for many forms of neurodegenerative diseases.

Metabolic analysis using stable isotopes is a powerful tool for probing enzyme reactions and transporter/carrier activities in an intact retina. Metabolites that differ in the number of isotopic atoms incorporated are called mass isotopomers. The distributions of mass isotopomers provide insights into the types of pathways in a tissue and how they are regulated. The distributions of isotopomers can be quantified by gas chromatography (GC) and/or liquid chromatography (LC) followed by mass spectrometry (MS).

We have used stable isotope methods to identify novel metabolic pathways in intact mouse retinas (Du, Cleghorn, Contreras, Lindsay, et al., 2013; Du, Cleghorn, Contreras, Linton, et al., 2013; Lindsay et al., 2014). We present our methods in detail including experiment design, retinal culture, sample preparation, mass spectrometry analysis, and examples of applications.

2. METHODS

A flow chart of the stable isotope metabolic analysis in retinas is shown in Fig. 2.

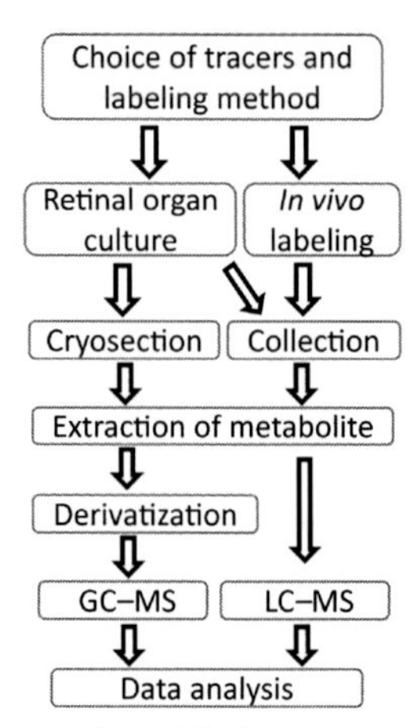

Figure 2 A flow chart of the steps in stable isotope experiments in retina.

2.1 Choice of Stable Isotope Tracers and Labeling Method

^{13}C, ^{2}H, ^{15}N, and ^{18}O are stable isotopes commonly used to study metabolism. Many tracers labeled with these stable isotopes at various positions are commercially available. Specific tracers can be chosen based on the types of biochemical reactions being studied. There are many reports on how to select tracers rationally for mass spectrometry. Here, we present a short list of tracers we used in retina or that will be potentially useful for future studies of retinal metabolism in Table 1. Some neuronal cell types might prefer certain types of tracers, e.g., ^{13}C acetate and ^{13}C glutamate are highly preferred by astrocytes in brain metabolism (Deelchand, Shestov, Koski, Ugurbil, & Henry, 2009; Pellerin & Magistretti, 1994). The types of tracers should be chosen to most effectively investigate the specific cell type and incubation conditions.

There are at least three methods to label metabolites based on the time of experiment: (1) Steady-state labeling, (2) Dynamic state (nonsteady state) labeling, and (3) Pulse-chase labeling. Steady-state labeling can take hours to days in mammalian culture depending on the pathways or types of cells and tissue. It is called steady state because the metabolic labeling pattern, concentration, and fluxes remain unchanged over time. Most flux analyses, calculations, and modeling assume that metabolites have reached steady state (Gianchandani, Chavali, & Papin, 2010). For retinas in *ex vivo* culture incubated with U-^{13}C glucose, glycolytic intermediates reach steady state after about 90 min, but intermediates in the tricarboxylic acid (TCA) cycle require about 4 h to reach steady state (Du, Cleghorn, Contreras, Linton, et al., 2013).

Table 1 A Short List of Stable Isotope Tracers in the Study of Retinal Metabolism

Tracers	Metabolic Pathways
^{13}C glucose	Glycolysis, TCA cycle, fatty acids, PPP pathway
^{13}C glutamine	TCA cycle, IDH carboxylation, fatty acids
^{13}C glutamate	Astrocyte metabolism, glutamine metabolism
^{13}C lactate	Gluconeogenesis, TCA cycle
^{13}C pyruvate	Pyruvate transport, TCA cycle
^{13}C palmitate	Fatty acid oxidation, ketone body metabolism
^{13}C glycine	One carbon metabolism, purine synthesis
^{13}C acetate	Astrocyte metabolism, TCA cycle
^{2}H glucose	Glucose metabolism, NAD(P)H reaction
^{15}N glutamine	Transamination, nucleotide synthesis
^{13}C ^{15}N ATP	Purine degradation
$^{2}H_2O$	Lipogenesis, gluconeogenesis
$H_2{}^{18}O$	ATP hydrolysis and biosynthesis, cGMP metabolism, glycolysis, glycogenolysis

Steady-state labeling facilitates calculating metabolic flux accurately; however, it takes a long time and one needs to make sure that there is no loss of cell viability or changes in cellular phenotypes. Steady-state labeling also obscures dynamic information about metabolic pathways (Fernie, Geigenberger, & Stitt, 2005; Noh & Wiechert, 2011; Wiechert & Noh, 2013). Multiple cycling produces labeling patterns in which it is difficult to distinguish metabolites from different pathways. For example, under steady-state conditions U-^{13}C glucose labeling, M3 malate can be produced either from the second turn of the TCA cycle or from the pyruvate carboxylase reaction. When dynamic labeling is used, i.e., labeling for only short time (1–5 min), the TCA metabolites have not started the second cycle. Under those conditions any M3 malate must be exclusively from carboxylation. Dynamic state labeling uses shorter incubation times with the tracer and can limit the labeling to specific pathways. However, in dynamic state labeling, differences in pool sizes should be considered for accurate calculation of metabolic flux. As the tracers have not reached equilibrium with unlabeled metabolic pools, pool sizes are inversely correlated to the rate of enrichment and can significantly affect flux calculation, especially, when comparing flux with different tissue or treatment that alters pool sizes. The pool sizes can be measured simultaneously with the same or different tracers (Gastaldelli, Coggan, & Wolfe, 1999; Noh et al., 2007). Pulse-chase labeling starts with addition of a labeled metabolite for a brief interval (pulse) and then changes to incubation with the same metabolite in unlabeled form (chase). This technique can be very useful for analyzing pathways where metabolites are exchanged between different cells types in a tissue (see details in Section 3.1).

2.2 Retinal Organ Culture

Mammalian retinal organ cultures provide a very useful model for identifying and characterizing metabolic pathways. Its strength is that it gives investigators the flexibility to design protocols that focus on the roles of specific fuels, metabolites, and metabolic pathways. Its weakness is that it may not accurately reflect *in vivo* metabolism. We use a defined Krebs-Ringer/HEPES/bicarbonate (KRB) medium to culture mouse or rat retinas. The KRB medium has several advantages over enriched medium: (1) it minimizes interference caused by metabolites added with serum and supplements; (2) It can be customized by addition of specific nutrients or changes of solvent, e.g., with commercial enriched medium, it is hard to use ^{18}O water as a tracer; and (3) It is inexpensive. We found that retinas can maintain physiological functions (light response and O_2 consumption) in this medium. In KRB containing 5 m*M* glucose, there is no appreciable cell death within 6 h (based on propidium iodide staining). We prepare KRB medium in stock concentration and filter it through a 0.22 μ*M* membrane (Table 2). The KRB is diluted to its final concentration with water. Nutrients or stable isotope tracers are added just before each experiment.

Table 2 The Formula of KRB Medium

Solute	Stock (m*M*)	Final (m*M*)	Sigma Cat.
NaCl	102.60	98.50	S5886
KCl	5.11	4.90	P5405
KH_2PO_4	1.25	1.20	P5655
$MgSO_4$	1.25	1.20	M1880
HEPES	20.84	20.00	H4034
$CaCl_2$	2.71	2.60	C7902
$NaHCO_3$	26.98	25.90	S5761

When ^{18}O water is used as a tracer, the components of the KRB are dissolved in 10% ^{18}O water. Light- or dark-adapted mice are euthanized by cervical dislocation. The eyes are enucleated and the retina is dissected out of the eye cup into cold HBSS under a dissecting microscope. For dark-adaptation experiments, the mice are dark-adapted for longer than 4 h. All operations including retina dissections are performed under infrared illumination with the investigator using night-vision goggles to see. Retinas are washed once with cold HBSS and then transferred into preincubated KRB medium containing nutrient and tracers of choice at 37 °C in a 5% CO_2 incubator. The entire operation from euthanasia to putting the culture dishes into the incubator should take less than 5 min to minimize hypoxia and photoreceptor cell death. Photoreceptors are very sensitive to hypoxia and nutrient deprivation. The block of blood supply for only 10 s could cause vision loss (Linsenmeier, 1990) and the availability of glucose is essential for the survival of photoreceptors (Chertov et al., 2011). Rat retinal isolation and culture are done in the same way as the mouse except the rat retina is cut into four pieces to increase nutrient accessibility.

2.3 Retinal Organ Culture for Cryosection

The unique laminated structure of retina (Fig. 1) allows the separation of different cell layers by cryosectioning (see Section 2.5.3). Our lab has used this technique to localize protein expression and metabolites in different cell layers (Du, Cleghorn, Contreras, Linton, et al., 2013; Lindsay et al., 2014; Linton et al., 2010). We have optimized this method for tracer experiments. Retinas should be flattened to ensure successful cryosectioning. A dissecting microscope is used to place the isolated retina in HBSS onto the grid side of a cellulose membrane disc (0.45 μm, 13 mm, gridded, EMD Millipore) with the photoreceptor side facing the cellulose. Cuts are made in the retina to relieve mechanical stress so that it can be flattened. To enhance attachment of the retina to the cellulose, the retina/cellulose disc is blotted quickly on filter paper and then returned to HBSS. We repeat this 3 × and then incubate the retina/cellulose discs in 35 m*M* culture dishes containing 2 ml tracers and KRB.

2.4 Tracing Retinal Metabolism *In Vivo*

Studying retina metabolism *in vivo* has unique advantages. The retinas are in their native physiological environment and they are not stressed by euthanasia, dissection, or changes in their environment. *In vivo* stable-isotope labeling has been used extensively in brain tissue together with NMR and mass spectrometry (Dobbins & Malloy, 2003; Kanamori, Kondrat, & Ross, 2003). However, to our knowledge there is no data available yet on stable-isotope labeling of retinas *in vivo*. Our preliminary experiments have shown that intraperitoneal injection of a bolus of U-^{13}C glucose (500 mg/Kg) enriched the glycolysis and TCA cycle intermediates (listed in Table 2) between 10% and 40% in 45 min in mouse retina. In this experiment, the retina was isolated quickly from the mouse and snap frozen in liquid nitrogen (Section 2.5.2). More *in vivo* studies will be needed to optimize this method to explore labeling of retinas through intravenous and intravitreal injection and infusion, and to determine the influence of permeability of retinal-blood barrier on the choice of tracers that can be used. However, some disadvantages in *in vivo* labeling should also be considered in experiment design: (1) Multiple and bigger pools compared with *ex vivo* culture can take longer to reach isotopic steady state; (2) Metabolism and metabolic exchange among tissues (e.g., Cori cycle between muscle and liver) can result in scrambling of mass isotopomers; (3) The patterns of mass isotopomers distribution (MID) and time to reach steady state can be influenced by the sites (intraperitoneal, tail vein, or intravitreal injections) of tracer administration.

2.5 Sample Collection and Preparation

Good sample preparation is the starting point for successful metabolite analysis. Since cellular metabolites have very rapid turnover, sample collection and preparation require the fastest possible quenching and most effective extraction of cellular metabolites. There are many reports on quenching and extraction methods. We found it is effective to quench the reaction with cold saline (Strumilo, 1995) and extract polar metabolites with a cold methanol/chloroform/water mixture. To obtain spatial resolution of the metabolic reaction, the retina can be sliced into sections to separate cell layers. For GC–MS, the extract metabolites need to be derivatized to render them volatile.

2.5.1 Collection and Preparation of Medium Sample

We collect KRB medium samples directly into EP tubes and centrifuge them at 14,000 rpm for 10 min. The supernatant is transferred and saved at −20 °C. Glucose and glutamine can be analyzed by LC–MS (see Section 2.5) or by commercial colorimetric analysis such as Amplex red assays (Life Sciences). Lactate, pyruvate, citrate, adenosine, and hypoxanthine are often released by mouse retinas, so they can be measured by GC–MS and LC–MS. We transfer 20–50 μl of medium into an insert (Agilent) inside a 1.5 EP tube and dry it for derivatization for GC–MS, or we transfer it into an EP tube and dry it for LC–MS. It is important to note that high salt concentration can suppress mass spectrometry signals. To test for this effect in LC–MS, we redissolved the dried sample in 2 × volume of mobile phase and compared it to dissolving the standard directly in the mobile phase. There is no significant difference in the signals we detected in most of our samples. However, if larger volumes of sample medium are used, the salt should be removed by ion exchange columns before analysis.

2.5.2 Collection and Preparation of Retinas

After labeling *ex vivo* or *in vivo*, the retina is transferred quickly into a 1.5 ml conical screw cap tube containing 1 ml ice cold 0.9% NaCl to quench metabolism. We often use a transfer pipette or wide-opening 1 ml pipette tip to transfer the retina to avoid damage. The tube is spun quickly to remove cold saline and immediately frozen in liquid nitrogen. In advance, we prepare extraction buffer (methanol:chloroform:H_2O (700:200:50)) and store it on ice. We add 140 μl of this extraction buffer together with 10 μl of 100 μ*M* methylsuccinate or norvaline into the tube. These are added as internal standards. We homogenize the retina with pestles powered by a cordless motor (Kimble-Kontes) for 15 s and leave the homogenate on ice for 15 min to precipitate proteins. The homogenate then is centrifuged at 14,000 rpm for 15 min at 4 °C and the supernatant containing extracted metabolites is transferred to a 250 μl glass insert placed inside a 1.5 ml EP tube. The sample can be dried for mass spectrometry or stored at −80 °C. The pellet is dissolved in 200 μl of 0.1 M NaOH overnight at 37 °C. Protein concentrations are determined by the BCA assay kit (Thermo Fisher Scientific). *Note: Be careful when taking the tube out of centrifuge to avoid loosening the pellet.*

2.5.3 Cryosection of Labeled Retina

The retina/cellulose disc (from Section 2.3) is picked up with forceps, being careful not to touch the retina. The disc is dipped into a dish containing cold 0.9% NaCl for 5 s to quench metabolism. We place the disc with its cellulose side down (the side without the retina) on filter paper to dry the liquid quickly. We mount the disc with the same side down, on a quarter of a glass slide prepared with a small drop of superglue on its top. The slide is then snap frozen under slight pressure on a drill press. In preparation for this, the head of a bolt that has been polished to a mirror finish is kept cold in liquid nitrogen. Just before flattening the retina we insert the bolt into the chuck of a drill press and tighten it in place. The head of the bolt is lowered quickly onto the retina and contact is made with slight pressure for ~5 s. After freezing, the retina/cellulose slide is stored on dry ice.

A second quarter of a glass slide then is wrapped in polytetrafluoroethylene (PTFE) tape. This is to prevent the retina from sticking to the slide. A ~0.5 mm plastic spacer is glued on each side of the PTFE tape. Then we sandwich the retina/cellulose between the slides. Binder clips are used to hold the sandwich together. The sandwiched retina then is placed on dry ice for 30 min and stored overnight at −20 °C.

We then cover a cryostat pedestal in optimal cutting temperature compound (OCT) and allow it to freeze completely. The OCT covered pedestal is mounted in the cryostat and 20 μm slices are made until the OCT is completely flat in one plane. We disassemble the retina sandwich and mount the retina/cellulose slide on the flat OCT pedestal. We use a drop of water applied to the top corners of the slide OCT interface and allow it to freeze in place.

We trim the flattest part of the retina with a scalpel blade to an area ~1 mm^2. After trimming, the cryostat blade is advanced in 20 μm steps until contact has been made with the trimmed retina. Sections are then cut at 50 μm until contact with the cellulose. At this thickness, the outer segment layer, photoreceptor cell body layer, inner retina layer, and ganglion cell layer are included in a total of 4–5 sections (Fig. 1). Each section is picked up with a plastic pipette tip and placed into a 1.5 ml tube containing extraction buffer on dry ice. The metabolites are extracted as described in Section 2.5.2. The pellet is dissolved in protein lysis buffer. The quality of the sectioning is monitored by immunoblotting with antibodies against photoreceptor or ganglion cell specific markers.

2.5.4 Derivatization for GC–MS

GC–MS requires all analytes to be volatilized. The polar nature of amino acids and organic acids requires derivatization before GC analysis. Depending on the metabolites of interest, there are a large variety of derivatization reagents and methods available for GC–MS. Here, we provide a protocol to derivatize keto groups with methoxyamine and to replace hydrogen groups in COOH, NH, and SH with volatile *N*-tert-butyldimethylsilyl-*N*-methyltrifluoroacetamide (TBDMS). The derivatives from this method are more stable and less moisture sensitive than most other types of derivatives. We use the following protocol:

1. Dry the samples (as prepared in Sections 2.5.1–2.5.3) in a lyophilizer (Labconco) or Speedvac without heat until they are completely dry.
2. Add 25 μl of freshly prepared methoxyamine hydrochloride (Sigma) in pyrimidine (20 mg/ml) to the dried whole retinal extract. For dried retinal sections or dried medium samples use only 10 μl. Gently tap the tubes with inserts inside 3–5 × to mix, then spin the samples down and close the lid tightly.
3. Incubate them at 37 °C for 90 min in an oven or heat block.
4. After incubation, quickly spin the tubes and transfer into the inserts, 25 μl for whole retinal extract, 10 μl for section extract or 10 μl for medium sample. Gently tap the tubes 3–5 × to mix, then spin the samples and close the lid tightly.
5. Incubate the tubes with samples inside at 70 °C for 30 min. Spin tubes and transfer inserts into GC–MS glass vials. For medium samples, the supernatant will be transferred to new inserts in glass vials. The sample should be clear after derivatization although they may be colored from drugs used in the incubations. If it is yellowish, the sample might be contaminated with proteins during preparation.

2.5.5 Sample Preparation for LC–MS

Samples to be analyzed by LC–MS do not need derivatization. The retinal samples are dried completely and suspended in 100 μl of mobile phase (40% of A and 60% of B) with vortex for 10 s. The samples are then passed through 0.45-μ*M* PVDF syringe filters directly into glass inserts in vials.

2.6 Analysis of Metabolites by GC–MS

GC–MS is used to measure metabolites including most of the intermediates in glycolysis and the TCA cycle (Table 3). We use an Agilent 7890/5975C GC/MS system (Agilent Technologies) with an Agilent HP-5MS column (30 m × 0.25 mm × 0.25 μm film) under electron ionization at 70 eV. Ultrahigh-purity helium is the carrier gas at a constant flow rate of 1 ml/min. One microliter of sample is injected in split-less mode by the auto sampler. The temperature gradient starts at 100 °C with a hold time of 4 min and then increases at a rate of 5 °C/min to 300 °C, where it is held for 5 min. The temperatures reset as follows: inlet 250 °C, transfer line 280 °C, ion source 230 °C, and quadrupole 150 °C. Mass spectra are collected from *m/z* 50–600 at a rate of 1.4 spectra/s after a 6.5-min solvent delay. Select ion

Table 3 Metabolites for GC–MS Analysis

Metabolite	M-57 Ion	OH/NH/SH +Keto	Metabolite Formula	Derivative Formula	SIM	RT (Min)
Metabolites in glycolysis						
3PG	585	4	$C_3H_3O_7P$	$Si_4C_{20}H_{51}$	585–588	35.81
GAP	484	3+1	$C_3H_4O_6P$	$Si_3C_{15}H_{39}N$	484–487	31.71
DHAP	484	3+1	$C_3H_4O_6P$	$Si_3C_{15}H_{39}N$	484–487	32.08
PEP	453	3	$C_3H_2O_6P$	$Si_3C_{14}H_{36}$	453–456	29.50
Pyruvate	174	1+1	$C_3H_3O_3$	SiC_3H_9N	174–177	8.50
Lactate	261	2	$C_3H_4O_3$	$Si_2C_8H_{21}$	261–264	14.39
Alanine	260	2	$C_3H_5NO_2$	$Si_2C_8H_{21}$	260–263	15.51
Serine	390	3	$C_3H_4NO_3$	$Si_3C_{14}H_{36}$	390–393	25.12
Metabolites in TCA cycle						
Citrate	591	4	$C_6H_4O_7$	$Si_4C_{20}H_{51}$	591–597	35.66
α-Ketoglutarate	346	2+1	$C_5H_4O_5$	$Si_2C_9H_{24}N$	346–351	25.57
Malate	419	3	$C_4H_3O_5$	$Si_3C_{14}H_{36}$	419–423	27.40
Succinate	289	2	$C_4H_4O_4$	$Si_2C_8H_{21}$	289–293	20.52
Fumarate	287	2	$C_4H_2O_4$	$Si_2C_8H_{21}$	287–291	21.14
Aspartate	418	3	$C_4H_4NO_4$	$Si_3C_{14}H_{36}$	418–422	28.19
Glutamate	432	3	$C_5H_6NO_4$	$Si_3C_{14}H_{36}$	432–437	30.26
Glutamine	431	3	$C_5H_7N_2O_3$	$Si_3C_{14}H_{36}$	431–436	32.67
Asparagine	417	3	$C_4H_5N_2O_3$	$Si_3C_{14}H_{36}$	417–421	30.73
2-HG	433	3	$C_5H_5O_5$	$Si_3C_{14}H_{36}$	433–438	29.22
NAA	460	3	$C_6H_6NO_5$	$Si_3C_{14}H_{36}$	460–466	30.93
3HBA	275	2+1	$C_4H_6O_3$	$Si_2C_9H_{24}N$	275–279	16.93
IS						
Methylsuccinate	303	2	$C_5O_4H_6$	$Si_2C_8H_{21}$	303	20.70
Norvaline	288	2	$C_5H_9NO_2$	$Si_2C_8H_{21}$	288	18.42

3PG, 3-phosphoglycerate; GAP, glyceraldehyde 3-phosphate; DHAP, dihydroxyacetone phosphate; PEP, phosphoenolpyruvic acid; IS, internal standard.

monitoring (SIM) records only selected ranges of *m*/*z* in expected retention time windows. This mode significantly improves the sensitivity and specificity of detection.

The TBDMS group adds 114 and methoxyamine adds 29 to the molecular weight. After ionization, the molecular weight 57 less than the derivatives (C4H9 (M-57)) is often the major ion in the spectra. Therefore, M-57 mass is selected to quantify the intensity of each metabolite. Table 3 lists the mass we monitor in the SIM mode for metabolites from glucose, glutamine, and ketone metabolism. The chromatograms are analyzed using Agilent Chemstation software. Abundances of the selected ions are extracted.

Natural abundance from stable isotopes and derivative agents interferes with the MID, and it must be corrected before MID analysis (Fernandez, Des Rosiers, Previs, David, & Brunengraber, 1996; Hellerstein & Neese, 1999). IsoCor, freely available software under an open-source license, provides a simple solution to correct both common naturally occurring isotopes (C, H, O, S, N, and Si) and the purity of the isotopic tracer (Millard, Letisse, Sokol, & Portais, 2012) (http://metasys.insa-toulouse.fr/software/isocor). The metabolite formula (the formula after derivatization) and derivative formula listed in Table 3 can be used in IsoCor for natural abundance correction. The fractional abundance of isotopomers are named as M0, M1, M2, ..., M*x*. M0 is the mass without labeling (M-57 ion in Table 3) and 1 to *x* represents the mass shift from the isotope labeling.

The concentration of each unlabeled metabolite is determined using an external calibration curve containing internal standard.

2.7 Analysis of Metabolites by LC–MS

Large and/or thermo-unstable organic molecules including sugars, phosphate, and nucleotides are particularly suitable for LC–MS. There are a large selection of methods and instruments available depending on the metabolites of interest. Labeled glucose, glutamine, and glutathione from ^{13}C glucose or ^{13}C glutamine in the medium and retina can be quantified directly by LC–MS. Analysis of the mass isotopomer distributions of nucleotides can estimate the rate of ATP production, cGMP degradation, and pathways in nucleotide degradation. Here, we present a protocol to measure metabolites listed in Table 4 by LC–MS. We use a Waters Xevo TQ Tandem mass spectrometer with a Waters ACQUITY system with UPLC. An ACQUITY UPLC BEH Amide analytic column (2.1 × 50 mm, 1.7 μm, Waters) is used for separation. The mobile phase is (A) water with

Table 4 Metabolites for LC–MS Analysis

Metabolite	Mode	Parent (*m/z*)	Daughter (*m/z*)	Dwell (s)	Cone (v)	Collision (v)
Glucose	Negative	179	89	0.025	18	8
Glutamine	Positive	147	84	0.025	14	16
GSSH	Positive	613	231	0.04	86	38
GSH	Positive	308	84	0.04	22	22
ATP	Positive	508	136	0.08	28	30
ADP	Negative	426	134	0.025	34	22
AMP	Negative	346	79	0.025	34	22
Adenosine	Positive	268	136	0.06	22	16
NAD	Positive	664	136	0.025	28	52
NADH	Positive	666	108	0.06	24	58
GTP	Negative	521	159	0.04	40	34
cGMP	Positive	346	135	0.06	30	44
GDP	Positive	444	135	0.06	20	60
GMP	Positive	364	135	0.06	20	46
Guanosine	Positive	284	152	0.06	14	16
Xanthine	Negative	151	108	0.025	32	14
Hypoxanthine	Negative	135	92	0.025	32	14

GSH, reduced glutathione; GSSH, oxidized glutathione; cGMP, cyclic GMP.

10 m*M* ammonium acetate (pH 8.9) and (B) acetonitrile/water (95/5) with 10 m*M* ammonium acetate (pH 8.9) (All solvents are LC–MS Optima grade from Fisher Scientific). The gradient elution is (1) 95–61% B in 6 min, (2) 61–44% B at 8 min, (3) 61–27% B at 8.2 min, and 27–95% B at 9 min. The column is reequilibrated with 95% B at the end of each run. The Flow rate for all gradient is 0.5 ml/min and the total run is 11 min. The injection volume for each sample is 5 μl. Mass spectrometer settings are shown in Table 4. Each transition includes a parent ion and fragmented daughter ion. Transitions for isotopomers can be set up based on the tracers. *A priori* knowledge of the labeled moiety on the parent and/or daughter ions is important. The formula and fragment pattern can be checked in public databases such as METLIN (http://metlin.scripps.edu/index.php) and Human Metabolome

Database (http://www.hmdb.ca/). For example, the daughter ion of ATP is the purine base. In ^{18}O water tracer experiments, the transitions of isotopomers of parent ions can be set at 508 (M0), 510 (M2), 512 (M4), and 514 (M6), while the daughter ion is 136 for all these transitions since purine base will not be labeled by ^{18}O. The chromatograms are analyzed by MassLynx (Waters). Corrections for natural abundance are made as described in the section on GC–MS.

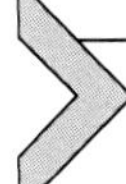

3. APPLICATIONS

3.1 Neuron–Glia Interaction

Müller glia interact closely with neurons to maintain metabolic homeostasis. Müller cells quickly undergo reactive gliosis after neuronal damage and similarly the damage of Müller cells can cause the death of photoreceptor neurons (Dyer & Cepko, 2000; Shen et al., 2013). This interdependence might be attributed to the intricate metabolite exchange between glia and neurons. It has been hypothesized that Müller glia supply glutamine and glucose to neurons while neurons might send lactate and glutamate to Müller glial cells. Stable-isotope labeling has been proved to be a useful tool to understand the metabolic interactions between glia and neurons in retina (Lindsay et al., 2014).

An example from our lab demonstrates that glia can use lactate and aspartate from neurons as surrogates to compensate for specific enzyme deficiencies in the Müller cells (Lindsay et al., 2014). We found that photoreceptors have abundant expression of both pyruvate kinase M2 isoform (PKM2) and aspartate glutamate carrier 1 (AGC1), but there is very little expression of these two enzymes or any other forms of pyruvate kinase in Müller glial cells. Glutamine synthetase (GS) is expressed exclusively in Müller glia in retinas, where it produces glutamine for neurons. To understand how Müller glia overcome their deficiency in Pyruvate Kinase and AGC1 in glutamine synthesis, we labeled retinas and isolated cells with different ^{13}C labeled fuels, and found that ^{13}C aspartate is the most efficient fuel for glial cells. Aspartate enhanced glycolysis, the TCA cycle, and glutamine synthesis in both intact retina and isolated glial cells. To test our hypothesis that aspartate transports oxidizing power and carbons from neurons to Müller cells in intact retinas, we designed a pulse-chase labeling strategy. This strategy is based on our results and *a priori* knowledge about glutamine/glutamate metabolism. The photoreceptors are the primary site of glutamine catabolism and Müller cells are the primary site of glutamine synthesis. Glutaminase mostly expresses in neurons and GS is exclusively in MCs. To test the strategy, we treated retinas for 5 min with 5 m*M* U-^{13}C glutamine (M5), followed by incubation with 5 m*M* unlabeled lactate. Unlabeled glutamine was not included in the chase, because the intense signal from the added glutamine would have obscured the isotopomer signals we intended to quantify. At various times, retinas were harvested and metabolites were extracted and analyzed by GC–MS as described in Section 2. Figure 3A tracks the flow of carbons in this model. Figure 3B (upper) shows that M5 Glutamine

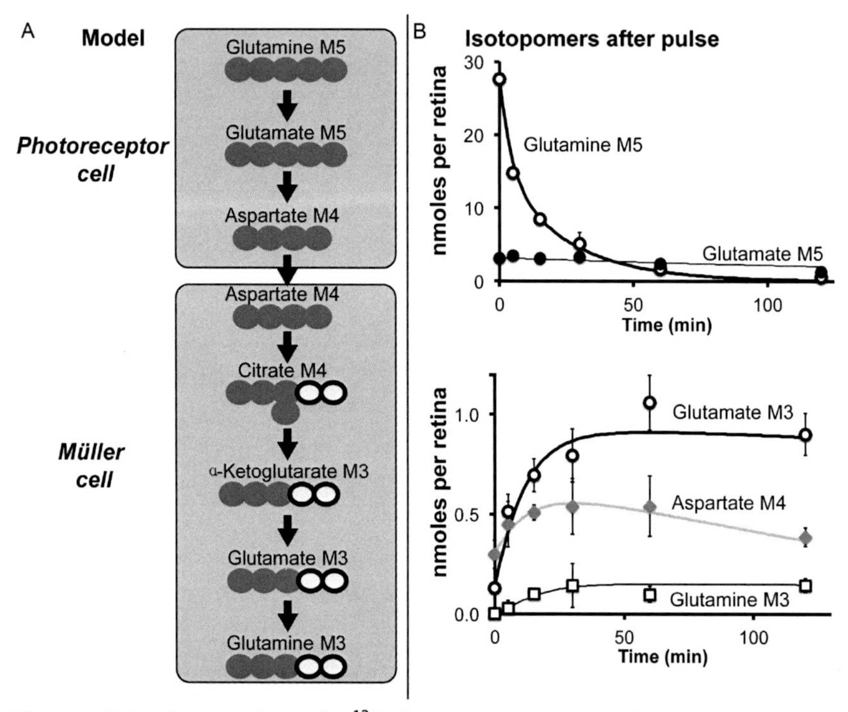

Figure 3 Pulse-chase analysis of U-^{13}C Gln in retina. Data are taken from the authors' previous study (Lindsay et al., 2014). (A) Schematic model for the role of aspartate as a carrier of oxidizing power between retinal neurons and glia. Red circles represent the ^{13}C carbons, and black circles represent the ^{12}C carbons. (B) ^{13}C labeling of Aspartate, Glutamate, and Glutamine from the pulse of U-^{13}C Glutamine. (Upper) The M5 Glutamine and Glutamate are derived directly from the pulse of 5 m*M* U-^{13}C Glutamine. After 5 min, the medium was changed to 5 m*M* unlabeled Lac with no added Gln. The retinas were subsequently harvested at the indicated times after the pulse. (Lower) The M4 Aspartate derived from oxidation of Glutamate via the TCA. The M3 Glutamate is made by further oxidation via citrate, and M3 Glutamine is made only in MCs by Glutamine synthetase ($n=6$). (See the color plate.)

converts rapidly to M5 Glutamate. The lingering M5 glutamate and M5 α-ketoglutarate are consistent with our previously reported finding that α-ketoglutarate in retinas is protected from oxidation in neurons. Figure 3B (lower) shows how M4 aspartate accumulates and then decays as M3 Glutamate and then M3 Glutamine accumulate. Conversion of M4 aspartate into M3 Glutamine confirms that the subsequent reactions (Fig. 3A) occur in MCs, because only MCs express GS. We further confirmed this model by blocking the transport of aspartate. The availability of more information on differential enzyme distribution, animal models of specific cell ablation, and models of conditional gene knockout will substantially facilitate further advance in understanding of metabolic interactions using stable isotope-based methodologies.

3.2 Retinal Metabolism in Light and Dark Adaption

Many studies have shown that both energy demand and energy production in retinas are greater in darkness than in light (Ames et al., 1992; Cornwall et al., 2003; Linsenmeier, 1986; Okawa et al., 2008; Wang et al., 1997). Photoreceptors are constitutively depolarized in darkness because ion channels in the outer segment of the photoreceptor are opened by their agonist, cyclic GMP. Ions leaking in through those channels in darkness must be removed from the cell via an ion pump that consumes ATP. Light activates cGMP hydrolysis to close channels and diminish ATP consumption but light also stimulates specific anabolic requirements, regeneration of visual pigment, and renewal of membranes. The molecular mechanisms by which retinas control the pace of their metabolism under different illumination conditions are still unknown. Energy consumption in retina has been evaluated by measuring O_2 consumption, production of CO_2 and lactate, and pumping of sodium ion (Okawa et al., 2008; Wang et al., 1997; Winkler, 1981). These experiments have limitations in providing mechanistically biochemical insights. Furthermore, besides energy consumption, little information is available on changes of nucleotide metabolism in light and darkness. This is important because large changes in cGMP metabolism could alter metabolism of other purines as well. Stable isotope tracers can be used with light- and dark-adapted retinas both *ex vivo* and *in vivo*. These tracers can monitor metabolic enzyme reactions directly by measuring the labeled metabolites. Cryosection of labeled retinal layers will provide information about the metabolic changes occurring in the retina.

3.3 Retinal Degenerative Diseases

Retinal degenerative diseases cause blindness in both young and old in a variety of diseases including age-related macular degeneration (AMD), retinitis pigmentosa, Leber's congenital amaurosis, Stargardt disease, and Usher Syndrome (Rejdak et al., 2012). Retinal degenerative diseases have been linked to mutations in energy metabolism and purine metabolism such as isocitrate dehydrogenase 3b, pyruvate dehydrogenase E1/2, nicotinamide mononucleotide adenylyltransferase, inosine monophosphate dehydrogenase, phosphodiesterase 6, and guanylyl cyclase (Aherne et al., 2004; Dizhoor, Boikov, & Olshevskaya, 1998; Farber & Lolley, 1974; Hartong et al., 2008; Maurer et al., 2010; Perrault et al., 2012; Taylor, Hurley, Van Epps, & Brockerhoff, 2004). Disruption of nutrient availability and metabolic regulation might play important roles in AMD and several models of retinitis pigmentosa (Punzo, Kornacker, & Cepko, 2009; Umino et al., 2006). The diseased neurons might reprogram metabolism and/or cause some sort of metabolic failure during degeneration. The stable isotope-based methodologies allow for kinetic analysis of metabolic reactions in diseased retinas that will help advance our understanding of these diseases. Because metabolites are very sensitive to environmental changes and cellular metabolic state, it is important to compare mutant animals with littermate controls with the same age. It also is important to include several time points spanning the full time course of the degeneration. Besides research in *ex vivo* labeling, the labeling of retinas *in vivo* will be critical to provide information on the integrity of the blood-retinal barriers and the efficiency of nutrient utilization.

4. SUMMARY

Retina is an excellent model for studying metabolism that occurs within a nonhomogeneous, multicellular tissue. Its laminar structure is suited for cryosection analysis with or without culture. Recent advances in metabolite imaging directly on tissue sections with mass spectrometry such as MALDI (matrix assisted laser desorption and ionization)-MS (Sun et al., 2014), SIMS (second ion mass spectrometry) (Wedlock et al., 2011), and NIMS (nanostructure-initiator mass spectrometry) (Greving, Patti, & Siuzdak, 2011) would help resolve the conundrum of metabolite compartmentalization and exchange among difference cells in the normal and diseased retinas. Vertebrate retinas share some features of tumor metabolism, such as Warburg effect and high expression of PKM2. They also have their own unique metabolic demands and pathways in light and darkness and in neurons and glial cells. Stable isotope-assisted methods have been successfully applied in metabolism research including retinal metabolism. The findings in cancer metabolism are useful to guide studies of retinal metabolism. Similarly, the heterogeneous nature of metabolism in retina with its different cell types might provide useful examples for tumor metabolism research. In this chapter, we have described methods we use to probe metabolism in intact retinas using stable isotopes. We first discussed how to design experiments, to choose tracers and labeling methods. We then detailed our protocols for use of retinal organ culture, cryosectioning of retinas, sample preparation, and analysis of samples by both GC–MS and LC–MS. Finally, we gave examples of our research and discussed the application of the methods to several areas of retinal metabolism research including neuron–glia interaction, light and dark adaptation, and degenerative diseases.

ACKNOWLEDGMENTS

Research on retinal metabolism in Hurley lab is funded by National Eye Institute Grants (EY06641 and EY023346). J.D. is supported by the Career Starter Award from the Knights Templar Eye Foundation.

REFERENCES

Aherne, A., Kennan, A., Kenna, P. F., McNally, N., Lloyd, D. G., Alberts, I. L., et al. (2004). On the molecular pathology of neurodegeneration in IMPDH1-based retinitis pigmentosa. *Human Molecular Genetics*, *13*(6), 641–650.

Ames, A., 3rd, Li, Y. Y., Heher, E. C., & Kimble, C. R. (1992). Energy metabolism of rabbit retina as related to function: High cost of Na+ transport. *The Journal of Neuroscience*, *12*(3), 840–853.

Bui, B. V., Kalloniatis, M., & Vingrys, A. J. (2004). Retinal function loss after monocarboxylate transport inhibition. *Investigative Ophthalmology & Visual Science*, *45*(2), 584–593.

Campbell, M., & Humphries, P. (2012). The blood-retina barrier: Tight junctions and barrier modulation. *Advances in Experimental Medicine and Biology*, *763*, 70–84.

Chertov, A. O., Holzhausen, L., Kuok, I. T., Couron, D., Parker, E., Linton, J. D., et al. (2011). Roles of glucose in photoreceptor survival. *The Journal of Biological Chemistry*, *286*(40), 34700–34711.

Cornwall, M. C., Tsina, E., Crouch, R. K., Wiggert, B., Chen, C., & Koutalos, Y. (2003). Regulation of the visual cycle: Retinol dehydrogenase and retinol fluorescence measurements in vertebrate retina. *Advances in Experimental Medicine and Biology*, *533*, 353–360.

Cunha-Vaz, J., Bernardes, R., & Lobo, C. (2011). Blood-retinal barrier. *European Journal of Ophthalmology*, *21*(Suppl. 6), S3–S9.

Deelchand, D. K., Shestov, A. A., Koski, D. M., Ugurbil, K., & Henry, P. G. (2009). Acetate transport and utilization in the rat brain. *Journal of Neurochemistry*, *109*(Suppl. 1), 46–54.

Dizhoor, A. M., Boikov, S. G., & Olshevskaya, E. V. (1998). Constitutive activation of photoreceptor guanylate cyclase by Y99C mutant of GCAP-1. Possible role in causing human autosomal dominant cone degeneration. *The Journal of Biological Chemistry*, *273*(28), 17311–17314.

Dobbins, R. L., & Malloy, C. R. (2003). Measuring in-vivo metabolism using nuclear magnetic resonance. *Current Opinion in Clinical Nutrition and Metabolic Care*, *6*(5), 501–509.

Dowling, J. E. (1970). Organization of vertebrate retinas. *Investigative Ophthalmology*, *9*(9), 655–680.

Du, J., Cleghorn, W. M., Contreras, L., Lindsay, K., Rountree, A. M., Chertov, A. O., et al. (2013). Inhibition of mitochondrial pyruvate transport by zaprinast causes massive accumulation of aspartate at the expense of glutamate in the retina. *The Journal of Biological Chemistry*, *288*(50), 36129–36140.

Du, J., Cleghorn, W., Contreras, L., Linton, J. D., Chan, G. C., Chertov, A. O., et al. (2013). Cytosolic reducing power preserves glutamate in retina. *Proceedings of the National Academy of Sciences of the United States of America*, *110*(46), 18501–18506.

Dyer, M. A., & Cepko, C. L. (2000). Control of Muller glial cell proliferation and activation following retinal injury. *Nature Neuroscience*, *3*(9), 873–880.

Farber, D. B., & Lolley, R. N. (1974). Cyclic guanosine monophosphate: Elevation in degenerating photoreceptor cells of the C3H mouse retina. *Science*, *186*(4162), 449–451.

Fernandez, C. A., Des Rosiers, C., Previs, S. F., David, F., & Brunengraber, H. (1996). Correction of 13C mass isotopomer distributions for natural stable isotope abundance. *Journal of Mass Spectrometry*, *31*(3), 255–262.

Fernie, A. R., Geigenberger, P., & Stitt, M. (2005). Flux an important, but neglected, component of functional genomics. *Current Opinion in Plant Biology*, *8*(2), 174–182.

Gastaldelli, A., Coggan, A. R., & Wolfe, R. R. (1999). Assessment of methods for improving tracer estimation of non-steady-state rate of appearance. *Journal of Applied Physiology*, *87*(5), 1813–1822.

Gianchandani, E. P., Chavali, A. K., & Papin, J. A. (2010). The application of flux balance analysis in systems biology. *Wiley Interdisciplinary Reviews. Systems Biology and Medicine*, *2*(3), 372–382.

Greving, M. P., Patti, G. J., & Siuzdak, G. (2011). Nanostructure-initiator mass spectrometry metabolite analysis and imaging. *Analytical Chemistry*, *83*(1), 2–7.

Hartong, D. T., Dange, M., McGee, T. L., Berson, E. L., Dryja, T. P., & Colman, R. F. (2008). Insights from retinitis pigmentosa into the roles of isocitrate dehydrogenases in the Krebs cycle. *Nature Genetics*, *40*(10), 1230–1234.

Hellerstein, M. K., & Neese, R. A. (1999). Mass isotopomer distribution analysis at eight years: Theoretical, analytic, and experimental considerations. *The American Journal of Physiology*, *276*(6 Pt. 1), E1146–E1170.

Jablonski, M. M., & Iannaccone, A. (2000). Targeted disruption of Muller cell metabolism induces photoreceptor dysmorphogenesis. *Glia*, *32*(2), 192–204.

Kanamori, K., Kondrat, R. W., & Ross, B. D. (2003). 13C enrichment of extracellular neurotransmitter glutamate in rat brain—Combined mass spectrometry and NMR studies of neurotransmitter turnover and uptake into glia in vivo. *Cellular and Molecular Biology (Noisy-le-Grand, France)*, *49*(5), 819–836.

Kang Derwent, J., & Linsenmeier, R. A. (2000). Effects of hypoxemia on the a- and b-waves of the electroretinogram in the cat retina. *Investigative Ophthalmology & Visual Science*, *41*(11), 3634–3642.

Lindsay, K. J., Du, J., Sloat, S. R., Contreras, L., Linton, J. D., Turner, S. J., et al. (2014). Pyruvate kinase and aspartate-glutamate carrier distributions reveal key metabolic links between neurons and glia in retina. *Proceedings of the National Academy of Sciences of the United States of America*, *111*(43), 15579–15584.

Linsenmeier, R. A. (1986). Effects of light and darkness on oxygen distribution and consumption in the cat retina. *The Journal of General Physiology*, *88*(4), 521–542.
Linsenmeier, R. A. (1990). Electrophysiological consequences of retinal hypoxia. *Graefe's Archive for Clinical and Experimental Ophthalmology*, *228*(2), 143–150.
Linton, J. D., Holzhausen, L. C., Babai, N., Song, H., Miyagishima, K. J., Stearns, G. W., et al. (2010). Flow of energy in the outer retina in darkness and in light. *Proceedings of the National Academy of Sciences of the United States of America*, *107*(19), 8599–8604.
Maurer, C. M., Schonthaler, H. B., Mueller, K. P., & Neuhauss, S. C. (2010). Distinct retinal deficits in a zebrafish pyruvate dehydrogenase-deficient mutant. *The Journal of Neuroscience*, *30*(36), 11962–11972.
Millard, P., Letisse, F., Sokol, S., & Portais, J. C. (2012). IsoCor: Correcting MS data in isotope labeling experiments. *Bioinformatics*, *28*(9), 1294–1296.
Newman, E., & Reichenbach, A. (1996). The Muller cell: A functional element of the retina. *Trends in Neurosciences*, *19*(8), 307–312.
Noh, K., Gronke, K., Luo, B., Takors, R., Oldiges, M., & Wiechert, W. (2007). Metabolic flux analysis at ultra short time scale: Isotopically non-stationary 13C labeling experiments. *Journal of Biotechnology*, *129*(2), 249–267.
Noh, K., & Wiechert, W. (2011). The benefits of being transient: Isotope-based metabolic flux analysis at the short time scale. *Applied Microbiology and Biotechnology*, *91*(5), 1247–1265.
Ochrietor, J. D., & Linser, P. J. (2004). 5A11/Basigin gene products are necessary for proper maturation and function of the retina. *Developmental Neuroscience*, *26*(5–6), 380–387.
Okawa, H., Sampath, A. P., Laughlin, S. B., & Fain, G. L. (2008). ATP consumption by mammalian rod photoreceptors in darkness and in light. *Current Biology*, *18*(24), 1917–1921.
Pellerin, L., & Magistretti, P. J. (1994). Glutamate uptake into astrocytes stimulates aerobic glycolysis: A mechanism coupling neuronal activity to glucose utilization. *Proceedings of the National Academy of Sciences of the United States of America*, *91*(22), 10625–10629.
Perrault, I., Hanein, S., Zanlonghi, X., Serre, V., Nicouleau, M., Defoort-Delhemmes, S., et al. (2012). Mutations in NMNAT1 cause Leber congenital amaurosis with early-onset severe macular and optic atrophy. *Nature Genetics*, *44*(9), 975–977.
Punzo, C., Kornacker, K., & Cepko, C. L. (2009). Stimulation of the insulin/mTOR pathway delays cone death in a mouse model of retinitis pigmentosa. *Nature Neuroscience*, *12*(1), 44–52.
Rejdak, R., Szkaradek, M., Czepita, M., Taslaq, W., Lewicka-Chomont, A., & Grieb, P. (2012). Retinal degenerative diseases—Mechanisms and perspectives of treatment. *Klinika Oczna*, *114*(4), 301–307.
Shen, W., Zhu, L., Lee, S. R., Chung, S. H., & Gillies, M. C. (2013). Involvement of NT3 and P75(NTR) in photoreceptor degeneration following selective Muller cell ablation. *Journal of Neuroinflammation*, *10*, 137.
Strumilo, E. (1995). Effect of Ca2+ on the activity of mitochondrial NADP-specific isocitrate dehydrogenase from rabbit adrenals. *Acta Biochimica Polonica*, *42*(3), 325–328.
Sun, N., Ly, A., Meding, S., Witting, M., Hauck, S. M., Ueffing, M., et al. (2014). High-resolution metabolite imaging of light and dark treated retina using MALDI-FTICR mass spectrometry. *Proteomics*, *14*(7–8), 913–923.
Taylor, M. R., Hurley, J. B., Van Epps, H. A., & Brockerhoff, S. E. (2004). A zebrafish model for pyruvate dehydrogenase deficiency: Rescue of neurological dysfunction and embryonic lethality using a ketogenic diet. *Proceedings of the National Academy of Sciences of the United States of America*, *101*(13), 4584–4589.

Umino, Y., Everhart, D., Solessio, E., Cusato, K., Pan, J. C., Nguyen, T. H., et al. (2006). Hypoglycemia leads to age-related loss of vision. *Proceedings of the National Academy of Sciences of the United States of America*, *103*(51), 19541–19545.

Wang, L., Kondo, M., & Bill, A. (1997). Glucose metabolism in cat outer retina. Effects of light and hyperoxia. *Investigative Ophthalmology & Visual Science*, *38*(1), 48–55.

Wassle, H., & Boycott, B. B. (1991). Functional architecture of the mammalian retina. *Physiological Reviews*, *71*(2), 447–480.

Wedlock, L. E., Kilburn, M. R., Cliff, J. B., Filgueira, L., Saunders, M., & Berners-Price, S. J. (2011). Visualising gold inside tumour cells following treatment with an antitumour gold(I) complex. *Metallomics*, *3*(9), 917–925.

Wiechert, W., & Noh, K. (2013). Isotopically non-stationary metabolic flux analysis: Complex yet highly informative. *Current Opinion in Biotechnology*, *24*(6), 979–986.

Winkler, B. S. (1981). Glycolytic and oxidative metabolism in relation to retinal function. *The Journal of General Physiology*, *77*(6), 667–692.

CHAPTER FIVE

Analysis of Cell Metabolism Using LC-MS and Isotope Tracers

Gillian M. Mackay[1], Liang Zheng, Niels J.F. van den Broek, Eyal Gottlieb[1]

Cancer Research UK Beatson Institute, Glasgow, UK

[1]Corresponding authors: e-mail address: g.mackay@beatson.gla.ac.uk; e.gottlieb@beatson.gla.ac.uk

Contents

Abstract

Here we discuss our methods to analyze small polar compounds involved in central carbon metabolism using LC-MS. Methods described include sample extraction procedures for cells and medium, as well as for plasma/serum, urine, CSF, and tissue samples. Different extraction solvents are assessed. Our methods for using ^{13}C stable isotope tracers to examine the kinetics and distributions of mass isotopologues of many metabolites are discussed. Quantification methods are described for ^{13}C stable isotope tracer experiments as well as for unlabeled experiments. These methods were applied in a

Methods in Enzymology, Volume 561
ISSN 0076-6879
http://dx.doi.org/10.1016/bs.mie.2015.05.016

fumarate hydratase deficient cell model to show how isotope tracing can demonstrate shifts in metabolic pathways and, together with metabolite exchange rates, can be used to gain insights into changes in cell metabolism.

Definitions

GC-MS Separation of compounds using gas chromatography (GC) and usually electron impact (EI) ionization produces a spectrum of different mass ions for each metabolite, detected by a mass spectrometer.

HILIC HPLC The retention mechanism for hydrophilic interaction liquid chromatography (HILIC) is a combination of partitioning, ion-exchange, and hydrogen bonding, suitable for the separation of polar compounds. Hydrophilic partitioning of polar metabolites occurs between a highly organic solvent (e.g., 80% acetonitrile/20% water) as the mobile phase and a partially immobilized aqueous layer formed on the column material (stationary phase).

Ion suppression Ion suppression of a metabolite refers to reduced detector response due to competition for ionization efficiency in the ionization source between the metabolite of interest and other compounds present. For LC-MS analysis, a limitation of ESI is that it is a competitive process. Ion suppression in LC-MS can occur when a compound in the matrix co-elutes from the HPLC with a metabolite and reduces the ionization of the metabolite in the source of the mass spectrometer, resulting in a lower abundance of metabolite detected.

^{13}C isotopologues ^{13}C isotopologues of a metabolite occur when the metabolite exists with different numbers of ^{13}C atoms present. When a carbon source such as glucose is labeled with ^{13}C atoms, many metabolites are produced which have different numbers of their C atoms labeled as ^{13}C atoms.

LC-MS For LC-MS analysis, compounds are separated on a high-performance liquid chromatography (HPLC) column, the solvent is removed and compounds are ionized usually by electrospray ionization (ESI), and the mass/charge ratio (m/z) of the resultant ions are detected in a mass spectrometer. With the mass spectrometer operating in full scan mode, mass spectra are produced over the specified mass range, as the compounds are eluted from the HPLC. The abundance of an individual ion is measured as peak area (or height) from an extracted ion chromatogram, within a predetermined accuracy of its exact mass.

LC-MS/MS An extension of LC-MS, where ions in the mass spectrometer are fragmented in a collision cell using inert gas. The mass spectrometer, depending on instrument, can isolate and fragment specific ions, the most abundant ions or all ions.

Matrix The background matrix comprises all compounds present in the extracted sample solution in addition to the metabolites of interest.

Metabolomics Metabolomics, as defined by Oliver Fiehn (Fiehn, 2002), is "a comprehensive analysis in which all the metabolites of a biological system are identified and quantified" which "reveals the metabolome of the biological system under study."

Reversed phase HPLC Compounds bind to the HPLC column by hydrophobic interactions, in the presence of a hydrophilic solvent (mobile phase) and are eluted from the column by a more hydrophobic solvent. The column packing material (stationary phase), is often silica based with a bonded phase attached, such as long hydrocarbon chains (C18, C8) or a phenyl or cyano group, to produce the hydrophobic retention.

Reversed phase HPLC with ion pairing With reversed phase chromatography, a compound (ion pair reagent) is added to the mobile phase which associates with the stationary phase by hydrophobic interaction and also associates with ions in the sample, enhancing retention of polar molecules.

^{13}C stable isotopes ^{13}C stable isotopes are nonradioactive isotopes which have been chemically modified to substitute ^{12}C atoms in the molecules with ^{13}C isotopes. All the ^{12}C atoms can be replaced by ^{13}C atoms or only some ^{12}C atoms at specific positions in the molecule. Each ^{12}C atom that is replaced with a ^{13}C atom has an increase in mass of 1.0034 Daltons. The natural abundance of ^{13}C atoms is 1.1% of all carbons present in a molecule.

1. INTRODUCTION

Much information about cell metabolism can be obtained by extracting biological samples in solvent and analyzing the extracts with an LC-MS metabolomics screen for many known metabolites. This can be applied to cells from cell culture, spent medium from cell culture, as well as plasma, urine, and tissues from animals or humans.

Of considerable importance in cell metabolism are the rates of reactions of the enzymes of glycolysis and the tricarboxylic acid (TCA) cycle and changing substrate and product concentrations. To understand the metabolism of these small compounds, the fate of stable isotope tracers (often ^{13}C-labeled glucose) can be determined by measuring isotopologues of many intracellular metabolites over time. The main carbon sources for the TCA cycle are glucose or glutamine, so these compounds are often added as uniformly fully labeled (U) ^{13}C compounds to the medium, replacing the original natural isotopes of glucose or glutamine. However, isotope tracing can also be done with ^{15}N compound labeling or compounds labeled with isotopes of other atoms.

Cell metabolism has become a key area of cancer research. Between 2000 and 2002, genes encoding succinate dehydrogenase (SDH) and fumarate hydratase (FH), two enzymes of the TCA cycle were identified as tumor suppressors (King, Selak, & Gottlieb, 2006). Cancer cells adapt to the loss of SDH or FH, suggesting that significant metabolic rewiring is occurring.

Targeted metabolomics and metabolite profiling can be used to show changes in metabolite levels in cancer cells. Here, we focus on the targeted methods we have established. However, we are also involved in untargeted approaches, where we use metabolite profiling to compare control and experimental cells and look for novel metabolic changes, by identifying compounds which show different abundances, often using their accurate mass and fragmentation spectra with LC-MS/MS.

We have established a relatively simple, quick method for sample extraction. We have identified over 300 metabolites in various samples on our LC-MS platform, by matching mass and retention time with commercial standard compounds. Metabolites include amino acids, organic acids, sugars, phosphates (glycolysis and pentose phosphate pathways), nucleotides, and cofactors (such as CoA, NADH). We routinely use stable isotope tracers in our work and have added isotopologues of many metabolites into our compound databases which are very useful for examining intracellular kinetics.

Often a comparison of metabolite peak areas is sufficient showing relative changes in metabolite abundances between samples. However, the concentrations of metabolites present can be important. We have developed methods for quantifying the concentrations of known metabolites in the samples often using ^{13}C-labeled commercial standards. We have quantified intracellular and extracellular metabolites over time and can calculate metabolite exchange rates for extracellular metabolites.

We will show some results of our stable isotope tracing using our FH-deficient cell model, to demonstrate shifts in metabolic pathways.

2. METHODS FOR SAMPLE EXTRACTION

Metabolites are extracted in a polar solvent (50% methanol, 30% acetonitrile, 20% water) and centrifuged to precipitate and remove any proteins present, in preparation for LC-MS analysis.

2.1 Adherent Cells Grown on Cell Culture Plates

Immediately prior to the end of the incubation period, the number of cells per sample is counted. A separate replicate sample is used for counting. At the end of the incubation, it is important to extract the samples as quickly as possible.

1. The medium is discarded. With very adherent cells, the medium can be poured from all the samples at the same time, or with less adherent cells, the medium can be quickly and carefully removed from each sample by aspiration.
2. Each sample is washed three times with chilled (0–4 °C) phosphate buffered saline (PBS). The excess PBS is aspirated from all the samples after the final wash, important as the PBS can interfere with the

chromatography. (For poorly adhering cells, fewer PBS washes may be required to ensure that all the cells remain and are not washed away with the PBS.)

3. The extraction solution used is 50% methanol, 30% acetonitrile, 20% water and should be stored in the freezer (−20 °C) before use. All solvents must be of high quality and free from impurities: methanol and acetonitrile must be HPLC or LC-MS grade (e.g., HiPerSolv Chromanorm, VWR, Lutterworth, Leics., UK) and the water should be from a Millipore Ultrapure Milli-Q purification system (Merck Millipore, Billerica, MA, USA) or equivalent. The volume of extraction solution required is calculated from the cell count and is usually in the range $1–2 \times 10^6$ cells/ml. Once the experimental conditions have been established a set volume is used for all samples. The added extraction solution must cover all the cells in the sample.
4. The cell metabolites are extracted from the samples by placing the cell culture plate on a rocking shaker at 0–4 °C for 5 min. The extraction solution from each well is then pipetted into a microcentrifuge tube and shaken in the Thermomixer (Thermomixer comfort, Eppendorf AG, Hamburg, Germany) at high speed of 1400 rpm at 0–4 °C for 10 min.
5. The microcentrifuge tubes are then centrifuged at $16,100 \times g$ for 10 min at 0–4 °C. The supernatants are transferred to glass HPLC vials and kept at −75 °C prior to LC-MS analysis.

2.2 Fresh or Spent Medium from Cell Culture

Medium from cell culture plates is diluted 20- to 50-fold with the same extraction solvent used to extract metabolites from cells.

1. In microcentrifuge tubes, 20 μl medium is added to cold 980 μl extraction solvent (50% methanol, 30% acetonitrile, 20% water), taken directly from −20 °C freezer.
2. To mix, the microcentrifuge tubes are shaken in the Thermomixer at high speed of 1400 rpm at 0–4 °C for 10 min.
3. The microcentrifuge tubes are centrifuged at $16,100 \times g$ for 10 min at 0–4 °C. The supernatants are transferred to glass HPLC vials and kept at −75 °C prior to LC-MS analysis.

2.3 Plasma, Urine, and Cerebrospinal Fluid

In a similar extraction procedure to medium extraction, plasma, urine, and cerebrospinal fluid (CSF) can be extracted.

1. Plasma, urine, and CSF are diluted 20- to 50-fold with cold extraction solvent taken from −20 °C freezer.
2. Samples are vortexed for 30 s to mix, but plasma samples are usually vortexed for longer.
3. The samples are centrifuged at 16,100 × *g* for 10 min at 0–4 °C. The supernatants are transferred to glass HPLC vials and kept at −75 °C prior to LC-MS analysis.

2.4 Soft Tissue (e.g., Liver, Brain) and Hard Tissue (e.g., Tumor, Lung)

Tissue is homogenized at 20–40 mg/ml with the same extraction solvent. Our homogenizer is a Precellys 24 (lysis and homogenization) with a cryolys advanced temperature controller (Stretton Scientific Ltd., Derbyshire, UK) and the tubes used for homogenizing samples are Precellys Lysing Kits (Stretton Scientific Ltd.), comprising prefilled tubes with beads (CK28R homogenizing tubes for hard tissue containing 2.8 mm ceramic beads and CK14 tubes for soft tissue containing 1.4 mm ceramic beads).

1. Frozen tissue is cut, weighed and the approximate 40 mg sample is added to the Precellys homogenization tube with beads.
2. The exact volume of extraction solvent is added to achieve 40 mg tissue/ml solvent.
3. Samples are homogenized at approximately 0 °C in the Precellys 24 homogenizer. (Homogenization at 3 × 30 s, with a 30 s gap between each of the three cycles.) If the tissue is not completely homogenized, this procedure can be repeated.
4. Homogenized samples are centrifuged in the Precellys tubes at 16,100 × *g* for 10 min at 0–4 °C. The supernatant is collected in a micro-centrifuge tube and centrifuged again.
5. The supernatants are transferred to glass HPLC vials and kept at −75 °C prior to LC-MS analysis.

2.5 Assessment of Different Extraction Solvents

For serum or plasma, precipitation of proteins with cold methanol is a simple and effective method for extraction of metabolites. For adherent cells in culture, 80% methanol has been used as the extraction solvent for human fibroblasts which is reported to capture free and macromolecule-bound intracellular metabolites (Bennett, Yuan, Kimball, & Rabinowitz, 2008). Mixtures of either methanol, acetonitrile and water, or acetonitrile and water, has been suggested for the extraction of water-soluble metabolites from cultures of *E. coli* (Rabinowitz & Kimball, 2007). These mixed extraction solvents were particularly effective for improved extraction of nucleotide triphosphates. Alternatively, Bruce and colleagues extracted metabolites in a range of solvents, dried the sample extracts under nitrogen, and then redissolved them all in the same solvent for analysis (Bruce et al., 2009). A chosen extraction solvent which gives a high recovery of metabolites can be used to extract the metabolites from the samples, and then after drying under nitrogen gas or lyophilizing, a different solvent more similar to the initial mobile phase of the LC-MS analysis can be used for resolubilizing. This method also has the advantage of redissolving in a smaller volume of solvent thus concentrating the samples, which may be useful for detecting low abundant metabolites. However, the additional step of drying the samples may introduce a source of error and more variability in the results.

A good extraction solvent should be polar as the majority of the small metabolites of interest are water soluble and an organic solvent is useful for stability of the metabolites. Another consideration is that with the HILIC chromatography we use, the initial mobile phase is 80% acetonitrile and 20% aqueous, and therefore it is beneficial if the sample solvent is of a similar polarity. We compared three different extraction solutions used to extract adherent HCT116 cells (human colon carcinoma cell line), in triplicate, grown for 24 h in standard DMEM medium (25 m*M* glucose) with 2 m*M* glutamine and 10% fetal bovine serum (FBS) added.

Figure 1 shows the differences in peak areas of several metabolites using the different solvents: 50% methanol/30% acetonitrile/20% water, 80% methanol/20% water, 75% acetonitrile/25% water.

Compared to the other solvents, 80% methanol gave the best recovery for many of the amino acids, except for cystine, which had the poorest recovery. Many of the organic acids, in particular, lactate and pyruvate, extracted with 80% methanol gave lower peak areas compared with the other extraction solvents. The 75% acetonitrile gave similar peak areas for most metabolites as the mixture of all three solvents. The 50% methanol, 30% acetonitrile, and 20% water is a good extraction solvent for the amino acids and organic acids, and although acetyl CoA, glyceraldehyde 3-phosphate, and glucose, peak areas are lower than with the 75% acetonitrile solvent, peaks are detectable and reproducible.

We examined 75% acetonitrile/25% water as an extraction solvent in more detail, but observed phase separation of the acetonitrile and aqueous layers with some media extracts when chilled, with most metabolites

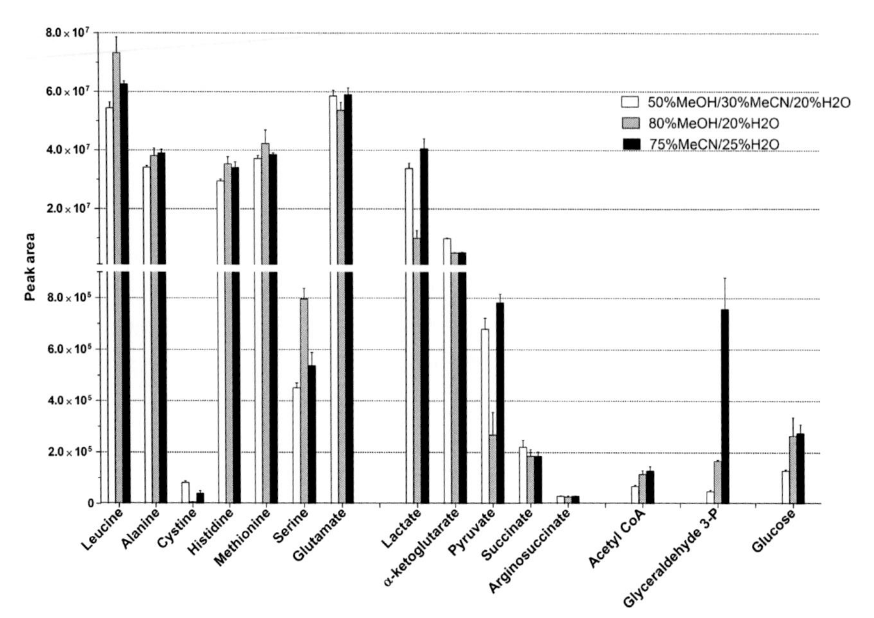

Figure 1 Peak areas of several metabolites extracted in three different solvents: 50% methanol/30% acetonitrile/20% water, 80% methanol/20% water, and 75% acetonitrile/25% water.

favoring the aqueous layer (data not shown). We hypothesized that this phase separation occurred due to high salt concentrations in the medium. The 50% methanol, 30% acetonitrile, and 20% water solvent has been used throughout this work.

As many metabolites have a rapid turnover, it is important to quench metabolism as quickly and as effectively as possible. Metabolism can be quenched by the addition of cold organic solvents. The solvents halt metabolism, by decreasing temperature and by denaturing enzymes (Bennett et al., 2008). In our hands, no observable metabolism of exogenously-added ^{13}C-labeled compounds is detected when cells are extracted in cold organic solutions. This indicates lack of enzymatic activity during metabolite extraction. Prior to the addition of the cold extraction solvent, cells are washed with chilled PBS. It was demonstrated that washing with cold isotonic saline at 0.5 °C stops the conversion of ATP to ADP and AMP, indicating metabolic arrest (Dietmair, Timmins, Gray, Nielsen, & Kromer, 2010). Indeed other researchers have used the adenylate energy charge (a parameter which describes the relationship between ATP, ADP, and AMP in cells) to demonstrate that the quenching procedure was maintaining ATP levels and thus inactivating cell metabolism (Sellick, Hansen, Stephens, Goodacre, & Dickson, 2011). With an established extraction procedure, the time taken for extraction of metabolites should be kept to a minimum and should be consistent between samples.

3. LC-MS METHODS

Mass spectrometry has been chosen to be our preferred method of detection for metabolomics, with high resolution and accurate *m/z* signals (mass/charge ratios). Compared with NMR, mass spectrometry provides higher resolution, increased sensitivity, and the capability to analyze highly complex mixtures (Dettmer, Aronov, & Hammock, 2007). With LC-MS, separation on the chromatography column enables retention time to be an additional parameter for distinguishing between metabolites. With HPLC separation, different metabolites are retained on the HPLC column for different lengths of time and thus are introduced into the mass spectrometer at different times, which reduces ion suppression and saturation effects. GC-MS can be useful for metabolomics analysis, but usually metabolites need to be derivatized to render them volatile and so involves more complex sample preparation. Compounds must also be thermostable for GC-MS analysis (Fiehn, Kopka, Trethewey, & Willmitzer, 2000). The GC column separates metabolites with high resolving power which can be useful to distinguish between metabolites and isomers, however, the electron impact (EI) ionization usually used in GC-MS produces fragment rich spectra which complicates the deconvolution of overlapping chromatographic peaks.

ThermoFisher Scientific Exactive and Q-Exactive mass spectrometers (Bremen, Germany) with the Orbitrap technology offer high resolution (>100,000) and accurate mass (routinely <2 ppm), and as such are extremely useful for targeted or untargeted metabolomics. High mass resolution separates metabolites from interfering peaks of similar masses in complex cellular samples. The accurate mass of the Exactive mass spectrometer improves our ability to identify unknown metabolites and also enables reliable analysis of specific isotopologues in samples made more complex with isotope labeling (Lu et al., 2010). The fast positive/negative polarity switching feature of these instruments (one positive and one negative scan within 1 s at the lowest resolution) enables many metabolites of different classes to be analyzed in the same run. For our cell culture experiments, the Q-Exactive is at least fivefold more sensitive than the Exactive system, due to its S-lens (replacing the tube lens and skimmer of the Exactive) and its quadrupole (not present in the Exactive) controlling the mass range being transmitted into the Orbitrap mass analyzer. The Q-Exactive has the additional fragmentation capability, which can be used to distinguish between metabolites with the same monoisotopic molecular ion (isobaric compounds).

The small molecules are ionized in the source of the mass spectrometer by electrospray ionization (ESI) after the HPLC separation. Singly charged positive or negative monoisotopic mass ions are formed. Positive ion mode on the mass spectrometer gives good responses for amino acids. However, many organic acids in the TCA pathway and phosphates in glycolysis and the pentose phosphate pathway, important in our analysis, ionize better in negative ion mode on the mass spectrometer. To optimize the formation of these negative ions, an HPLC mobile phase with a pH above the pK_a of each of these acidic metabolites is required.

Zhang et al. compared various HPLC modes of separation: coupling reversed phase, aqueous normal phase, and hydrophilic interaction chromatography (HILIC) for LC-MS analysis of metabolites in urine (Zhang, Creek, Barrett, Blackburn, & Watson, 2012). The two HILIC columns assessed were packed with a stationary phase where the zwitterionic sulfobetaine group is covalently bonded to either porous silica (ZIC-HILIC column) or porous polymer beads (ZIC-pHILIC column) (Wade, Garrard, & Fahey, 2007). This permanent hydrophilic zwitterion produces weak electrostatic interactions between polar analytes and the neutral zwitterions on the column. The unique polymeric beads of the ZIC-pHILIC column allow this column to be used at a high pH (pH 9.2, maximum pH 10 for this column), compared to the ZIC-HILIC column (used at pH 2.8, maximum pH 8 for this column). The comparison of the four columns, running two mixtures of standards, showed that organic acids and phosphates had a much better performance on the ZIC-pHILIC column. Performance was based on repeatability of retention time and peak area between replicates and visual inspection of the peak shapes. The range of classes of metabolites examined included amino acids, nucleosides, phosphates, organic acids, and sugars. The ZIC-pHILIC column gave the broadest coverage of metabolites with an acceptable performance in each of these compound classes, compared with the other three columns. The more basic amino acids and amines did not perform well on the ZIC-pHILIC at high pH, but showed good symmetrical sharp peaks with the ZIC-HILIC column at low pH.

As discussed by Zhang et al., we also chose to use the ZIC-pHILIC column (SeQuant, VWR, Lutterworth, Leics., UK) as our standard column to measure a broad range of metabolites of different classes and the ZIC-HILIC column (SeQuant, VWR), when we were interested in specific metabolites, such as cysteine, cystine, glutathione, and fumarate (Zhang et al., 2012). Cysteine and glutathione, with their free thiol groups, are less reactive at the low pH mobile phase used with the ZIC-HILIC column. In addition, fumarate is more accurately quantified with the ZIC-HILIC column at low pH. Although peak shape for fumarate is better with the ZIC-pHILIC column, a significant proportion of malate is converted to fumarate by in-source fragmentation in the mass spectrometer under the high pH conditions of the mobile phase, and as these two acids co-elute on this column, fumarate cannot be accurately quantified.

Others have used different columns for LC-MS metabolomics analysis. Lu et al. (2010) used a reversed phase LC method, with a C18 column and tributylamine as the ion-pairing reagent. This group also used the Exactive mass spectrometer, but only in negative ion mode, as the tributylamine causes ion suppression in positive ion mode. A broad range of water soluble metabolites in core metabolism were successfully detected in cellular extracts from yeast. Buescher and colleagues used a similar UPLC method of ion pairing with tributylamine on a T3 endcapped reversed phase column (Buescher, Moco, Sauer, & Zamboni, 2010). The small particle size of 1.8 μm of the UPLC column resulted in good separation of isomers, such as citric acid and isocitric acid, and hexose phosphate isomers. Yanes et al. compared three C18 columns with unique properties to retain polar compounds and demonstrated that a multimodal column combining reversed phase separation with cation and anion exchange was suitable for most of the 31 model compounds of different polarities (Yanes, Tautenhahn, Patti, & Siuzdak, 2011). However, the ionic strength of the mobile phase had to be increased to elute a few compounds which had been completely retained on the column.

3.1 LC-MS Analysis with ZIC-pHILIC Chromatography

We use ThermoFisher Scientific Accela 600 pump and autosampler with our Exactive mass spectrometer and ThermoFisher Scientific Dionex UltiMate 3000 pump and autosampler with the Q-Exactive. For our standard targeted metabolites screen, the HPLC system consists of a SeQuant ZIC-pHILIC column (vwr), 150 mm × 2.1 mm, 5 μm, preceded with a guard column SeQuant ZIC-pHILIC, 20 mm × 2.1 mm. The aqueous mobile phase solvent is 20 m*M* ammonium carbonate, adjusted to pH 9.4 with 0.1% ammonium hydroxide solution (25%) and the organic mobile

Table 1 ZIC-pHILIC LC Gradient

Time (min)	20 m*M* $(NH_4)_2CO_3$ (pH 9.4) (%)	Acetonitrile (%)	Flow Rate (μl/min)
0	20	80	200
2	20	80	200
17	80	20	200
17.1	20	80	200
19.1	20	80	400
22.1	20	80	400
22.2	20	80	200

phase is 100% acetonitrile. A linear gradient from 80% organic to 80% aqueous is run over a 15 min period, followed by an equilibration back to the starting conditions (Table 1). The flow rate used is 200 μl/min and the column temperature is 45 °C. Samples are maintained at 4 °C in a chilled autosampler, prior to injection. The run time is 22.2 min. 5 μl of sample is injected for the Exactive mass spectrometer but for the more sensitive Q-Exactive, only 2 μl of sample is required.

3.2 LC-MS Analysis with ZIC-HILIC Chromatography

For our alternative HILIC method, the HPLC system consists of a SeQuant ZIC-HILIC column, 150 mm × 4.6 mm, 5 μm, preceded with a guard column SeQuant ZIC-HILIC, 20 mm × 2.1 mm. The aqueous mobile phase solvent is 0.1% formic acid in water and the organic mobile phase is 0.1% formic acid in acetonitrile. The mobile phase gradient is described in Table 2. The flow rate used is 300 μl/min and the column temperature is maintained at 30 °C. Samples are maintained at 4 °C in a chilled autosampler, prior to injection. The run time is 45 min. 5 μl of sample is injected for the Exactive mass spectrometer detection.

3.3 LC-MS Analysis with Exactive and Q-Exactive Mass Spectrometers

For many targeted analyses, the Exactive mass spectrometer is used. Metabolites separated on the HPLC column undergo ESI in the (heated) HESI source (typical settings shown in Table 3) and are detected across a mass range of 75 to 1000 *m/z*. The resolution we use for a full scan on the

Table 2 ZIC-HILIC LC Gradient

Time (min)	0.1% Formic Acid in Water (%)	0.1% Formic Acid in Acetonitrile (%)	Flow Rate (μl/min)
0	20	80	300
12	50	50	300
26	50	50	300
28	80	20	300
36	80	20	300
37	20	80	300
45	20	80	300

Table 3 Typical MS Settings for Full MS

	Exactive		Q-Exactive	
Scan range (m/z)	75–1000		75–1000	
Resolution	25,000		35,000	
AGC target	1×10^6		1×10^6	
HESI	**Positive Ion**	**Negative Ion**	**Positive Ion**	**Negative Ion**
Sheath gas flow rate	50	50	25	25
Aux gas flow rate	10	10	15	15
Spray voltage (kV)	4.5	3.5	4.25	3.25
Capillary temperature (°C)	275	275	275	275
Capillary voltage (V)	40	40	n/a	n/a
Tube lens voltage (V)	95	95	n/a	n/a
Skimmer voltage (V)	18	18	n/a	n/a
S-lens RF level	n/a	n/a	50	50
Heater temperature (°C)	30	30	50	50

n/a = not applicable.

Exactive mass spectrometer is 25,000 (at 200 m/z), scan rate 4 Hz and on the Q-Exactive is 35,000 (at 200 m/z), scan rate 7 Hz (Table 3). Although these mass spectrometers offer much higher resolutions, for targeted metabolomics these resolutions were chosen to obtain enough data points for the peak areas to be accurately quantified. The automatic gain control (AGC) regulates the ion injection time and the number of ions (the AGC target) entering the Orbitrap. If the Orbitrap is overloaded with ions, it distorts the electric field and causes space-charge effects. We chose a mid-range AGC target of 1×10^6 ions for our full scans where resolution is high enough to resolve isotope peaks, and the scan rate allows enough scans/peak to accurately quantify peak areas. Polarity switching is used to enable both positive and negative ions to be determined in the same run. Lock masses are used and the mass accuracy for all metabolites is below 5 ppm (routinely below 2 ppm). Data is acquired with Thermo Xcalibur software.

4. LC-MS DATA ANALYSIS

For targeted metabolomics, commercial standards of all metabolites detected had been previously analyzed with both of these LC-MS methods. For each metabolite, HPLC retention time was recorded as was the ionization mode which produced the higher response. Together with the exact masses of the singly charged ions, we built our own compound database for LC-MS analysis for each HPLC method. The ZIC-pHILIC database for our general metabolites screen has over 100 metabolites. This information was collected in ThermoFisher Scientific TraceFinder software, where raw sample data files are analyzed.

Metabolites are identified by accurate mass and retention time. For each metabolite, the peak area of the extracted ion chromatogram at the appropriate retention time is determined. The relative comparison of metabolite peak areas between samples is often sufficient to show metabolic changes, but metabolites can be quantified if required.

5. USE OF STABLE ISOTOPE TRACERS

Carbon supply to the TCA cycle is mainly provided by glucose or glutamine. To demonstrate changes in TCA metabolism and in related metabolic pathways, stable isotope tracing with U-$^{13}C_6$-glucose or U-$^{13}C_5$-glutamine as the carbon source has proved extremely useful. The metabolic nutrients U-$^{13}C_6$-glucose (or U-$^{13}C_5$-glutamine) are added in cell culture medium, replacing all natural isotopes of glucose (or glutamine). Buescher et al. has discussed in detail the importance of labeling of metabolites at steady state, as well as examining the kinetics for the rate of the isotope labeling (Buescher et al., 2015). These stable isotope tracers are from

Cambridge Isotope Laboratories (Tewksbury, MA, USA). We can determine the isotopologues of many intracellular and extracellular metabolites over time with our LC-MS methods. At steady state, the proportion of labeled metabolite is important, as well as the difference in overall abundance of all the isotopologues of each metabolite added together.

In a typical cell culture experiment with U-$^{13}C_6$-glucose or U-$^{13}C_5$-glutamine isotope tracing, 2×10^5 cells are plated in six-well plates in medium supplemented with 5–25 m*M* glucose, 0–1 m*M* pyruvate, 0.5–2 m*M* glutamine, and 1–10% FBS, depending on the requirements for the individual experiment. After 24 h, the medium is replaced with the same medium without natural isotopes of glucose or glutamine, supplemented with U-$^{13}C_6$-glucose or U-$^{13}C_5$-glutamine. The cells are cultured for different times until steady state labeling pattern of the studied metabolites is reached (typically 24 h, but can be 48–72 h). A minimum of triplicate samples are required for each time point for statistical analysis.

5.1 Detection of Isotopologues

Each isotopologue of a compound has a different mass from the others: with ^{13}C labeling the mass difference between isotopologues is 1.0034 Daltons. The HPLC shows each isotopologue of a compound elutes at the same retention time. The peak areas from the extracted ion chromatograms at the exact mass of each isotopologue are proportional to their concentrations.

When the stable isotope tracer is added to the medium, the sum of all peak areas (or concentrations) of all the isotopologues, of each compound measured, will remain the same, but over time, some isotopologues will increase as the labeled glucose (or glutamine) is consumed. The ^{13}C-labeling pattern of the urea cycle metabolite, argininosuccinate derived from U-$^{13}C_6$-glucose, U-$^{13}C_5$-glutamine, and U-$^{13}C_6$-arginine is shown in Fig. 2.

When isotope distribution is calculated, it is sometimes necessary to correct for the contribution of natural stable isotopes (such as ^{13}C, which is 1.1% of all carbons). The way to deduct the natural contribution of stable isotopes for each isotopologue is nicely described by Buescher et al. (2015).

With the LC-MS analysis described here, the different isotopologues are distinguished by their different masses, but there is no information about the position of the stable isotope within each metabolite. Fragmentation of the carbon backbone of metabolites (in a mass spectrometer such as the Q-Exactive) will produce multiple fragment ions, and may provide information about the location of the ^{13}C atom in the molecule. Nevertheless,

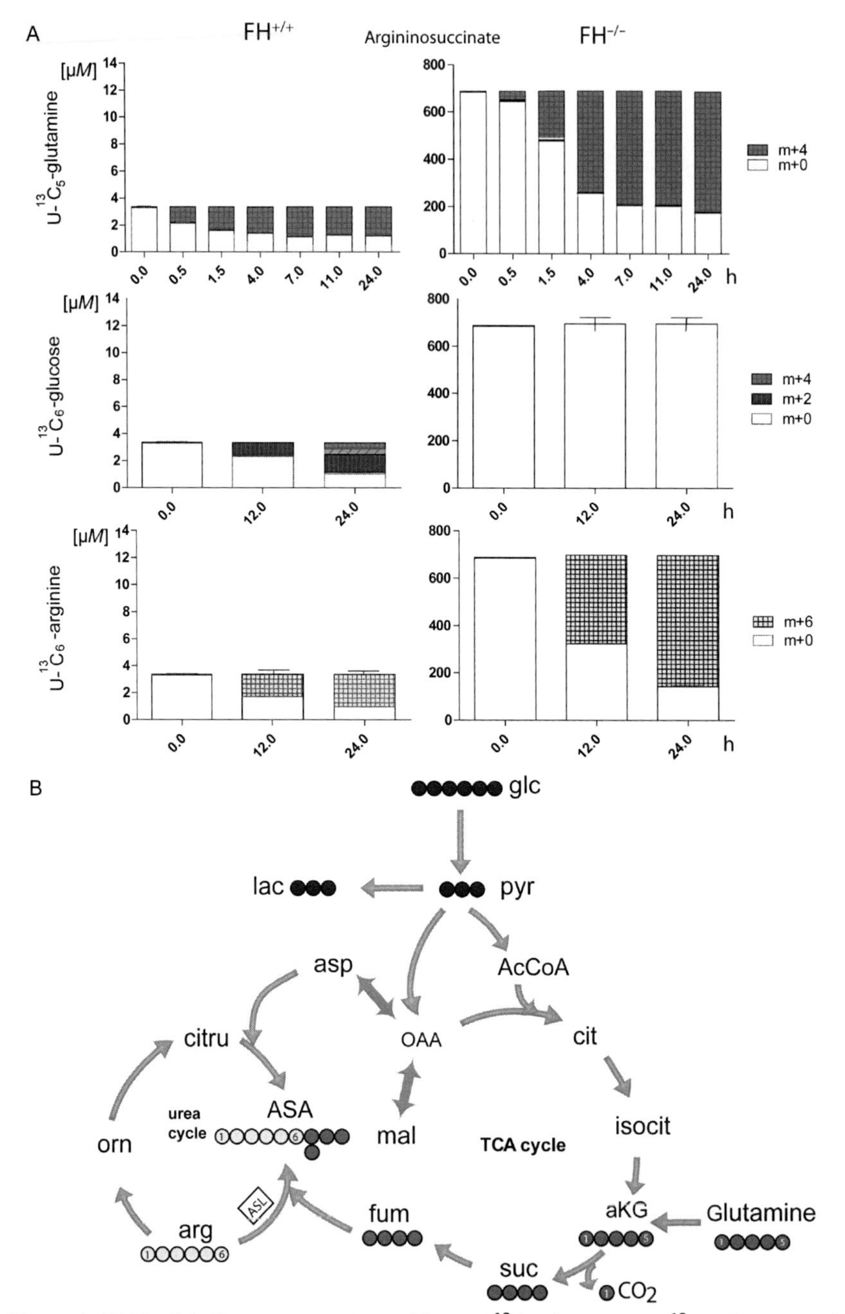

Figure 2 (A) The labeling patterns derived from U-$^{13}C_5$-glutamine, U-$^{13}C_6$-glucose, and U-$^{13}C_6$-arginine over time and the concentrations of the urea cycle metabolite, argininosuccinate (ASA) in FH-proficient or deficient cells. (B) The labeling patterns derived from U-$^{13}C_5$-glutamine, U-$^{13}C_6$-glucose, and U-$^{13}C_6$-arginine demonstrated an unexpected synthesis of ASA from arginine and fumarate by the reverse activity of ASL. (See the color plate.)

even without knowledge of the position of the ^{13}C labeling in the molecules, conclusions can typically be drawn on nutrient contribution and pathway activities (Buescher et al., 2015).

6. NORMALIZATION

Cell and media extracts used for LC-MS analysis must be normalized before comparisons can be made. This can be done either by cell number or protein content of the cells. Urine metabolites are usually normalized using the ratio of the peak area of the studied metabolite to the peak area of creatinine.

Cell number can be calculated from extra samples incubated under the same conditions. At the experimental time point, each sample is washed once with PBS, trypsinized, resuspended in a CASYton solution and counted with a CASY cell counter.

Because the metabolite extraction procedure precipitates protein on the plate, protein content of the cells can be measured by the Lowry assay on the same plate, after removal of the metabolites in extraction solvent.

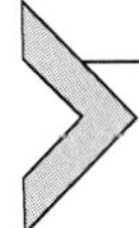

7. QUANTIFICATION OF METABOLITES IN CELLS AND MEDIUM

With LC-MS analysis, there is a linear response between concentration of a metabolite and peak area, within a range of concentrations. The peak area for very low concentrations will not be accurate enough to be within the linear range, and the peak areas for high concentrations show saturation effects, with the concentration increasing but the peak area not increasing proportionally. The presence of other compounds in the sample co-eluting with the metabolite of interest (the matrix effect), can change the linear range: for example, the ionization efficiency of the metabolite can decrease with increasing concentration. For quantification of metabolites, it is important to keep calibration standards and experimental samples within the linear range and to use an extract of cells or medium to prepare calibration standards rather than preparing standards in solvent.

For accurate quantification, external calibration curves, using ^{13}C-labeled metabolites spiked into unlabeled cells or medium, are used to quantify both intracellular and extracellular metabolites. The ^{13}C-labeled compounds are purchased commercially from Cambridge Isotope Laboratories. A stock solution containing a mixture of all the ^{13}C-labeled standards for quantification is prepared and spiked at a minimum of six different concentrations into unlabeled cell extract or unlabeled medium. The peak areas are used to prepare a calibration curve (Fig. 3A). The concentrations of the spiking solutions vary between metabolites and are chosen so that all peak areas for the experimental samples fall within the range of the calibration curve peak areas. It is easiest to prepare one stock solution of mixed metabolites at different concentrations and then dilute to prepare several calibration solutions. One of the key considerations for quantification is that the samples are diluted to the same extent as the spiked calibration standards. Another important criterion is that the medium used for calibration standards has exactly the same composition as that used for the samples, with the only difference being that the isotope tracer metabolite is labeled.

The data analysis for quantification of samples by this method may have to be done separately from the data processing software, if the software does not recognize that a labeled compound is linked to the unlabeled and all the isotopologues of this compound.

An alternative quantification method, which is used if ^{13}C-labeled standards are not available, is the standard addition method (Luo, Groenke, Takors, Wandrey, & Oldiges, 2007). An unlabeled standard compound (with natural isotopes) can be used to spike into unlabeled extracts in a similar way, resulting in a calibration curve for a range of concentrations above the concentration in the sample (Fig. 3B). This calibration curve can be extrapolated and the concentration of metabolite in experimental samples can be calculated from peak areas using the regression equation of the calibration curve.

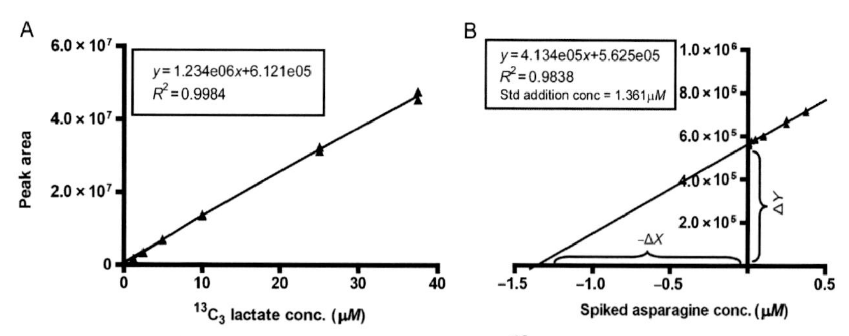

Figure 3 (A) Calibration curve for lactate, using $^{13}C_3$-lactate spiked into medium. (B) Standard addition calibration curve for asparagine (^{12}C, unlabeled) spiked into cell extract. ΔY represents the peak area derived from asparagine in the cell extract, and $-\Delta X$ equals the concentration of asparagine in the cell extract.

The issue with the standard addition method is that the peak areas being measured in the samples is often below the original calibration curve. There needs to be some confidence that the samples and the calibration curves are all within the linear range of metabolite response. To best achieve this, we recommend that all six concentrations of each metabolite used for the calibration curve are close in concentration to the sample being spiked, but different enough to give a robust linear calibration curve. It is useful that some software, such as ThermoFisher Scientific TraceFinder, calculates the concentration of metabolites in the samples, where the calibration standards are indicated as standard addition calibrants.

Others have used other methods. It is a great concept to fully label all the metabolites of interest for quantification in the cells and then use this with unlabeled spiking compounds, which are easily available and inexpensive (Bennett et al., 2008). The main potential issue is that all the metabolites of interest may not be fully labeled, and this has to be addressed by complex calculations.

7.1 Quantification of Intracellular Metabolites

A sample of unlabeled control cell extract for quantification is required. However, a volume of approximately 3 ml is required, so several wells may need to be extracted to achieve this total volume.

7.1.1 Intracellular Quantification Method

The method describes intracellular quantification using isotope tracing when U-$^{13}C_6$-glucose has been added to the medium for 24 h:

1. Control and experimental cells in culture, are prepared in triplicate, with seven extra wells of control cells for preparing calibration curves. Medium (with serum, glutamine, and any other additives) added to control cells for quantification has exactly the same composition as the control and experimental cells, except that the glucose is fully labeled in the experimental samples.
2. At end of the experiment, at 24 h, cells are extracted in extraction solvent, at the usual concentration of 1×10^6 cells/ml.
3. After mixing and centrifugation, as described earlier, the supernatants from the control cells for calibration are combined.
4. A stock solution containing the U-^{13}C metabolites to be quantified is prepared and diluted to six concentrations, with 10- to 50-fold difference in concentration between the lowest and highest.

5. 180 μl unlabeled cell extract is spiked with 20 μl of one of the six calibration solutions or with 20 μl extraction solvent, as a blank. Calibration solutions are prepared in duplicate.
6. 180 μl of all the control and experimental cell extracts are then diluted by adding 20 μl extraction solvent, so all cell extracts, including the calibrations standards, are diluted to contain the same concentration of cell extract.
7. The spiked calibration samples are run before and after running the samples on the LC-MS and are usually run increasing in concentration, to prevent any carryover.
8. A linear calibration curve is prepared using the duplicate peak areas from the labeled standards against the known concentrations.
9. The equation of the linear calibration curve is used to calculate the concentrations of each isotopologue in the samples. An alternative approach is to sum the peak areas of all isotopologues, calculate the total concentration from this total peak area for this metabolite from the calibration curve, and calculate concentrations of each isotopologue from the proportion of each peak area.
10. Concentrations are then normalized to cell number or protein content.

For a single batch of intracellular samples to be quantified, this method is ideal. However, we can also use concentrations calculated here for quantification of future batches of the same cells grown in the same medium, with at least one of the same experimental time points. For this, we recommend that the above calibration procedure should be run with a batch of samples on two separate occasions. For example, cell extracts were quantified in unlabeled medium and the mean concentration of citrate in these cells was calculated to be 3 nmol/10^6 cells at 24 h, with a mean peak area of 2.5×10^6. In a later similar experiment while using ^{13}C-glucose in the medium and examining different time points including 24 h, the sum of the peak areas of all the isotopologues of citrate at 24 h is equivalent to 3 nmol/10^6 cells (although total peak area has changed), and each isotopologue at each time point can now be quantified by direct proportion of each peak area. Alternatively, a cold extract can be run with the new experiment, which has a known concentration and therefore the peak area for this sample can be used to calculate concentrations from peak areas in the new batch.

The highest concentration may be too high for good linearity, which is observed if the calibration curve starts to curve at the high concentration. If, when this concentration is removed from the regression line, the sample peak areas are still within the concentration range of the other calibration standards, the peak areas for the highest concentrations can be removed. A peak area, which is obviously an outlier, can also be removed from the calibration curve. The most important aspect is that the there are enough points on the calibration curve near the peak areas of the samples to calculate concentration accurately.

7.2 Quantification of Extracellular Metabolites

We are interested in the cells' ability to consume and secrete metabolites from the medium. We commonly measure concentrations of glucose, glutamine, glutamate, lactate, pyruvate, and alanine in medium. Medium incubated without cells is compared with spent medium from cells and we calculate the cells' uptake and secretion of metabolites and metabolite exchange rates.

Samples are quantified using external calibration curves prepared using unlabeled medium extracts, spiked with known concentrations of ^{13}C-labeled standards. For accurate quantification of abundant metabolites in medium, medium samples are diluted 500-fold. The same method is used for isotope tracing experiments and for unlabeled experiments.

7.2.1 Extracellular Quantification Method

The method below describes the method using isotope tracing with U-$^{13}C_6$-glucose:

1. Six replicate wells for each sample and six replicate wells of medium only are required. These are incubated with medium containing U-$^{13}C_6$-glucose for the experimental time, often 24 h.
2. At the end of the experiment, all the samples are extracted as described earlier, using a 50-fold dilution of the medium in extraction solvent. Supernatants are collected after centrifugation.
3. To prepare calibration standards, the same medium, but with unlabeled glucose present, is extracted in the same way.
4. A stock solution containing the U-^{13}C-labeled metabolites to be quantified is prepared. This is diluted into six concentrations, with a 10- to 50-fold difference in concentration between the lowest and highest.
5. Calibration solutions are prepared in duplicate: To 160 μl extraction solvent, 20 μl unlabeled medium extract (1 in 50 dilution) is added and spiked with 20 μl of one of the six calibration solutions or with 20 μl blank extraction solvent. The final dilution is 1 in 500.

6. All the sample extracts are diluted a further 10-fold, to produce 1 in 500 medium dilution.
7. The spiked calibration samples are run before and after running the samples on the LC-MS and are usually run increasing in concentration.
8. A linear calibration curve is prepared using the duplicate peak areas from the calibration standards against the known concentrations.
9. The equation of the linear calibration curve is used to calculate the concentrations of each isotopologue in the sample. An alternative approach is to sum the peak areas of all isotopologues, calculate the total concentration from this total peak area for this metabolite from the calibration curve, and calculate concentrations of each isotopologue from the proportion of each peak area.
10. To calculate the concentration in the medium, the concentrations from the calibration curve are multiplied by 500, as the medium was diluted 500-fold. The concentration of medium per well in nmol can be calculated, from the volume of medium per well.
11. The change in spent medium concentrations is calculated. The mean amount (nmol) per well for the medium incubated without cells is subtracted from the amount per well for each replicate of spent medium.
12. The metabolite exchange rate (nmol/million cells/h) or (nmol/μg protein/h) is calculated by dividing the change in nmol calculated above, by the average cell number (or protein content) over the time of the experiment and by the experiment time. The average cell number or protein content is calculated by the following equation: $X{*}D{*}(2\hat{}(T/D)-1)/(T{*}\ln(2))$, where X = number of cells/protein content at time 0, D = doubling time of the cells, T = experimental time. This accounts for the exponential growth of cells.

8. STABLE ISOTOPE TRACING WITH FUMARATE HYDRATASE-DEFICIENT CELL MODEL

Using stable isotope tracing with U-$^{13}C_6$-glucose, U-$^{13}C_5$-glutamine, or U-$^{13}C_6$-arginine added to the medium, in mouse FH-deficient or proficient cell lines, we revealed metabolic changes upon the loss of FH. There was a large increase in the concentrations of the urea cycle metabolite, argininosuccinate in FH-deficient cells (Fig. 2A). The labeling patterns derived from U-$^{13}C_5$-glutamine showed that the argininosuccinate was not produced by the expected metabolism from citrulline and aspartate (three ^{13}C would be labeled via aspartate), but from arginine and fumarate as four

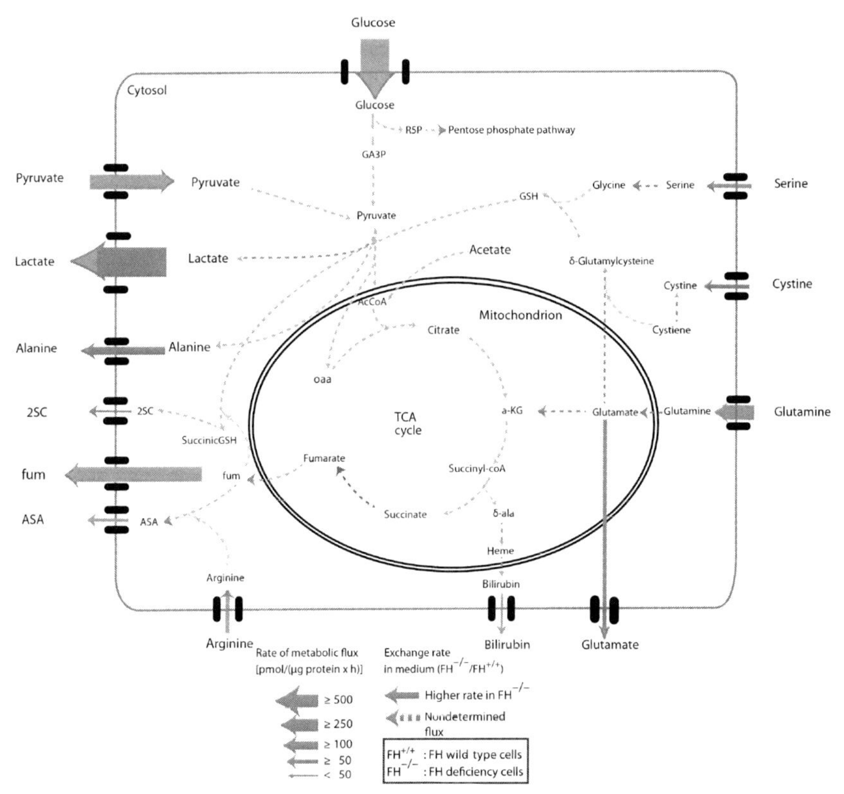

Figure 4 For the FH-deficient model, the metabolite exchange rates in medium demonstrate the rate of metabolic flux and also show several new mechanisms that enable cells to survive without a functional TCA cycle.

^{13}C were labeled from fumarate (Fig. 2B). This suggested that the enzyme, argininosuccinate lyase (ASL) is working in reverse. To confirm this observation, tracing with U-$^{13}C_6$-arginine produced argininosuccinate with all six ^{13}C labeled from the U-$^{13}C_6$-arginine. These results demonstrate the use of stable isotope tracing (corrected for carbon natural isotope abundance) to show shifts in metabolic reactions (Zheng et al., 2013).

In Fig. 4, the metabolite exchange rates between cells and medium, for each genotype, were measured and presented in a diagram that best describes the metabolic flux changes observed in FH-deficient cells compared with control cells. Several new mechanisms to enable cells to survive without a functional TCA cycle have been discovered (Fig. 4): (1) The loss of FH in mouse kidney cells caused massive accumulation of fumarate, slowed down respiration and increased glycolysis, which is indicated by the substantial uptake of glucose and pyruvate and secretion of lactate and alanine (Frezza et al., 2011). (2) A metabolic pathway beginning with glutamine uptake and ending with bilirubin secretion involving the biosynthesis and degradation of heme has been identified and demonstrated to be essential for FH-deficient cells (Frezza et al., 2011). (3) The moderate reactivity of the double bond of fumarate interacts with the thiol group of cysteine to form the succination of reduced glutathione (GSH) to succinicGSH (Zheng et al., 2015). This molecule can be degraded and secreted to the medium as 2-(succinyl)-cysteine (2SC). The succination caused the depletion of GSH and increased demand of cystine, thus decreased the NADPH pool and enhanced redox stress which led to cellular senescence (Zheng et al., 2015). At the same time, as discussed above, the high level of fumarate reversed ASL activity in the urea cycle and produced argininosuccinate which was secreted to the medium (Zheng et al., 2013).

9. FUTURE

For targeted analysis, we are currently developing methods on the Q-Exactive using fragmentation, where product ions of the fragmentation can be used to confirm peak identification. The confirming ions have to be acquired using commercial standards for the compounds of interest and are added to ThermoFisher Scientific TraceFinder software for data analysis and can be stored in a fragmentation library. Also for targeted analysis, a selected ion monitoring (SIM) can be applied to focus on a few target metabolites. This results in a better signal to noise ratio, therefore being more sensitive than applying a full scan.

To be able to do a true untargeted analysis for metabolites, using HILIC chromatography and a system with fragmentation capability, we need to consider multiple difficulties. Reproducibility is fortunately reliable when using a (p)HILIC column. Nonuniform fragmentation is the biggest problem for untargeted metabolomics. Unfortunately, the monoisotopic mass is not a unique identifier for most compounds. Target compounds can be identified by using a known retention time, but this cannot be applied for the unknown compounds. For these we would need to acquire a unique spectrum. This does not seem to be straightforward for the wide range of structurally different compounds. Setting up a proper data dependent fragmentation is therefore quite challenging. Larger metabolites fragment more easily at lower fragmentation energy than smaller ones. Small metabolites need higher fragmentation energy to create a unique spectrum. Setting up a range of fragmentation energies (or ramping) to be applied per compound will help to get a better and unique spectrum. These spectra can then be used to identify compounds in further data analysis, or compared with online fragmentation databases like MZcloud (http://www.mzcloud.org, Highchem LLC) or Metlin (http://metlin.scripps.edu, Scripps center for Metabolomics).

Although there are multiple software packages available, the application can be challenging. Examples are the freely available online tool XCMS online (Scripps) or commercially available as Sieve (ThermoFisher Scientific) or Progenesis QI (Nonlinear Dynamics). As mentioned before, identifying metabolites with help of their unique fragmentation pattern is difficult. Not all software is optimized yet to incorporate this identifier to accurately identify the compounds. Software like Progenesis QI is able to use these unique spectra by comparing the spectra of the compounds to an *in silico* fragmented database of choice. It also combines additional compound properties such as retention time lists to more accurately identify compounds of interest.

We usually use fully labeled ^{13}C compounds for isotope tracing experiments. However, in some cases, specific partially labeled tracers can be of use. One notable example being 1,6-^{13}C-glucose, where the first and sixth carbons can be traced into $^{13}C_1$-lactate (m + 1) or to $^{13}C_0$-lactate (m + 0) as an indication for the branching of glucose into the pentose phosphate pathway where the first carbon is oxidized to CO_2.

REFERENCES

Bennett, B. D., Yuan, J., Kimball, E. H., & Rabinowitz, J. D. (2008). Absolute quantitation of intracellular metabolite concentrations by an isotope ratio-based approach. *Nature Protocols*, *3*(8), 1299–1311.

Bruce, S. J., Tavazzi, I., Parisod, V., Rezzi, S., Kochhar, S., & Guy, P. A. (2009). Investigation of human blood plasma sample preparation for performing metabolomics using ultrahigh performance liquid chromatography/mass spectrometry. *Analytical Chemistry*, *81*(9), 3285–3296.

Buescher, J. M., Antoniewicz, M. R., Boros, L. G., Burgess, S. C., Brunengraber, H., Clish, C. B., et al. (2015). A roadmap for interpreting C metabolite labeling patterns from cells. *Current Opinion in Biotechnology*, *34C*, 189–201.

Buescher, J. M., Moco, S., Sauer, U., & Zamboni, N. (2010). Ultrahigh performance liquid chromatography-tandem mass spectrometry method for fast and robust quantification of anionic and aromatic metabolites. *Analytical Chemistry*, *82*(11), 4403–4412.

Dettmer, K., Aronov, P. A., & Hammock, B. D. (2007). Mass spectrometry-based metabolomics. *Mass Spectrometry Reviews*, *26*(1), 51–78.

Dietmair, S., Timmins, N. E., Gray, P. P., Nielsen, L. K., & Kromer, J. O. (2010). Towards quantitative metabolomics of mammalian cells: Development of a metabolite extraction protocol. *Analytical Biochemistry*, *404*(2), 155–164.

Fiehn, O. (2002). Metabolomics—The link between genotypes and phenotypes. *Plant Molecular Biology, 48*(1-2), 155–171.

Fiehn, O., Kopka, J., Trethewey, R. N., & Willmitzer, L. (2000). Identification of uncommon plant metabolites based on calculation of elemental compositions using gas chromatography and quadrupole mass spectrometry. *Analytical Chemistry, 72*(15), 3573–3580.

Frezza, C., Zheng, L., Folger, O., Rajagopalan, K. N., MacKenzie, E. D., Jerby, L., et al. (2011). Haem oxygenase is synthetically lethal with the tumour suppressor fumarate hydratase. *Nature, 477*(7363), 225–228.

King, A., Selak, M. A., & Gottlieb, E. (2006). Succinate dehydrogenase and fumarate hydratase: Linking mitochondrial dysfunction and cancer. *Oncogene, 25*(34), 4675–4682.

Lu, W., Clasquin, M. F., Melamud, E., Amador-Noguez, D., Caudy, A. A., & Rabinowitz, J. D. (2010). Metabolomic analysis via reversed-phase ion-pairing liquid chromatography coupled to a stand alone orbitrap mass spectrometer. *Analytical Chemistry, 82*(8), 3212–3221.

Luo, B., Groenke, K., Takors, R., Wandrey, C., & Oldiges, M. (2007). Simultaneous determination of multiple intracellular metabolites in glycolysis, pentose phosphate pathway and tricarboxylic acid cycle by liquid chromatography-mass spectrometry. *Journal of Chromatography A, 1147*(2), 153–164.

Rabinowitz, J. D., & Kimball, E. (2007). Acidic acetonitrile for cellular metabolome extraction from Escherichia coli. *Analytical Chemistry, 79*(16), 6167–6173.

Sellick, C. A., Hansen, R., Stephens, G. M., Goodacre, R., & Dickson, A. J. (2011). Metabolite extraction from suspension-cultured mammalian cells for global metabolite profiling. *Nature Protocols, 6*(8), 1241–1249.

Wade, K. L., Garrard, I. J., & Fahey, J. W. (2007). Improved hydrophilic interaction chromatography method for the identification and quantification of glucosinolates. *Journal of Chromatography A, 1154*(1-2), 469–472.

Yanes, O., Tautenhahn, R., Patti, G. J., & Siuzdak, G. (2011). Expanding coverage of the metabolome for global metabolite profiling. *Analytical Chemistry, 83*(6), 2152–2161.

Zhang, T., Creek, D. J., Barrett, M. P., Blackburn, G., & Watson, D. G. (2012). Evaluation of coupling reversed phase, aqueous normal phase, and hydrophilic interaction liquid chromatography with Orbitrap mass spectrometry for metabolomic studies of human urine. *Analytical Chemistry, 84*(4), 1994–2001.

Zheng, L., Cardaci, S., Jerby, L., MacKenzie, E. D., Sciacovelli, M., Johnson, T. I., et al. (2015). Fumarate induces redox-dependent senescence by modifying glutathione metabolism. *Nature Communications, 6*, 6001.

Zheng, L., Mackenzie, E. D., Karim, S. A., Hedley, A., Blyth, K., Kalna, G., et al. (2013). Reversed argininosuccinate lyase activity in fumarate hydratase-deficient cancer cells. *Cancer & Metabolism, 1*(1), 12.

CHAPTER SIX

Analysis of Fatty Acid Metabolism Using Stable Isotope Tracers and Mass Spectrometry

Sergey Tumanov, Vinay Bulusu, Jurre J. Kamphorst[1]
Cancer Research UK Beatson Institute & Institute of Cancer Sciences, University of Glasgow, Glasgow, United Kingdom
[1]Corresponding author: e-mail address: jurre.kamphorst@glasgow.ac.uk

Contents

Abstract

Cells can synthesize fatty acids by ligating multiple acetyl units from acetyl-CoA. This is followed by desaturation and elongation reactions to produce a variety of fatty acids required for proper cellular functioning. Alternatively, exogenous lipid sources can contribute to cellular fatty acid pools. Here, we present a method based on incorporation of ^{13}C-carbon from labeled substrates into fatty acids and subsequent mass spectrometry analysis. The resulting labeling patterns can be used to determine (1) ^{13}C-enrichment of lipogenic acetyl-CoA, (2) the relative contributions of synthesis and uptake, and (3) absolute fatty acid fluxes. We begin by providing a background and general principles regarding the use of stable isotopes to study fatty acid metabolism. We then proceed with detailing procedures for sample preparation and both GC-MS and LC-MS

Methods in Enzymology, Volume 561
ISSN 0076-6879
http://dx.doi.org/10.1016/bs.mie.2015.05.017

analysis of isotope incorporation. Finally, we discuss the interpretation of the resulting fatty acid-labeling patterns.

1. INTRODUCTION

Fatty acids play important roles in cells as structural elements of membranes, energy storage, and signaling molecules (Quehenberger, Armando, & Dennis, 2011). Deregulation of fatty acid metabolism has been associated with some of the most prevalent diseases, including obesity and cancer (Baenke, Peck, Miess, & Schulze, 2013; Khandekar, Cohen, & Spiegelman, 2011; Zaidi et al., 2014). Given its prominent role in both normal cellular functioning and disease, it is not surprising that a variety of approaches utilizing isotopes to "trace" fatty acid metabolism have been developed over the years. One of the earliest approaches involved the use of radiolabeled substrates such as ^{14}C-acetate to measure isotopic enrichment in the lipid fraction. Incorporation into fatty acids can be determined by performing liquid scintillation counting directly on lipid extracts resulting from Bligh & Dyer or Folch extraction procedures (Daniels et al., 2014). The advantage of this approach is that it requires minimal sample preparation and no mass spectrometry. The downside, however, is that it is not specific to particular fatty acids and therefore does not provide information on isotope enrichment per molecule. Hence, it cannot be used to determine the contribution of individual fatty acid metabolic reactions.

Increased specificity can be obtained by combining stable isotope tracers with mass spectrometry analysis. For example, ^2H_2O can be used to study lipogenesis in cultured cells or animals; the ^{2}H is incorporated into fatty acids during *de novo* synthesis and the degree of incorporation is proportional to the rate of biosynthesis (Herath et al., 2014). Recent advances in analytical chemistry approaches facilitate the analysis of ^{2}H incorporation into individual fatty acids (Herath et al., 2014).

As an alternative to ^{2}H, ^{13}C-labeled precursors that are metabolized to acetyl-CoA (AcCoA) can be used as tracers. One of the primary carbon sources for AcCoA production is glucose. Glucose-derived pyruvate enters the mitochondrion and is metabolized to AcCoA by the pyruvate dehydrogenase complex (Fig. 1). AcCoA is then used to produce citrate, which is transported into the cytosol and metabolized to cytosolic (lipogenic) AcCoA which can be used for fatty acid synthesis and elongation. In addition to

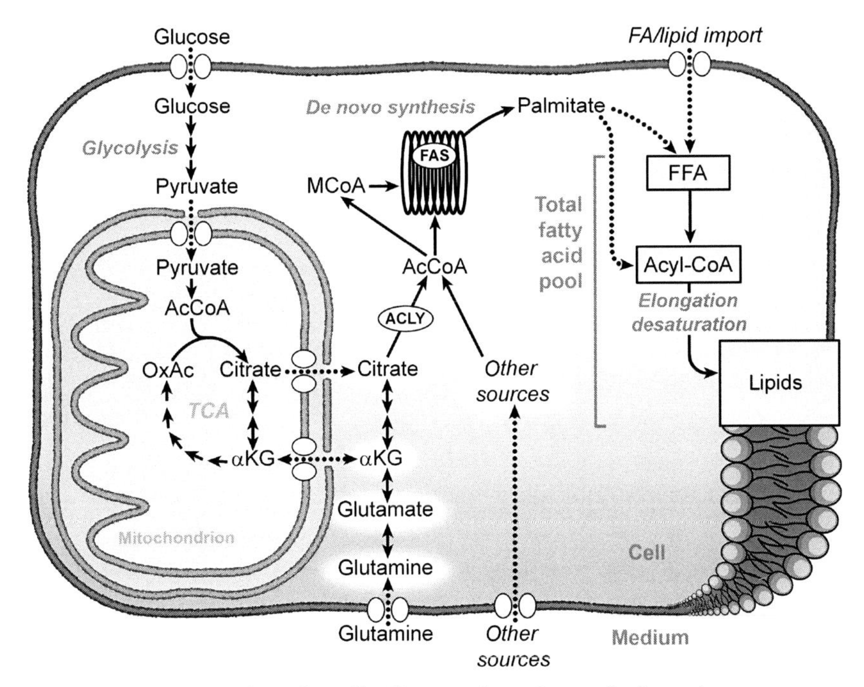

Figure 1 Fatty acid synthesis by cells. Glucose, glutamine, and other substrates are precursors for the production of cytosolic acetyl-CoA. Acetyl-CoA in turn is the two-carbon donor for fatty acid synthesis. The resulting palmitate is subjected to elongation and desaturation reactions to produce a variety of fatty acids that are required for proper cellular functioning. Abbreviations: AcCoA, acetyl-CoA; OxAc, oxaloacetate; αKG, α-ketoglutarate; ACLY, ATP citrate lyase; MCoA, malonyl-CoA; FAS, fatty acid synthase; FFA, free (nonesterified) fatty acid. (See the color plate.)

glucose, other substrates like glutamine (through glutaminolysis or reductive carboxylation) and acetate may be important carbon donors for AcCoA and their ^{13}C-labeled forms are good tracers. These nutrients are especially important in conditions when entry of glucose-carbon into the TCA cycle is limited, such as during hypoxia (Comerford et al., 2014; Kamphorst, Chung, Fan, & Rabinowitz, 2014; Mashimo et al., 2014; Schug et al., 2015). Fatty acids and certain amino acids may also contribute to the cytosolic AcCoA pool, depending on cell type and specific conditions (Shyh-Chang et al., 2013).

AcCoA serves as the substrate for *de novo* fatty acid synthesis. Fatty acids are produced by fatty acid synthase through successive ligations of the two-carbon acetyl units of AcCoA. Thus, any ^{13}C-substrate that metabolizes to cytosolic AcCoA will consequently label fatty acids. It is important to note that beyond fatty acid synthase, numerous elongation (again involving AcCoA as substrate) and desaturation reactions take place to produce a variety of fatty acids that are required for proper cell functioning. Additionally, cells have the capacity to take up fatty acids from exogenous sources (Kamphorst et al., 2013; Okajima, 2002; Sato et al., 2014). A combination of all these events is what forms the cellular fatty acid pools and each of these reactions uniquely affects ^{13}C-incorporation in any given fatty acid. Mass spectrometry analysis of fatty acid ^{13}C-labeling patterns facilitates the elucidation of each of these reactions' contributions. Thus, the use of ^{13}C-labeled precursors in combination with mass spectrometry analysis is a powerful method for detailed analysis of fatty acid metabolism (Bederman et al., 2004).

Over recent years, we have developed novel approaches to quantify fatty acid metabolism in complex biological systems. Here, we describe how fatty acid ^{13}C-labeling can be used to study three central aspects of fatty acid metabolism. First, we show how fatty acid ^{13}C-labeling patterns can be used to determine how much carbon the various substrates (glucose, glutamine, acetate, etc.) contribute to lipogenic AcCoA production. Crucially, determination of AcCoA ^{13}C-enrichment from the fatty acid-labeling distributions enables selective investigation of the cytosolic AcCoA pool, rather than the entire (including mitochondrial) cellular pool. Second, we demonstrate how isotopic steady-state-labeling patterns can be used to determine the relative contributions of cellular synthesis (*de novo* synthesis, elongation, and desaturation) and uptake from exogenous sources to fatty acid pools. Third, we show how turnover rates and absolute fatty acid fluxes can be quantified using experiments that monitor ^{13}C incorporation over time. Successful interrogation of fatty acid metabolism requires high quality data. Because of this, we first discuss ^{13}C-isotope tracing in cell culture, sample preparation, and mass spectrometry analysis, and include practical considerations.

2. LABELING AND EXTRACTION OF FATTY ACIDS IN CULTURED CELLS

2.1 General Considerations

The vast majority of fatty acids in cells are incorporated into complex lipids, and only a small pool of free (nonesterified) fatty acids exists (Fig. 1). Free palmitate is a direct product of the fatty acid synthase enzyme, and its small pool may facilitate rapid labeling and hence assessment of *de novo* lipogenesis pathway activity. However, using free fatty acids to measure degree of labeling has inherent drawbacks. From a practical perspective, we and others have found that the low cellular concentration of free palmitate in combination with a relatively high background of "contaminating" free fatty acids (most notably palmitate and stearate, see later) obtained during sample preparation, often complicate analysis (Lee, Tumanov, Villas-Bôas, Montgomery, & Birch, 2015). From a biochemical perspective it has not been established what reactions contribute to generation of the free cellular palmitate pool (i.e., what percentage is coming from fatty acid synthase or lipases). Additionally, it is unknown if the palmitate generated by fatty acid synthase can be directly channeled to palmitoyl-CoA and subsequently lipids. This would prevent the newly generated palmitate from freely mixing with the existing free palmitate pool, and hence would make analysis of the pool unrepresentative of actual fatty acid metabolic activity. For these reasons, we rely instead on a sample preparation approach involving (1) extraction and (2) hydrolysis (saponification) of cellular lipids, thereby generating free fatty acids, followed by (3) liquid chromatography-mass spectrometry (LC-MS) analysis of this saponified lipid fraction or gas chromatography-mass spectrometry (GC-MS) profiling of re-esterified (methylated) fatty acids (FAMEs). While not all complex lipids are amenable to hydrolysis (e.g., sphingolipids), the samples generated this way represent the vast majority of the fatty acids that were originally present in the cell in either esterified (phospholipids, neutral lipids, etc.) or free form (free fatty acids). This approach has the advantage that the fatty acid concentrations in these samples relative to background are much higher than for free fatty acids alone. This relative enrichment facilitates more reliable measurements. Additionally, as isotope tracing studies are often designed to answer questions regarding to how cells produce biomass, we consider "total" fatty acid pools to be more representative of fatty acid metabolic activity than free fatty acids whose origin is poorly understood.

2.2 Contamination in Fatty Acid Analysis

As noted earlier, contaminant fatty acids are a concern for mass spectrometry analysis. These contaminants can lead to significant errors in the quantitation of fatty acid concentrations and in the determination of the unlabeled peak (M^0). Saturated free fatty acids, especially palmitic acid (C16:0) and stearic acid (C18:0), are common contaminants in pipette tips, microcentrifuge tubes, and other plastic consumables (Lee et al., 2015). Their levels increase over time in reagent solutions due to exposure to the laboratory environment and pipette tips. To alleviate this, glassware should be used (rather than plastic) and solvents should be taken from clean stocks frequently (preferably once a week). It is also good practice to include blank samples in each experiment. These blanks, which we call the "procedure blanks," are dummy samples that are subjected to the entire sample preparation protocol. This provides a good strategy for assessing the contribution of fatty acids from solvents and consumables to the measured fatty acid levels.

2.3 Incubating Cells with Stable Isotope Precursors

Standard methods for mammalian cell culture involve incubating cells in medium containing glucose, amino acids, salts, and vitamins. Additionally, serum is added as a source of growth factors and essential fatty acids. The medium composition can be changed such that the unlabeled glucose and/or other AcCoA precursors are replaced with uniformly labeled ^{13}C-analogs. Culturing cells in this "labeling" medium will cause labeling of fatty acids to occur. We describe here culturing and labeling procedures for adherent proliferating mammalian cancer cells. These procedures may need to be adapted to the cell type under investigation.

1. Cells are maintained in Dulbecco's Modified Eagle's Medium (DMEM) containing 10–25 m*M* glucose, 2 m*M* glutamine, and 10% fetal bovine serum (FBS) at 37 °C and 5% CO_2. Cells are routinely passaged when they reach 80% confluence.
2. A day before initiating labeling split the cells and reseed them in six-well plates at such a density that they reach 80% confluence by the end of the experiment. Incubate the cells overnight before changing to medium containing ^{13}C-labeled substrates.
3. Prepare the labeling medium from DMEM powder with no glucose and glutamine, such that unlabeled or ^{13}C-labeled substrates can be used from the same stock. This powder should not contain pyruvate, a glycolytic intermediate from which AcCoA can be produced, resulting in underestimation of the contribution of glucose-derived carbon to fatty acid production. Supplement medium with 10% dialyzed FBS (dFBS).
4. Incubate the cells with the labeling medium for four doubling times or more for isotopic steady-state-labeling data. Variable durations (for example, 0, 6, 12, 24, 48, and 72 h) can be used for dynamic labeling experiments. At the termination of the labeling experiment harvest

the cells for fatty acid extraction (see later). At this time, cell density should be around 80% confluence.

5. To assess cell proliferation over the course of the labeling experiment, seed a separate set of wells with the same number of cells as the wells used for the experiment. We find that either measuring total (packed) cell volume (VoluPAC, Sartorius AG) or protein concentration is representative for biomass production.

2.4 Sample Preparation for Mass Spectrometry Analysis

All solvents used for mass spectrometry analysis must be LC-MS grade. As previously mentioned, it is necessary to check background levels of palmitate and stearate for each experiment using procedure blanks. The following protocols apply to GC-MS and LC-MS. The initial quenching and extraction procedures of cellular lipids are the same for both types of analysis. The methods only differ in the final steps, with fatty acid methyl esters (FAMEs) generated for GC-MS analysis and free fatty acids for LC-MS analysis (Fig. 2). We have obtained good results with both LC-MS and GC-MS analyses, so the analytical approach can be based simply on the type of mass spectrometer available. The indicated volumes are for each well of a six-well plate with $\sim 2 \times 10^6$ cells/well. The volumes can be adapted for culture dishes with different dimensions.

2.4.1 Quenching and Extraction of Lipids

The first step in the protocol, the quenching and extraction of cellular lipids, is the same for GC-MS and LC-MS. After removing the medium the cells are first washed with phosphate buffered saline (PBS) to remove serum lipids. The metabolic activity of the cells is then quenched by addition of methanol/PBS (1:1) at −20 °C. This serves to minimize potential artifact formation during sample preparation. After scraping the cells into a glass tube, chloroform is added to selectively extract the lipids from the cells.

1. Remove medium from the dishes using an aspirator and wash twice with 1 mL of PBS at room temperature.
2. Add 0.75 mL of −20 °C methanol/PBS (1:1) to quench metabolism and place plate on ice. Gently swirl the plate such that the solution covers the whole surface area of the dish.
3. Store at −20 °C for 10 min.
4. Scrape cells on ice using a cell scraper. To maximize recovery, scrape the cells toward one side of the dish and collect as much volume as possible into glass tubes. Keep the tube on ice and proceed to the next dish.

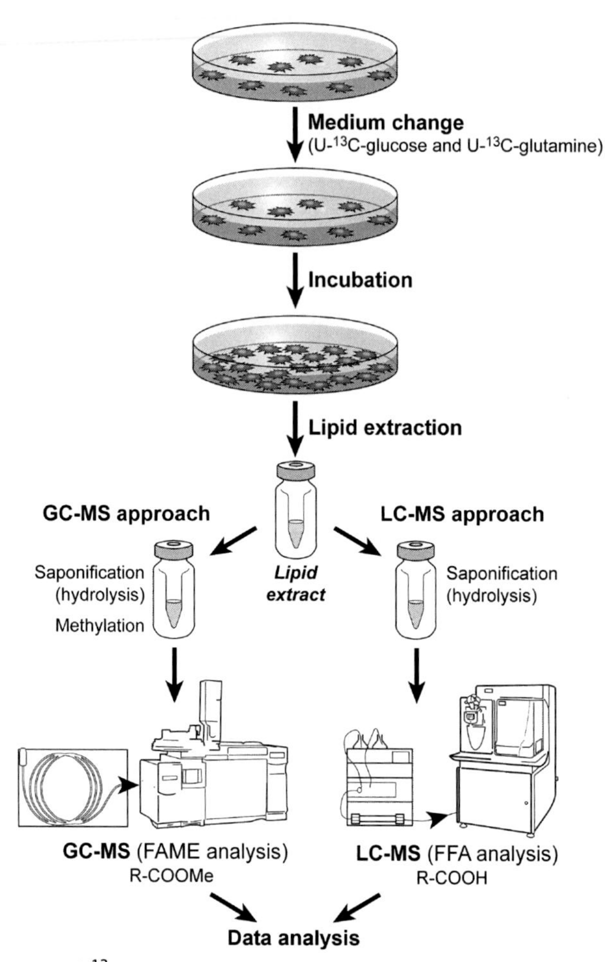

Figure 2 Fatty acid ^{13}C-isotope incorporation in cultured cells. Cells are incubated in medium containing U-^{13}C-glucose and U-^{13}C-glutamine. Then a whole-cell lipid extraction is performed followed by hydrolysis and methylation (GC-MS) or hydrolysis only (LC-MS) of fatty acids, followed by mass spectrometry analysis. Abbreviations: GC-MS, gas chromatography-mass spectrometry; LC-MS, liquid chromatography-mass spectrometry. (See the color plate.)

5. Add 0.5 mL ice cold chloroform ($CHCl_3$) to the tube and vortex for 1 min.
6. Spin down at up to 500 × *g* in a centrifuge for 10 min (note that higher *g*-force will break the glass).
7. Remove the lower chloroform layer and collect in a glass vial with a screw cap using a 500 μL glass Hamilton Syringe. Care should be taken

that the needle does not touch the protein precipitate. For this, the tube should be tilted toward one side and the organic layer can then be approached from the side. It is important to wash the needle three times with chloroform between samples to clean it from the previous sample.

8. Dry the lipid extracts under nitrogen gas.

2.4.2 Sample Preparation for Gas Chromatography-Mass Spectrometry

Analysis by GC-MS necessitates that the fatty acids are modified such that they can enter the gas phase. This is achieved by methylation of the carboxylic acid group, resulting in FAMEs. Formation of FAMEs occurs when heating samples in excess methanol and HCl as a catalyst.

9. After drying the lipid extracts add to each vial in the following order:
   - **I.** 80 μL toluene.
   - **II.** 600 μL methanol.
   - **III.** 120 μL methanolic-HCl.

   To make methanolic-HCL, slowly (to avoid splashing) add 9.7 mL concentrated HCl to 40.3 mL methanol. Methanolic-HCL should be stored at 4 °C and should only be used for a maximum of 1 month.
10. Vortex and incubate at 100 °C for 60 min. The caps of the vials should be tightened to prevent evaporation. It is advisable to tighten caps every ~20 min. This step will hydrolyze complex lipids and will generate FAMEs from these hydrolyzed, as well as free fatty acids.
11. Samples are then brought to room temperature. To each sample, add in the following order:
    - **I.** 400 μL water.
    - **II.** 300 μL hexane, vortex.
12. Extract the top hexane phase into the GC-MS vials and analyze.

2.4.3 Sample Preparation for Liquid Chromatography-Mass Spectrometry

Liquid chromatography-based separation of fatty acids does not require methylation of the carboxylic acid group. In fact, the electrospray ionization (ESI) that follows the separation is unable to ionize FAMEs. Instead, for analysis by LC-MS, esterified fatty acids are best treated with a base for saponification (hydrolysis) to nonesterified fatty acids with a free carboxylic acid group.

9. After drying, to each vial add 1 mL of 90% methanol in water containing 0.3 *M* potassium hydroxide (KOH).
10. Vortex and incubate at 80 °C for 60 min.

11. Samples are brought to room temperature. To each sample, add
    I. 100 μL formic acid.
    II. 800 μL hexane, vortex.
12. Extract the top hexane phase into a LC-MS vial and then re-extract sample with 700 μL hexane and pool.
13. Dry under nitrogen.
14. Reconstitute each vial in 200 μL methanol.
15. Transfer to LC-MS vials with a glass insert and analyze.

3. MASS SPECTROMETRY ANALYSIS OF FATTY ACID LABELING

3.1 GC-MS Analysis

For analyses of FAMEs, we use a GC-triple quadrupole mass spectrometer, but other systems like a GC-TOF may work equally well. Likewise, many types of columns have been described for the analysis of fatty acids. In our laboratory, we use an Agilent 7890B GC system coupled with an Agilent 7000 Triple Quadrupole GC/MS system. We use a Phenomenex ZB-1701 column (30 m × 0.25 mm × 0.25 μm). The GC-MS parameters are adapted from Smart, Aggio, Houtte, and Villas-Boas (2010). This method calls for 1 μL of sample to be injected in splitless mode, with an inlet temperature of 280 °C and a helium flow rate of 1 mL/min. The GC oven program is as described in Table 1.

For general fatty acid profiling the mass spectrometer is operated at 70 eV (EI), scanning with a mass range of 50–550 *m/z*. For analysis of isotope labeling, we set the source voltage 35 eV (EI), to reduce fragmentation of the parent molecule and hence increase its signal and that of its isotopologs. For example, for analysis of palmitate labeling, we set the mass range from

Table 1 GC Temperature Program for FAME Analysis

Start Temperature (°C)	Ramp (°C/min)	End Temperature (°C)	Hold Time (min)
45	–	45	2
45	9	180	5
180	40	240	11.5
240	40	280	2

265 to 295 *m/z* to exclude the more abundant fragmentation products but to include the parent molecule with a molecular weight of 270 and the entire labeling pattern (see Fig. 3). This range can be modified to measure labeling patterns of other fatty acids as they elute. Normally, we can observe palmitate and oleate, which have the highest abundances, together with other low abundant fatty acids (palmitoleic, stearic, docosanoic, docosenoic acids). Our GC program enables chromatographic separation of all fatty acids commonly present in the sample. For analysis of the relative intensities of the isotopologs, we use Agilent Mass Hunter B.06.00 software to measure peak intensities.

3.2 LC-MS Analysis

We routinely analyze negatively charged free (i.e., underivatized) fatty acids generated by lipid hydrolysis with high resolution orbitrap mass spectrometry (Kamphorst, Fan, Lu, White, & Rabinowitz, 2011) and envision that TOF, and perhaps other types of mass spectrometers may also be used successfully for this type of analysis. The ESI used to introduce the fatty acids into the mass spectrometer does not cause fragmentation. Therefore, only the intact molecular ion and its isotopologs are analyzed (Fig. 3). Our current setup is a Q Exactive orbitrap mass spectrometer coupled to a Dionex UltiMate 3000 LC system (Thermo Scientific). The LC settings we use are 2 μL of sample is injected onto a 1.9 μm particle 100 × 2.1 mm id Thermo Hypersil GOLD column (Thermo Scientific) which is heated to 60 °C. A binary gradient solvent system of (A) water/acetonitrile (40:60, v/v) with 10 m*M* ammonium formate and (B) acetonitrile/isopropanol (10:90, v/v) with 10 m*M* ammonium formate is used. Sample analysis is performed using a linear gradient from 30% to 60% B over 6 min followed by raising gradient to 100% B within next 0.5 min and holding at this composition for another 1.5 min. Thereafter, the column is equilibrated with the initial conditions for another 2 min. The flow rate we use is 0.3 mL/min. Fatty acids are analyzed in negative ionization mode. The electrospray and mass spec settings are as follows: spray voltage 3.5 kV, capillary temperature 300 °C, sheath gas flow rate 25 (arbitrary units), and auxiliary gas flow rate 15 (arbitrary units). The mass spec resolution is set to 70,000, automatic gain control is set at 1×10^6 with a maximum injection time of 100 ms and the scan range is 240–400 *m/z* for MS mode. Xcalibur 2.2 and MAVEN are used for data processing (Melamud, Vastag, & Rabinowitz, 2010).

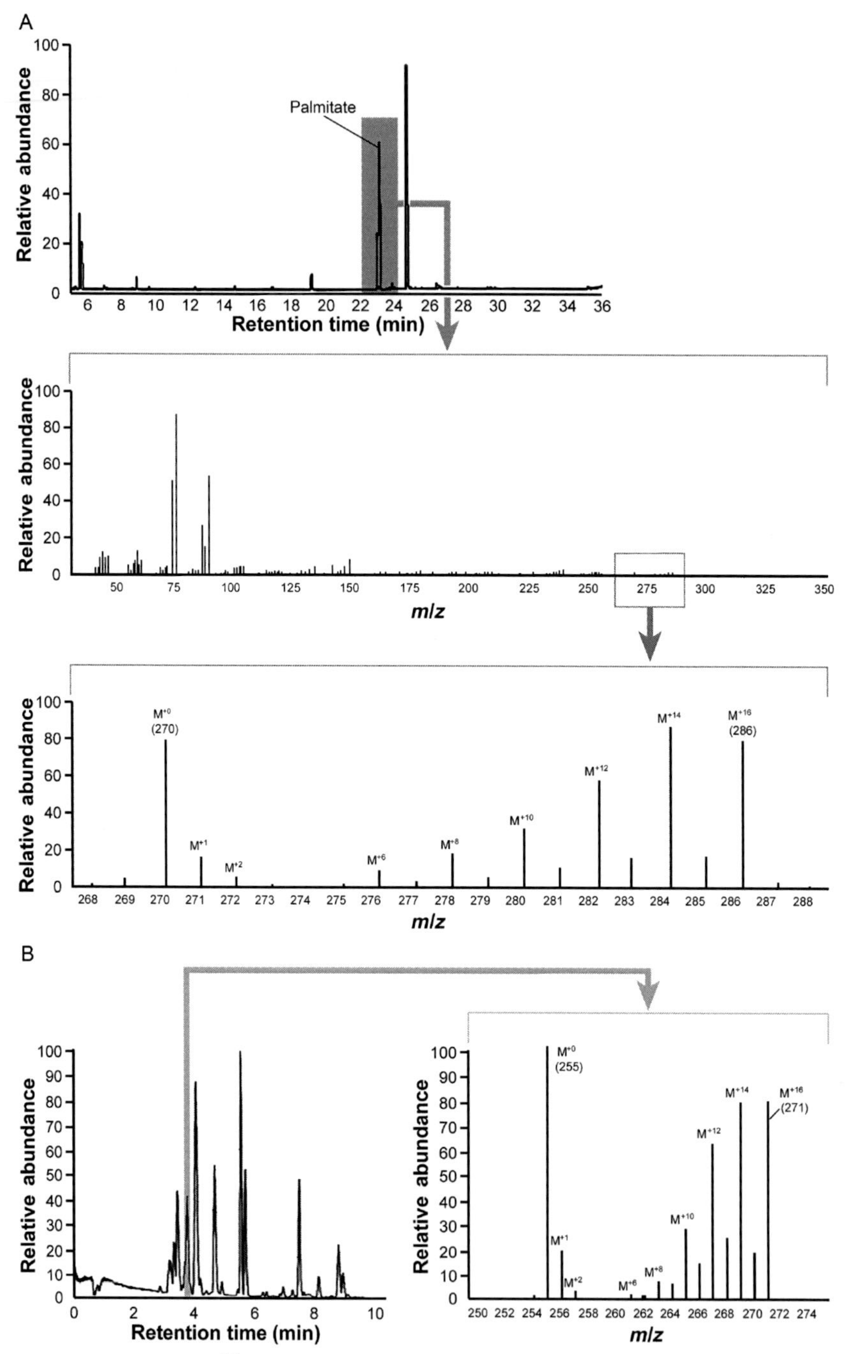

Figure 3 Analysis of ^{13}C fatty acid labeling by (A) GC-MS and (B) LC-MS. Palmitate (C16:0) is shown as example. With GC-MS extensive fragmentation occurs because of the electron impact (EI) ionization. Nevertheless, fatty acid labeling can reliably be determined (see inset). Electrospray ionization (ESI), used with LC-MS, does not cause fragmentation and allows for the convenient measurement of the molecular ion and its isotopologs.

4. FATTY ACID-LABELING DATA ANALYSIS AND INTERPRETATION

4.1 Correcting for ^{13}C Natural Abundance

The experimentally determined fatty acid-labeling distributions from cells cultured in the presence of U-^{13}C-glucose and U-^{13}C-glutamine (Fig. 3) reveal significant fractions with an odd number of ^{13}C atoms. However, based on known metabolic pathways, the production of the acetyl moiety of AcCoA from fully ^{13}C-labeled glucose, glutamine, or other precursor will only result in both carbons being ^{13}C labeled, or not labeled. The observed odd number labeling results from the natural occurrence of ^{13}C, whose fraction is significant in fatty acids. Correcting for natural abundance of ^{13}C is necessary for quantitative interpretation. Indeed, after correction the odd numbered ^{13}C-labeled fractions practically disappear (Fig. 4). An algorithm for natural abundance correction is described elsewhere (Yuan, Bennett, & Rabinowitz, 2008) and can be performed using the freely available computer program MAVEN (Clasquin, Melamud, & Rabinowitz, 2012; Melamud et al., 2010).

4.2 Determining the Carbon Sources of Lipogenic Acetyl-CoA

Culturing cancer cells in the presence of abundant U-^{13}C-glucose, U-^{13}C-glutamine, and oxygen, will result in fatty acid-labeling patterns with a significant fully ^{13}C-labeled fraction (see Fig. 4 with palmitate as example where M^{+16} is fully ^{13}C labeled). This is consistent with glucose and glutamine being the most important carbon sources for AcCoA in these conditions. Nevertheless, the fatty acid pools also contain partially labeled forms (M^{+14}, M^{+12}, etc.). These partially labeled isotopologs occur because AcCoA labeling from U^{13}C-glucose and U^{13}C-glutamine is not complete (i.e., other carbon sources also contribute to AcCoA production). As fatty acid synthesis is essentially a ligation of two-carbon acetyl units with both carbons either ^{13}C or ^{12}C, the distribution of labeling reflects a binomial distribution (Hellerstein & Neese, 1992; Kharroubi, Masterson, Aldaghlas, Kennedy, & Kelleher, 1992). For example, palmitate is the result of eight ligations of acetyl units and the fractional contribution of each labeling form (M^{+2}, M^{+4},..., M^{+16}) for any percentage of AcCoA labeling (p) can be calculated by (Kamphorst et al., 2014):

$$\text{Fraction palmitate M} + 2x = \binom{8}{x} p^x (1-p)^{8-x} \tag{1}$$

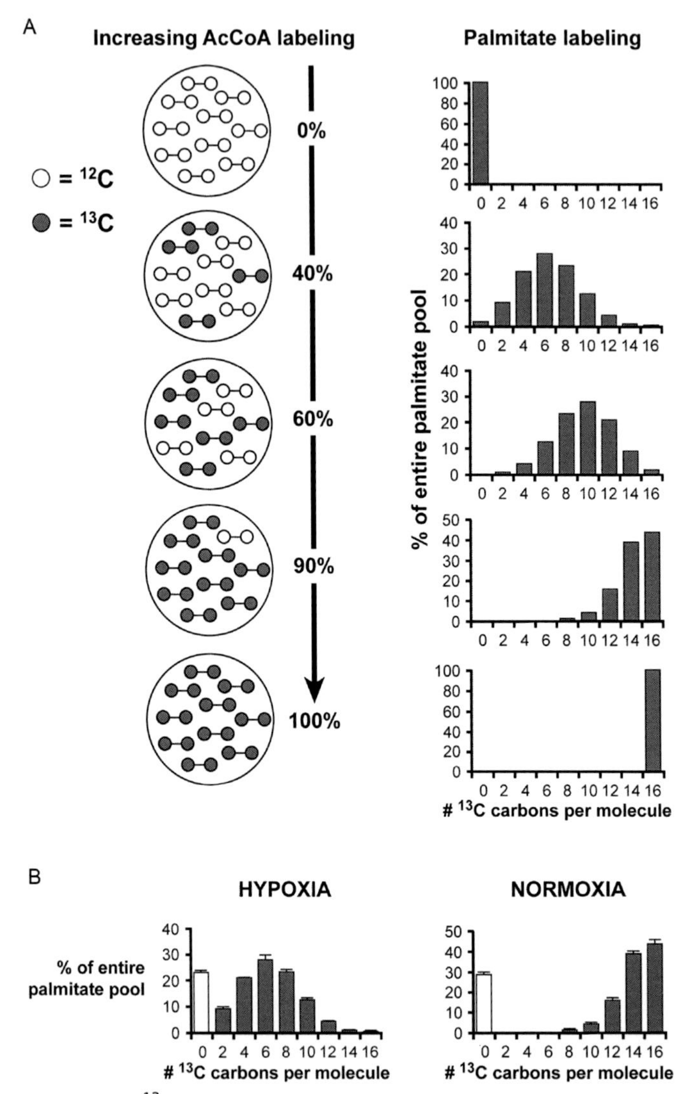

Figure 4 Percentage ^{13}C-labeling of cytosolic (lipogenic) acetyl-CoA can be quantified from palmitate labeling. (A) Increasing $^{13}C_2$-acetyl-CoA labeling from U-^{13}C-glucose, U-^{13}C-glutamine, and/or other sources shifts palmitate labeling pattern to the right. These are computed labeling distributions based on binomial distribution (Eq. 1) and indicated labeling percentages of $^{13}C_2$-acetyl-CoA. (B) Experimentally determined steady-state ^{13}C-labeling distributions of palmitate from MDA-MB-468 cells cultured in the presence of U-^{13}C-glucose and U-^{13}C-glutamine in normoxic and hypoxic (1% O_2) conditions. Cytosolic $^{13}C_2$-acetyl-CoA labeling can be quantified from these experimental labeling patterns by performing least-square fitting to the theoretical distribution for a range of acetyl-CoA ^{13}C-enrichments. In this experiment, we observed reduced acetyl-CoA labeling from glucose and glutamine in hypoxia and later identified acetate as an important additional acetyl-CoA carbon source. For (B), data are means $\pm$ SD, $n=3$.

The fractional labeling of AcCoA used for *de novo* lipogenesis can be determined from experimental labeling patterns by least-square fitting to computed distributions (using Eq. 1) for a wide range of AcCoA labeling fractions (Fig. 4). For the above equation, we typically disregard M^0 as we use a combination of U-^{13}C-glucose, U-^{13}C-glutamine, and/or U-^{13}C-acetate, which leads to such extensive labeling of AcCoA that there is no overlap with M^0. Determining the AcCoA labeling from fatty acid labeling, rather than from directly measured AcCoA, has two advantages. First, fatty acids are more stable and abundant than AcCoA, making for facile analysis by either LC-MS or GC-MS. Second, direct analysis of AcCoA by mass spec will provide information for the total cellular pool, whereas fatty acid labeling represents the cytosolic (i.e., lipogenic) pool only.

As a demonstration of the utility of the approach described here, we provide an example study from a recent publication (Kamphorst et al., 2014). In this study, we determined lipogenic AcCoA labeling in cultured cancer cells from U-^{13}C-glucose and U-^{13}C-glutamine in both normoxic (atmospheric) and hypoxic (1% O_2) conditions. We conducted this experiment to see if hypoxia-induced changes in central carbon metabolism would affect AcCoA production. This study showed a reduced contribution overall from U-^{13}C-glucose and U-^{13}C-glutamine to AcCoA production, as well as a 20–50% contribution (depending on cell line) from an unknown carbon source (Fig. 4B). Using this method, we were able to identify acetate as the main additional source, next to glucose and glutamine, for lipogenic AcCoA production in hypoxic cancer cells (Comerford et al., 2014; Kamphorst et al., 2014; Mashimo et al., 2014; Schug et al., 2015). This conversion from acetate to AcCoA occurs in the cytosol. Thus, AcCoA production is compartmentalized, and the role of acetate in fatty acid production can be better determined from fatty acid-labeling patterns than from whole-cell (including mitochondrial) AcCoA measurements.

4.3 Differentiating Between *De Novo* Lipogenesis and Import

Mammalian cells either produce nonessential fatty acids *de novo* or derive them from an exogenous source. For cultured cells, this source of fatty acids is the serum that is added to the medium. The relative contribution of both *de novo* synthesis and uptake of exogenous fatty acids can be determined from fatty acid-labeling patterns from cells grown in the presence of ^{13}C-labeled precursors. For this, the fatty acid labeling should meet two criteria. First, the labeling should be at isotopic steady state (the labeling should not increase further with longer incubation). For most (cancer) cell lines we worked with thus far isotopic steady state for C16–C18 fatty acids is obtained after 48–72 h of incubation with ^{13}C-labeled substrate(s). This steady state could be compromised by depletion of serum lipids and therefore regular medium changes may be required depending on the conditions. Second, the ^{13}C-labeling of fatty acid precursor AcCoA should be to such degree that the distribution of labeled isotopologs does not overlap with the unlabeled (M^0) peak. Culturing cancer cells in medium containing both U-^{13}C-glucose and U-^{13}C-glutamine will shift the isotopic envelope well away from the M^0 peak.

When the criteria of (1) isotopic state and (2) sufficient AcCoA labeling have been met, the relative contribution of *de novo* synthesis and uptake to a particular fatty acid pool can be determined. For saturated fatty acids, the contribution from uptake is simply the fraction of M^0 to all isotopologs (Fig. 5). Unsaturated fatty acids, such as oleate (C18:1), are produced from a desaturation reaction, which results in no addition of carbon. Therefore, the M^0 of unsaturated fatty acids is derived from both import of exogenous fatty acids and from the M^0 of the desaturation reaction precursor (C18:0 for C18:1). To formalize this, with C18:1 as an example, the cellular pool is the result of desaturation of C18:0 by stearoyl-CoA desaturase 1 (SCD1) (D) and import from the serum (I), where fractional import equals 1-D. The fractional contribution of D, in turn, can be derived by comparing L (the fraction of a given fatty acid that is labeled, i.e., 1-fraction M^0) of C18:0 and C18:1 (Kamphorst et al., 2013).

$$L_{\mathrm{C18:1}} = (D)(L_{\mathrm{C18:0}}) \tag{2}$$

$$D = L_{\mathrm{cell\ C18:1}} / L_{\mathrm{cell\ C18:0}} \tag{3}$$

4.4 Calculating Fatty Acid Fluxes

Beyond percentage labeling of AcCoA and the relative contribution of *de novo* synthesis and import, fatty acid-labeling data can also be used to determine absolute fluxes. One approach, termed kinetic flux profiling, relies on time-course labeling data and the cellular concentration of the fatty acid of interest (Yuan et al., 2008). The flux can be determined by fitting a differential equation to the unlabeled fraction of a fatty acid. For example, for palmitate (C16:0), which is produced from *de novo* synthesis (f_1) and uptake (f_2), the change in the unlabeled pool C16:0^{U} over time is described by the differential equation:

$$\frac{\mathrm{d}}{\mathrm{d}t}\mathrm{C16:0^U} = f_1(1-p)^8 + f_2 - f_X\frac{\mathrm{C16:0^U}}{\mathrm{C16:0^T}} \approx f_2 - f_X\frac{\mathrm{C16:0^U}}{\mathrm{C16:0^T}} \tag{4}$$

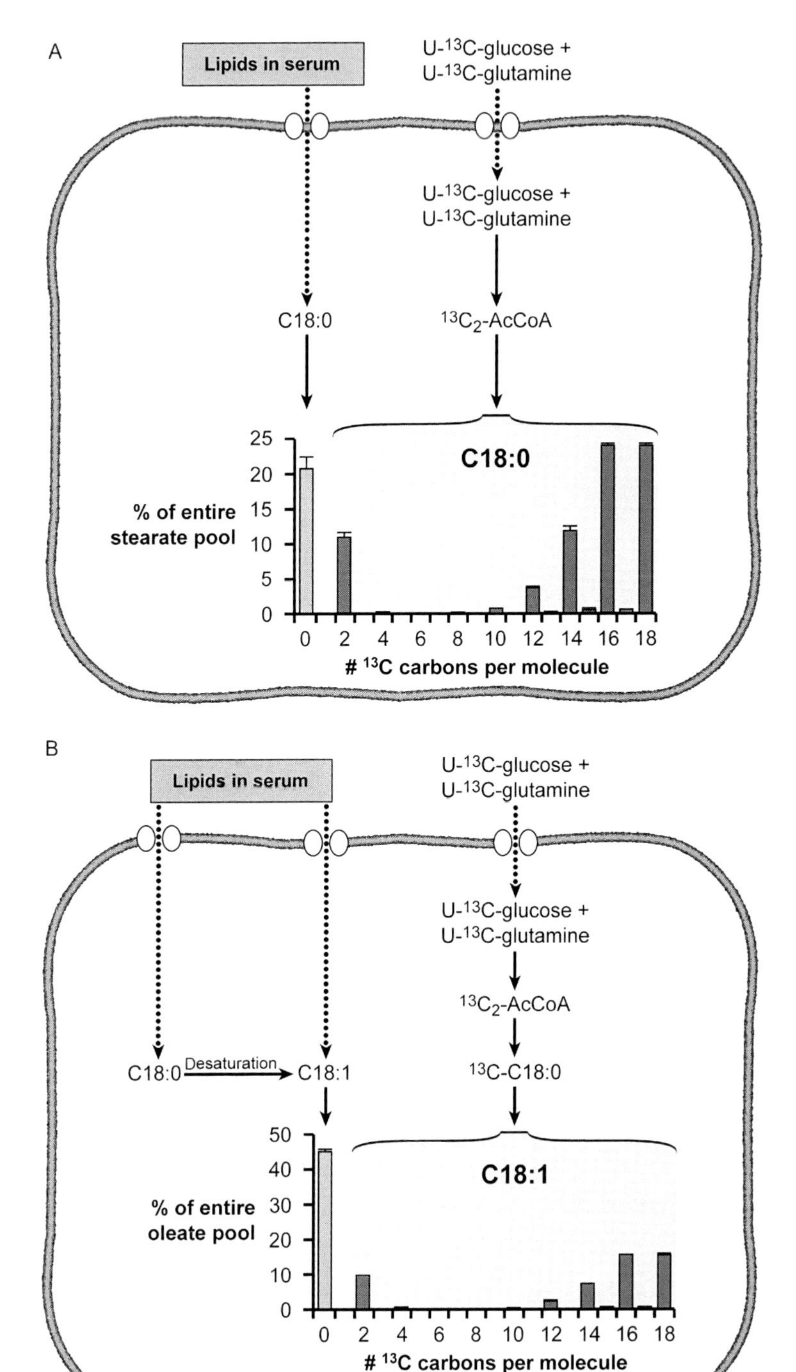

Figure 5 Determining the contribution from synthesis and import to cellular fatty acid pools from steady-state fatty acid-labeling patterns. Stearate (C18:0) and oleate (C18:1) are shown as examples. For saturated fatty acids, the fractional contribution from import equals the fraction of M^0. As desaturation of C18:0 to C18:1 does not introduce carbon, the M^0 of C18:1 is a mixture of the originally unlabeled C18:0 and imported C18:1. Their contribution can be determined using Eqs. (2) and (3).

Where C16:0^T is the total palmitate pool, $f_x = f_1 + f_2$, and p is the fraction of lipogenic AcCoA that is labeled. Like mentioned previously when using U-^{13}C-glucose, alone or in combination with other ^{13}C-labeled substrates, that fraction is sufficiently large and hence $(1-p)^8$ small enough to be omitted. The use of Eq. (4) rests on the assumption that the labeling of lipogenic precursor AcCoA reaches steady state very fast relative to palmitate. Indeed, we find that AcCoA steady-state labeling is reached in a matter of hours, whereas for palmitate it takes days. As Eq. (4) shows, the calculation of absolute fatty acid fluxes requires quantification of the cellular fatty acid pool. As the sample preparation described here is a two-step process (lipid extraction from cells, saponification, and derivatization) simply spiking in free fatty acids or FAME standards during sample preparation does not account for (1) losses during extraction and differences in extraction efficiencies between lipid classes, and (2) incomplete saponification and derivatization. Instead, a reasonable approach is to spike in known amounts of a set of commercially available phospholipids and triglycerides containing odd-chain fatty acids directly prior to quenching. The resulting free or esterified odd-chain fatty acids during the sample preparation can conveniently acts as standards, as they are nonnaturally occurring. The amount of standards to add should be determined empirically and potential differences in response factors between the standards and the fatty acids of interest should be evaluated.

5. SUMMARY AND PERSPECTIVE

We have discussed the use of ^{13}C-labeled substrates in combination with mass spectrometry for studying fatty acid metabolism. ^{13}C-carbon tracing provides valuable information concerning fatty acid metabolic reactions that cannot be obtained otherwise. It enables elucidation of the contribution of various carbon sources (glucose, glutamine, acetate, others) to fatty acid synthesis, and relative and absolute carbon fluxes can be calculated with this technique. U-^{13}C-glucose and U-^{13}C-glutamine labeling studies have been performed in mammals (Maher et al., 2012; Yang et al., 2009). While here the focus was on cell culture systems, we see no technical limitations for obtaining fatty acid-labeling data from *in vivo* experiments. One potential complicating factor is that ^{13}C-labeled glucose (or other ^{13}C-labeled substrates) may not distribute evenly across tissues (Previs et al., 2014). This should be carefully evaluated, for one by directly analyzing glucose ^{13}C enrichment in (parts of) the tissues of interest. As an alternative, the use of $^{2}H_2O$ may be more suitable in some cases.

One aspect that may affect labeling patterns of longer chain fatty acids was not discussed here; elongases may perform multiple rounds of acetyl addition prior to releasing a fatty acid. This is termed "channeling" and may cause a considerable fraction of a fatty acid intermediate not to mix with the cellular pool (Guillou, Zadravec, Martin, & Jacobsson, 2010; Jakobsson, Westerberg, & Jacobsson, 2006). This may create a considerable mismatch between computed and observed labeling patterns when only single acetyl additions are considered. For example, a 20-carbon saturated fatty acid (C20:0) can be synthesized from C18:0 via a single elongation step, or directly from C16:0 via two rounds of elongation. In the latter case, the labeling of the intermediate C18:0 will not mix with its cellular pool and there will be no M^0 contribution from serum-derived C18:0. We are currently developing an algorithm that accounts for possible channeling reactions and preliminary results indicate that it indeed occurs for some fatty acids.

The methods described here for indirect determination of AcCoA labeling, quantification of fatty acid synthesis versus uptake, and determination of fatty acid fluxes provide powerful tools that will continue to be used for better understanding complex biological systems. At the same time, we realize that lipids constitute a diverse and complex range of molecular structures and that novel technological innovation will be important for improved quantitative understanding of associated metabolic reactions. Recently, developed lipidomics methods in combination with a variety of stable isotope tracers (^{13}C-choline, ^{13}C-fatty acids) are very promising in this regard (Castro-Perez et al., 2011; McLaren et al., 2011; Wakelam, Pettitt, & Postle, 2007).

REFERENCES

Baenke, F., Peck, B., Miess, H., & Schulze, A. (2013). Hooked on fat: The role of lipid synthesis in cancer metabolism and tumour development. *Disease Models & Mechanisms*, *6*, 1353–1363.

Bederman, I. R., Reszko, A. E., Kasumov, T., David, F., Wasserman, D. H., Kelleher, J. K., et al. (2004). Zonation of labeling of lipogenic acetyl-CoA across the liver: Implications for studies of lipogenesis by mass isotopomer analysis. *Journal of Biological Chemistry*, *279*, 43207–43216.

Castro-Perez, J. M., Roddy, T. P., Nibbering, N. M. M., Shah, V., McLaren, D. G., Previs, S. F., et al. (2011). Localization of fatty acyl and double bond positions in phosphatidylcholines using a dual stage CID fragmentation coupled with ion mobility mass spectrometry. *Journal of the American Society for Mass Spectrometry*, *22*(9), 1552–1567.

Clasquin, M. F., Melamud, E., & Rabinowitz, J. D. (2012). LC-MS data processing with MAVEN: A metabolomic analysis and visualization engine. *Current Protocols in Bioinformatics*, *29*, 997–1003. http://dx.doi.org/10.1002/0471250953.bi1411s37.

Comerford, S. A., Huang, Z., Du, X., Wang, Y., Cai, L., Witkiewicz, A. K., et al. (2014). Acetate dependence of tumors. *Cell*, *159*, 1591–1602.

Daniels, V. W., Smans, K., Royaux, I., Chypre, M., Swinnen, J. V., & Zaidi, N. (2014). Cancer cells differentially activate and thrive on *de novo* lipid synthesis pathways in a low-lipid environment. *PLoS One*, *9*(9), e106913.

Guillou, H., Zadravec, D., Martin, P. G. P., & Jacobsson, A. (2010). The key roles of elongases and desaturases in mammalian fatty acid metabolism: Insights from transgenic mice. *Progress in Lipid Research*, *49*(2), 186–199.

Hellerstein, M. K., & Neese, R. A. (1992). Mass isotopomer distribution analysis: a technique for measuring biosynthesis and turnover of polymers. *American Journal of Physiology*, *263*, E988–E1001.

Herath, K. B., Zhong, W., Yang, J., Mahsut, A., Rohm, R. J., Shah, V., et al. (2014). Determination of low levels of ^{2}H-labeling using high-resolution mass spectrometry: Application in studies of lipid flux and beyond. *Rapid Communications in Mass Spectrometry*, *28*, 239–244.

Jakobsson, A., Westerberg, R., & Jacobsson, A. (2006). Fatty acid elongases in mammals: Their regulation and roles in metabolism. *Progress in Lipid Research*, *45*(3), 237–249.

Kamphorst, J. J., Chung, M. K., Fan, J., & Rabinowitz, J. D. (2014). Quantitative analysis of acetyl-CoA production in hypoxic cancer cells reveals substantial contribution from acetate. *Cancer & Metabolism*, *2*, 23.

Kamphorst, J. J., Cross, J. R., Fan, J., de Stanchina, E., Mathew, R., White, E. P., et al. (2013). Hypoxic and Ras-transformed cells support growth by scavenging unsaturated fatty acids from lysophospholipids. *Proceedings of the National Academy of Sciences of the United States of America*, *110*(22), 8882–8887.

Kamphorst, J. J., Fan, J., Lu, W., White, E., & Rabinowitz, J. D. (2011). Liquid-chromatography-high resolution mass spectrometry analysis of fatty acid metabolism. *Analytical Chemistry*, *83*, 9114–9122.

Khandekar, M. J., Cohen, P., & Spiegelman, B. M. (2011). Molecular mechanisms of cancer development in obesity. *Nature Reviews Cancer*, *11*, 886–895.

Kharroubi, A. T., Masterson, T. M., Aldaghlas, T. A., Kennedy, K. A., & Kelleher, J. K. (1992). Isotopomer spectral analysis of triglyceride fatty acid synthesis in 3T3-L1 cells. *American Journal of Physiology*, *263*, E667–E675.

Lee, T. W., Tumanov, S., Villas-Bôas, S. G., Montgomery, J. M., & Birch, N. P. (2015). Chemicals eluting from disposable plastic syringes and syringe filters alter neurite growth, axogenesis and the microtubule cytoskeleton in cultured hippocampal neurons. *Journal of Neurochemistry*, *133*, 53–65. http://dx.doi.org/10.1111/jnc.13009.

Maher, E. A., Marin-Valencia, I., Bachoo, R. M., Mashimo, T., Raisanen, J., Hatanpaa, K. J., et al. (2012). Metabolism of [U-^{13}C]-glucose in human brain tumors *in vivo*. *NMR in Biomedicine*, *25*(11), 1234–1244.

Mashimo, T., Pichumani, K., Vemireddy, V., Hatanpaa, K. J., Singh, D. K., Sirasanagandla, S., et al. (2014). Acetate is a bioenergetic substrate for human glioblastoma and brain metastases. *Cell*, *159*(7), 1603–1614.

McLaren, D. G., He, T., Wang, S. P., Mendoza, V., Rosa, R., Gagen, K., et al. (2011). The use of stable-isotopically labeled oleic acid to interrogate lipid assembly *in vivo*: Assessing pharmacological effects in preclinical species. *Journal of Lipid Research*, *52*(6), 1150–1161.

Melamud, E., Vastag, L., & Rabinowitz, J. D. (2010). Metabolic analysis and visualization engine for LC-MS data. *Analytical Chemistry*, *82*, 9818–9826.

Okajima, F. (2002). Plasma lipoproteins behave as carriers of extracellular sphingosine 1-phosphate: Is this an atherogenic mediator or an anti-atherogenic mediator? *Biochimica et Biophysica Acta—Molecular and Cell Biology of Lipids*, *1582*(1-3), 132–137.

Previs, S. F., McLaren, D. G., Wang, S. P., Stout, S. J., Zhou, H., Herath, K., et al. (2014). New methodologies for studying lipid synthesis and turnover: Looking backwards to enable moving forwards. *Biochimica et Biophysica Acta*, *1842*(3), 402–413.

Quehenberger, O., Armando, A. M., & Dennis, E. A. (2011). High sensitivity quantitative lipidomics analysis of fatty acids in biological samples by gas chromatography–mass spectrometry. *Biochimica et Biophysica Acta*, *1811*, 648–656.

Sato, K., Tobo, M., Mogi, C., Murata, N., Kotake, M., Kuwabara, A., et al. (2014). Lipoprotein-associated lysolipids are differentially involved in high-density lipoprotein and its oxidized form-induced neurite remodeling in PC12 cells. *Neurochemistry International*, *68*, 38–47.

Schug, Z. T., Peck, B., Jones, D. T., Zhang, Q., Grosskurth, S., Alam, I. S., et al. (2015). Acetyl-CoA synthetase 2 promotes acetate utilization and maintains cancer cell growth under metabolic stress. *Cancer Cell*, *27*(1), 57–71.

Shyh-Chang, N., Locasale, J. W., Lyssiotis, C. A., Zheng, Y., Teo, R. Y., Ratanasirintrawoot, S., et al. (2013). Influence of threonine metabolism on S-adenosyl-methionine and histone methylation. *Science*, *339*(6116), 222–226.

Smart, K. F., Aggio, R. B. M., Houtte, J. R. V., & Villas-Boas, S. G. (2010). Analytical platform for metabolome analysis of microbial cells using methyl chloroformate derivatisation followed by gas chromatography-mass spectrometry. *Nature Protocols*, *5*, 1–21.

Wakelam, M. J. O., Pettitt, T. R., & Postle, A. D. (2007). Lipidomic analysis of signaling pathways. *Methods in Enzymology*, *432*, 233–246.

Yang, C., Sudderth, J., Dang, T., Bachoo, R. G., McDonald, J. G., & DeBerardinis, R. J. (2009). Glioblastoma cells require glutamate dehydrogenase to survive impairments of glucose metabolism or *Akt* signalling. *Cancer Research*, *69*(20), 7986–7993.

Yuan, J., Bennett, B. D., & Rabinowitz, J. D. (2008). Kinetic flux profiling for quantitation of cellular metabolic fluxes. *Nature Protocols*, *3*, 1328–1340.

Zaidi, N., Lupien, L., Kuemmerle, N. B., Kinlaw, W. B., Swinnen, J. V., & Smans, K. (2014). Lipogenesis and lipolysis: The pathways exploited by the cancer cells to acquire fatty acids. *Progress in Lipid Research*, *52*(4), 585–589.

CHAPTER SEVEN

Dynamic Proteomics: *In Vivo* Proteome-Wide Measurement of Protein Kinetics Using Metabolic Labeling

W.E. Holmes*, T.E. Angel*, K.W. Li*, M.K. Hellerstein*,†,1

*KineMed Inc., Emeryville, California, USA

†Department of Nutritional Sciences and Toxicology, University of California, Berkeley, Berkeley, California, USA

[1]Corresponding author e-mail address: MHellerstein@kinemed.com

Contents

Methods in Enzymology, Volume 561
ISSN 0076-6879
http://dx.doi.org/10.1016/bs.mie.2015.05.018

Abstract

Control of biosynthetic and catabolic rates of polymers, including proteins, stands at the center of phenotype, physiologic adaptation, and disease pathogenesis. Advances in stable isotope-labeling concepts and mass spectrometric instrumentation now allow accurate *in vivo* measurement of protein synthesis and turnover rates, both for targeted proteins and for unbiased screening across the proteome. We describe here the underlying principles and operational protocols for measuring protein dynamics, focusing on metabolic labeling with ^2H_2O (heavy water) combined with tandem mass spectrometric analysis of mass isotopomer abundances in trypsin-generated peptides. The core principles of combinatorial analysis (mass isotopomer distribution analysis or MIDA) are reviewed in detail, including practical advantages, limitations, and technical procedures to ensure optimal kinetic results. Technical factors include heavy water labeling protocols, optimal duration of labeling, clean up and simplification of sample matrices, accurate quantitation of mass isotopomer abundances in peptides, criteria for adequacy of mass spectrometric abundance measurements, and calculation algorithms. Some applications are described, including the noninvasive "virtual biopsy" strategy for measuring molecular flux rates in tissues through measurements in body fluids. In addition, application of heavy water labeling to measure flux lipidomics is noted.

In summary, the combination of stable isotope labeling, particularly from ^2H_2O, with tandem mass spectrometric analysis of mass isotopomer abundances in peptides, provides a powerful approach for characterizing the dynamics of proteins across the global proteome. Many applications in research and clinical medicine have been achieved and many others can be envisioned.

1. INTRODUCTION

1.1 Importance of Kinetic Measurements (Rate-Based Metrics) in Biochemical Systems

Flux rates of molecules through functionally important metabolic pathways are at the center of phenotype, physiologic adaptation, and disease pathogenesis (Hellerstein, 2003, 2004; Turner & Hellerstein, 2005). Kinetic processes can involve flux rates at the level of intermediary metabolites, proteins and other macromolecules, nucleic acids, or cells. The measurement of molecular flux rates, however, requires different analytic technologies and experimental approaches than are used for static measurements of concentrations or chemical composition in living systems. Measurements of molecular dynamics *in vivo* require the addition of a metabolic tag, or label, that perturbs the biologic system in a time-dependent manner.

There are two basic classes of kinetic measurements in complex living systems—targeted and unbiased (or screening). Here, we will focus on *in vivo* approaches that allow both targeted and unbiased kinetic measurements of protein dynamics, and therefore can be used at the level of the whole molecular network (e.g., proteome dynamics) or individual proteins. We will also briefly mention flux metabolomics/lipidomics. Our focus will be on describing underlying concepts and the technical procedures involved, experimental design considerations, interpretation of data, potential pitfalls, and strategies to ensure the most reliable kinetic results. An important element of this discussion will be on the practical aspects of applying these approaches to humans, for application to clinical medicine and drug development.

Characterization of the proteins present in a proteome provides a foundation for better understanding the complexities inherent in biology on a system-wide level. The proteome is spatially, temporally, and chemically dynamic. Mass spectrometry (MS)-based proteomics technologies and approaches allow for highly multiplexed, high-throughput characterization and quantification of hundreds to thousands of proteins in a biological sample providing a direct measure of the active components of the biological system under investigation (Angel et al., 2012; Baker et al., 2012; Zhang,

Fonslow, Shan, Baek, & Yates, 2013). The application of MS-based technologies for quantification of proteins present in a single sample has had a transformative effect on analytical protein biochemistry and has been widely explored for clinical applications in characterizing disease and personalizing medicine. Knowledge of proteins that are present and their relative concentrations, however, has not been sufficient for identification of diagnostic and prognostic biomarkers in many cases and has fallen short in facilitating an understanding of the often very complex biology under investigation (Hawkridge & Muddiman, 2009).

1.2 Tools for Measuring Fluxes Differ from Tools for Measuring Static Concentrations

Fluxes differ from static measurements in the same way that motion pictures differ from snapshots: the dimension of time is included. The tools for measuring biochemical dynamics are fundamentally different from the tools for measuring static abundances. Fluxes are usually measured using isotopes, because isotope-labeling studies generate asymmetry in the dimension of time—the isotope is not present, then it is. This feature allows the dimension of time to be introduced and thereby allows kinetic processes to be measured.

Measuring protein flux rates throughout a proteome enables a more complete understanding of the biological system under investigation. The synthesis, breakdown, transport, secretion, and storage of proteins had not been readily directly measurable, however, through traditional proteomic techniques. The combination of metabolic labeling with stable isotopes and MS-based proteomics constitutes a powerful pairing of approaches for the unbiased interrogation of proteome-wide protein dynamics in complex biological systems (Claydon & Beynon, 2012; Claydon, Thom, Hurst, & Beynon, 2012; Klionsky et al., 2012; Li et al., 2012; Price, Holmes, et al., 2012; Price, Khambatta, et al., 2012).

1.3 Functional Role of Dynamic Processes, Including Protein Kinetics, in Disease

In a general sense, it should be appreciated that the static levels of components of biochemical networks typically have limited functional significance or explanatory power over the initiation, progression, severity or reversal of disease, or the processes and adaptations that characterizes normal physiology and good health. The static levels of the components of any complex system, particularly when considered in isolation but even when taken as an ensemble, do not have intrinsic functional significance in the dynamic steady state that characterizes the living world. Rather, it is the dynamic flow of molecules through integrated, functionally important processes and pathways that underlies true phenotype in biology. Proteins stand at the center of biologic phenotype and adaptation, so these concepts are very relevant to the measurements of proteins within a proteome.

For example, no single expression level of a gene, protein concentration, metabolite pattern, cell type, etc., is likely to explain the progression of complex chronic diseases such as tissue fibrosis in lung, liver, kidney, or skin. But the dynamic process of collagen turnover (synthesis and breakdown) is very likely to explain accumulation of collagen in the extracellular matrix (fibrogenesis) to a high degree. Whereas a single gene or protein may exhibit a few percent control strength in the complex web that leads to fibrosis, the explanatory sufficiency of altered collagen synthesis and breakdown *a priori* has a higher likelihood of capturing the progression or reversal of tissue fibrosis (Decaris et al., 2014; Gardner et al., 2007).

1.4 Proteins as Metabolic Substrates

Proteins are generally discussed for their catalytic activities or structural functions—that is, their role in facilitating and controlling metabolic processes and providing physical structure in the organism. But proteins are also metabolic substrates in a variety of ways and have a rich dynamic life after leaving the ribosome. The production and degradation of proteins are catalyzed by other proteins and are dependent on a wide range of functional activities, from kinases, deacetylases, serine proteases, lysosomal or autophagosomal hydrolases, ubiquitin ligases, glycosyl transferases, and hydroxylases to microtubule-mediated vesicular transport proteins, endoplasmic reticulum chaperones, receptor-mediated endocytosis, and cell membrane transporters.

The systems involved in the metabolism of proteins represent a higher level in the hierarchy of metabolic control. A protein-metabolizing system can modulate a wide variety of functional processes, and disturbances in protein metabolic systems may play causal roles in many diseases. All of the following protein metabolic processes are of potential interest in health and disease, yet the dynamics of these systems are typically not well understood, particularly in humans.

1.4.1 Transport, Folding, Targeting, Translocation, and Secretion

Proteins are folded, sorted, and transported to different parts of the cell or are secreted from cells by membrane-associated systems. The Golgi and endoplasmic reticulum, vesicular transport along microtubule-mediated, motor-driven pathways, for example, are required for proper localization and function for many proteins. Kinetic measurements have revealed dysfunction of protein cargo transport systems in neurodegenerative diseases, for example (Fanara et al., 2012).

1.4.2 Posttranslational Modifications

A number of modifications occur on proteins after translation. Protein posttranslational modifications can be either static or dynamic and include glycosylation, limited proteolysis, hydroxylation (proline), carboxylation (glutamic acid), methylation, and proteolytic activation (e.g., amplifying cascades). Nonenzymatic modifications, such as glycation of lysines, can also be important. A different type of covalent modification of proteins is the transient conjugation reactions that are part of regulatory cycles. Phosphorylation/dephosphorylation by kinases and phosphatase is the classic example, but acetylation/deacetylation, palmitoylation, and O-glycosylation, to name a few, have recently gained considerable attention (Choudhary, Weinert, Nishida, Verdin, & Mann, 2014). The kinetics of acetylation/deacetylation of histones and other proteins can be measured and carry important information about gene regulation and signaling pathways, for example (Evertts et al., 2013).

1.4.3 Assembly into Organelles and Functional Complexes

Proteins can reside as parts of larger assemblages, such as subcellular organelles or multiprotein complexes. These assemblages may then be metabolized as units—e.g., mitophagy and mitochondrial biogenesis link the turnover of an organelle consisting of hundreds of proteins that may not otherwise share functional or regulatory features and are informative when measured as a group.

1.4.4 Proteolysis and Degradation

Dysfunction in protein processing resulting in the accumulation of misfolded proteins as well as reduced degradation and clearance, for example, is associated with many diseases, including Alzheimer's disease, Parkinson's disease, type 2 diabetes mellitus, cystic fibrosis, and many others

(Mawuenyega et al., 2010; Valastyan & Lindquist, 2014). The development of proteasome inhibitors for cancer treatment takes advantage of the toxicity of accumulated proteins. The operational efficiency of protein-degrading systems such as ubiquitin-mediated proteasome activity, lysosomal proteases, and autophagosomes is best represented by proteolytic flux, but this has not been directly measurable, and reliance on static markers such as autophagic protein intermediates is recognized to be problematic (Klionsky et al., 2012).

1.4.5 Aggregation/Deaggregation

Several very important structural proteins undergo noncovalent polymerization/depolymerization reactions *in vivo*, such as actin, tubulin, myosin, and many others. Proteins, such as collagen, assemble in more permanent polymeric structures. A number of important disease states are characterized by pathologic aggregates in cells or in tissues. Amyloid-β in Alzheimer disease, Huntington aggregates in Huntington's disease, synuclein aggregates in Parkinson's disease, and prion aggregates in Kuru and scrapie disease are examples (Jucker & Walker, 2013). The dynamics of these aggregation/deaggregation or precursor protein turnover processes may be critical to the pathogenesis, natural history, and treatments for these conditions, and in principle can be measured *in vivo* (Mawuenyega, Kasten, Sigurdson, & Bateman, 2013; Mawuenyega et al., 2010; Shankaran et al., 2014).

1.4.6 Endocytosis and Recycling

Extracellular proteins can be taken up into cells by the process of receptor-mediated endocytosis. Classic examples of disorders of endocytosis or receptor metabolism in human disease, such as the LDL receptor in familial hypercholesterolemia (Brown & Goldstein, 1986), would undoubtedly be joined by many others if tools were available for routine measurement.

1.5 Dynamic Proteomics

The approach that we refer to here as dynamic proteomics (Decaris et al., 2014; Price, Holmes, et al., 2012; Price, Khambatta, et al., 2012) addresses unmet analytical needs for proteome-wide quantification of protein fluxes (synthesis, degradation, transport, modification, etc.). Dynamic proteomics is based on *in vivo* stable isotope labeling coupled with MS analysis of biological molecules, and comprises the measurement of protein synthesis rates across a proteome in biologic systems following the administration *in vivo* of a stable isotope tracer (e.g., ^{2}H, ^{13}C, ^{15}N) (Fig. 1). Label

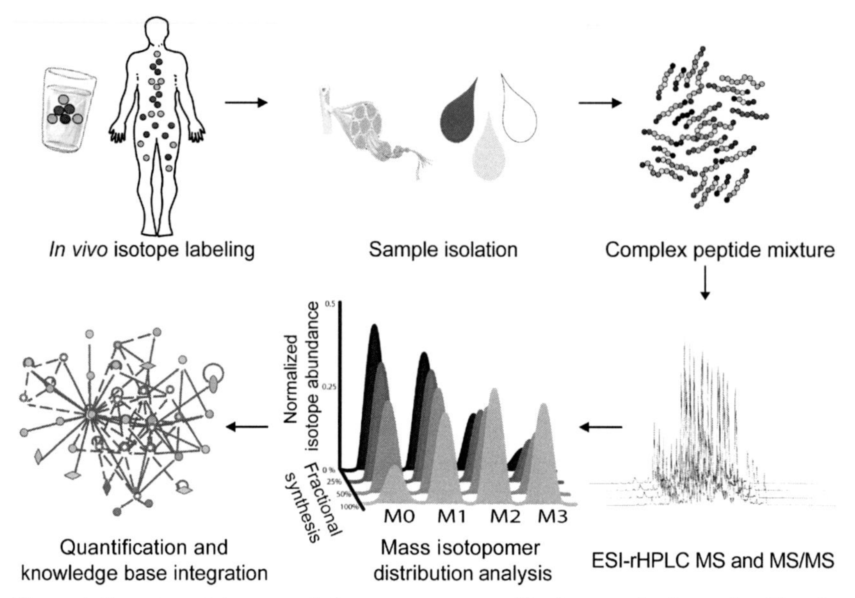

Figure 1 Experimental approach for measurement of *in vivo* protein dynamics. (See the color plate.)

incorporation into newly synthesized proteins is analyzed in proteolytic digests, using reversed-phase liquid chromatography coupled to tandem mass spectrometry (rHPLC–MS/MS) for peptide identification and rHLPC–MS for measurement of stable isotope enrichment. Quantification of the mass isotopomer distributions in peptides provides a means to estimate the fractional synthesis rate (FSR) of individual proteins over the labeling period from measured changes in isotopomer abundance of peptides (Hellerstein & Neese, 1999). Unlike traditional static proteomic techniques, this strategy provides information regarding which proteins are actively synthesized, degraded, transported, or otherwise dynamically altered during a defined time period.

Quantification of relative changes in protein pool size is routinely achieved by label-free quantitation or by addition of exogenous stable isotope-labeled standards, combined with MS-based proteomics (Ong et al., 2002). These quantitative proteomics approaches are often applied in cross-sectional studies capturing a snapshot of relative protein abundance profiles, but do not directly capture the dynamics of the biologic system. Incorporation of metabolic labeling with stable isotopes *in vivo* allows for the measurement of biological transients or flux, providing insights into the essential dynamic properties of a living system (Claydon & Beynon, 2012; Claydon et al., 2012; Klionsky et al., 2012; Li et al., 2012; Price, Holmes, et al., 2012; Price, Khambatta, et al., 2012).

The dynamic proteomics approach has a number of technical and operational advantages over static proteomics approaches, including the fact that changes in isotopomer abundance are not sensitive to differences in protein yield or recovery during sample processing. As such, these methods provide robust internally normalized analytical measurements. Applications of these techniques to preclinical and clinical studies will be discussed next.

2. FUNDAMENTALS: STABLE ISOTOPE-LABELING APPROACHES FOR MEASURING PROTEIN SYNTHESIS AND OTHER POLYMERIZATION BIOSYNTHESIS RATES

2.1 General Principles

The assembly and disassembly of polymers synthesized from repeating monomeric units is a central theme in biology. Such polymers may be as simple as fatty acids synthesized from acetyl-CoA units or as complex as proteins synthesized from amino acids or DNA made from nucleotides. Biological polymers may be homonuclear (defined as containing subunits that are identical), as in fatty acids, or heteronuclear (defined as containing more than one type of subunit), as in proteins or polynucleotides. Despite the importance of polymers in the chemistry of living systems, techniques for determining their rates of synthesis or breakdown have historically been unsatisfactory (Dietschy & Brown, 1974; Hetenyi, 1982; Srere, 1994). As a consequence, fields as wide ranging as lipid biosynthesis, protein metabolism, carbohydrate metabolic regulation, and control of cell proliferation have been severely limited.

Mass isotopomer distribution analysis (MIDA) is a technique based on analysis of combinatorial probabilities and the labeling patterns in intact polymers that provides a fundamental "equation for biosynthesis" (Fig. 2). MIDA was first presented as a systematic approach to polymerization biosynthesis over 20 years ago (Hellerstein, Christiansen, et al., 1991; Hellerstein, Kletke, Kaempfer, Wu, & Shackleton, 1991; Hellerstein & Neese, 1999; Neese et al., 1995; Papageorgopoulos, Caldwell, Shackleton, Schweingrubber, & Hellerstein, 1999). We will review here the theoretical and practical factors that must be taken into account if MIDA and related techniques are to be optimally applied to measure synthesis rates of proteins across the proteome.

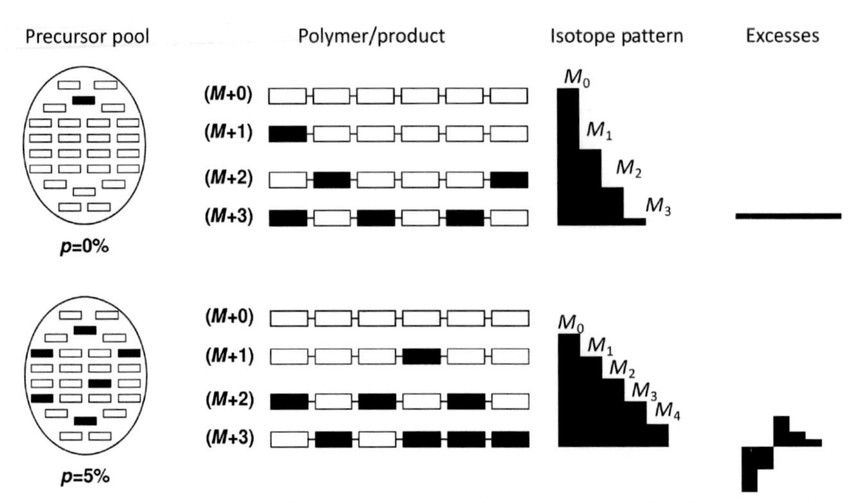

Figure 2 Stable isotope label incorporation in polymers at low and high p with resulting isotopomer relative abundances and excesses.

2.2 Terms and Rate Calculations for Dynamic Proteomics

p = measured precursor pool enrichment expressed as percent

n = number of nonlabile C – H positions per peptide exhibiting active metabolic incorporation of hydrogen (H) or deuterium (^{2}H or D) from body water

$$\%M_0 = \frac{M_0}{M_0 + M_1 + M_2 + M_3} \times 100$$

$\%M_{0\,(\text{baseline})}$ calculated by mass isotope distribution analysis (MIDA)

$$\text{EM}_{0t} = \text{EM}_0 \text{ at time } t = \%M_{0t\,(\text{measured})} - \%M_{0\,(\text{baseline})}$$

$$\text{EM}_0^* = \text{asymptotic EM}_0$$
$$= \text{max EM}_0 \text{ at specified } p \text{ (in newly synthesized peptide)}$$

$$f = \text{fractional synthesis at time } t$$
$$= \frac{\text{EM}_{0t}}{\text{EM}_0^*} \text{ (unitless, expressed as fraction or percent new)}$$

$$k = \text{FSR} = \frac{-\ln(1-f)}{t} \text{ (rate per unit time)}$$

FSR is calculated using the exponential rise to plateau equation of ($f = 1 - e^{-kt}$), where the rate constant k is synonymous with FSR, f is the measured fractional synthesis over the duration of label (t in days).

2.3 Definitions

The following definitions will be used here (Hellerstein & Neese, 1992).

Isotopes. Atoms with the same number of protons and hence of the same element but with different numbers of neutrons (e.g., H vs. D).

Exact mass. The mass calculated by summing the exact masses of all the isotopes in the formula of a molecule (e.g., 32.04847 for CH_3NHD).

Nominal mass. The integer mass obtained by rounding the exact mass of a molecule.

Isotopomers. Isotopic isomers or species that have identical elemental compositions but are constitutionally and/or stereochemically isomeric because of isotopic substitution, as for CH_3NH_2, CH_3NHD, and CH_2DNH_2.

Isotopologues. Isotopic homologues or molecular species that have identical elemental and chemical compositions but differ in isotopic content (e.g., CH_3NH_2 vs. CH_3NHD). Isotopologues are defined by their isotopic composition; therefore, each isotopologue has a unique exact mass but may not have a unique structure. An isotopologue usually comprises of a family of isotopic isomers (isotopomers) that differ by the location of the isotopes on the molecule (e.g., CH_3NHD and CH_2DNH_2 are the same isotopologue but are different isotopomers).

Mass isotopomer. A family of isotopic isomers that is grouped on the basis of nominal mass rather than isotopic composition. A mass isotopomer may comprise molecules of different isotopic compositions, unlike an isotopologue (e.g., CH_3NHD, $^{13}CH_3NH_2$, $CH_3{}^{15}NH_2$ are part of the same mass isotopomer but are different isotopologues). In operational terms, a mass isotopomer is a family of isotopologues that is not resolved by a mass spectrometer. For quadrupole mass spectrometers, this typically means that mass isotopomers are families of isotopologues that share a nominal mass. Thus, the isotopologues CH_3NH_2 and CH_3NHD differ in nominal mass and are distinguished as being different mass isotopomers, but the isotopologues CH_3NHD, CH_2DNH_2, $^{13}CH_3NH_2$, and $CH_3{}^{15}NH_2$ are all of the same nominal mass and hence are the same mass isotopomers. Each mass isotopomer is therefore typically composed of more than one isotopologue and is comprised of molecular species that differ in exact mass. The distinction between isotopologues and mass isotopomers is useful in practice, because all individual isotopologues are not resolved using quadrupole mass spectrometers and may not be resolved even by using mass spectrometers that produce higher mass resolution, so that for the interpretation of low-resolution mass spectrometric data one must consider the abundances of the mass isotopomers rather than isotopologues. The mass isotopomer lowest in mass is represented as M_0; for most organic molecules, this is the species containing all ^{12}C, ^{1}H, ^{16}O, ^{14}N, and the like. Other mass isotopomers are distinguished by their mass differences from M_0 (M_1, M_2, etc.). For a given mass isotopomer, the location or position of isotopes within the molecule is not specified and may vary (i.e., "positional isotopomers" are not distinguished).

Mass isotopomer pattern. A histogram of the abundances of the mass isotopomers of a molecule. Traditionally, the pattern is presented as percent relative abundances, where all of the abundances are normalized to that of the most abundant mass isotopomer; the most abundant isotopomer is said to be 100%. The preferred form for applications involving probability analysis, such as MIDA, however, is proportion or fractional abundance, where the fraction that each species contributes to the total abundance is used. The term isotope pattern is sometimes used in place of mass isotopomer pattern, although technically the former term applies only to the abundance pattern of isotopes in an element.

Monoisotopic mass. The exact mass of the molecular species that contains all ^{1}H, ^{12}C, ^{14}N, ^{16}O, ^{32}S, and the like. For isotopologues composed of C, H, N, O, P, S, F, Cl, Br, and I, the isotopic composition of the isotopologue with the lowest mass is unique and unambiguous, because the most abundant isotopes of these elements are also the lowest in mass. The monoisotopic mass is abbreviated as M_0, and the masses of other mass isotopomers are identified by their mass differences from M_0 (M_1, M_2, etc.).

Fractional abundances. The abundances of individual isotopes (for elements) or mass isotopomers (for molecules) given as the fraction of the total abundance represented by that particular isotope or mass isotopomer. This is distinguished from relative abundance, wherein the most abundant species is given the value 100 and all other species are normalized relative to 100 and expressed as percent relative abundance.

Isotopically perturbed. The state of an element or molecule that results from the explicit incorporation of an element or molecule with a distribution of isotopes that differs from the distribution found in nature, whether a naturally less abundant isotope is present in excess (enriched) or in deficit (depleted).

Monomer. A chemical unit that combines during the synthesis of a polymer and that is present two or more times in the polymer.

Polymer. A molecule synthesized from and containing two or more repeats of a monomer.

2.4 Biological Basis of Polymerization Biosynthesis Measurements

In a biological system, polymers that are newly synthesized mix into a pool that also contains preexisting polymer molecules. The goal of an isotope incorporation study is to quantify the fraction of molecules in the mixture that were newly synthesized during the label incorporation period (i.e., "what's new") and the rate at which the total pool of polymers is turning over. To determine the newly synthesized fraction (f) present in the mixture, one must first establish exactly the content of label in the population of newly synthesized polymers. Dilution of this labeled population by the population of preexisting, unlabeled molecules can then be determined, according to the precursor–product relationship (Hellerstein & Neese, 1992; Lee, Bergner, & Guo, 1992; Wolfe, 1984; Zilversmit, 1960).

Because there exists no purely physical technique for identifying in a mixed population of molecules which ones are new and which are not, the major practical difficulty has been establishing how much label is contained in the newly synthesized population of molecules. No classical extraction technique can reveal where different molecules in a population came from or how long they have been present. The biochemistry of the precursor–product relationship provides a possible solution, however, because the precursor pool of subunits in a cell has a physical reality and can in principle be characterized for its isotopic content (Hellerstein & Neese, 1999).

Serious problems arise, however, when surrogate monomer pools are used to represent the isotopic content of the true precursor pool (p): e.g., plasma amino acids or free intracellular amino acids to represent the tRNA-amino acid precursor pool for protein synthesis. Complicating factors deriving from subcellular or intracellular biochemical organization have been shown to affect every class of polymer so far examined in detail, including proteins (Waterlow, Garlick, & Millward, 1978), lipids, carbohydrates, and nucleic acids (Hentze, 1991).

2.5 The "Equation for Biosynthesis": Combinatorial Probabilities Are a Solution to the Precursor–Product Problem

MIDA is based on combinatorial probabilities (Hellerstein & Neese, 1992). Polymerization biosynthesis can be conceptualized as a combinatorial process, with monomeric subunits from a precursor pool combining into polymeric assemblages (Fig. 2). If the monomeric subunits are of more than one distinctive type, i.e., labeled and unlabeled with a stable isotope, then the population of assembled polymers will not be of uniform isotopic composition. The polymers will exist as distinguishable species containing varying numbers of the different types of isotopically labeled subunits.

Some species will include no labeled subunits, some will include one labeled subunit, some will contain two labeled subunits, and so on (Fig. 2). The relative proportion of each species of polymer is determined by and can be calculated from the binomial or multinomial expansion. The binomial expansion contains two variables, the number of subunits in the collection (n) and the probability (p) of each subunit being of a particular type. Because the number of subunits in a biological polymer is typically constant and known, the sole factor determining the relative proportions of each polymeric combination (i.e., the quantitative distribution of mass isotopomers) is p, the labeling probability in the precursor pool. The combinations of labeled and unlabeled subunits in the polymer population can be represented for mathematical analysis as the distribution of mass isotopomers. The population of intact polymeric assemblages therefore contains information about the precursor pool that is not available by analysis of the monomeric units in isolation. This is the central insight on which MIDA is based (Hellerstein & Neese, 1992). Because each mass isotopomeric distribution is uniquely determined by p at any given n, each distribution is characteristic of and capable of revealing the unique value of the precursor pool from which it was assembled. The distribution is, moreover, immutable; it is a fingerprint that will persist throughout the lifetime of the population, as long as there is no biological discrimination (no isotope effect) between species of the polymer and no remodeling of the core polymer after its original assembly. The combinatorial probability stable isotope-labeling approach can be applied to analysis of metabolic flux rates by use of different mathematic models, e.g., the isotopomer spectral analysis model developed by Kelleher and colleagues (Antoniewicz et al., 2007; Hiller, Metallo, & Stephanopoulos, 2011; Kelleher et al., 1994).

Mixing of a population of polymers assembled from a precursor pool of labeling probability p with a population of polymers assembled from an unlabeled precursor pool is what happens in a biological system when a labeling experiment is performed: newly synthesized polymers from the labeled pool mix with polymers that were present before the experiment began (i.e., there is dilution of the labeled polymer pool). A key mathematical feature of MIDA is that the relationships among those polymeric species that contain labeled subunits (the internal pattern among isotopomers) are unchanged by dilution from an unlabeled population of polymers.

2.6 Central Conceptual Features of Combinatorial Analysis (MIDA) Summarized

The first rule of MIDA is that there must be combinations or repeated subunits possible in the molecule analyzed. At least two repeats of a probabilistically identical subunit must be present. Metabolic pathways involving other kinds of chemical transformations but no polymerization are therefore not amenable to this combinatorial approach. Polymers studied must also be analyzed intact, or with at least two subunits present, because it is the distribution of isotopomeric species in a polymer that carries the essential information. Any maneuver that reduces the population to monomeric homogeneity, such as combustion to carbon dioxide for isotope ratio measurements or hydrolysis to monomeric subunits before analysis, results in the loss of the combinatorial information and precludes application of MIDA.

The second rule of MIDA is that subpopulations of molecules must be analytically distinguishable and quantifiable. Indeed, it is the measurable variations within a population of polymers that carry the information crucial for MIDA. The implicit notion is that subpopulations of monomeric precursors (some isotopically perturbed, some natural abundance) are present and result in identifiable subpopulations of polymeric products (some newly synthesized containing characteristic distributions of isotopomeric species and some preexisting containing a characteristic natural abundance distribution of isotopomeric species).

The analytic modality must therefore be capable of discriminating among different polymeric subpopulations. This is why radioisotopic methods cannot be used: specific activity is measured from the total counts and total mass of material present, treated as a uniform population; it is why average mass measurements by MS also cannot be used: a "centroid" average mass collapses all of the population variability in the polymer pool into a single value.

The third essential concept underlying MIDA is that dilution of the monomeric (precursor) and polymeric (product) pools affects abundance distributions differently. Both sources of dilution can alter the relative proportion of polymeric species containing no labeled subunits versus labeled subunits, but only dilution in the precursor pool can alter the internal quantitative relationships among labeled species (Hellerstein & Neese, 1992). It is this differential effect on "amount" (proportion of the polymer population containing any labeled subunits) versus "pattern" (relationships within the population of labeled polymers) that allows independent calculation of p and f, respectively.

2.7 Advantages of the Combinatorial Analysis (MIDA) Approach for Protein Synthesis

The combinatorial analysis, or MIDA, approach has a number of fundamental advances, in theory and in practice that apply to the measurement of protein synthesis and turnover rates *in vivo*.

- The isotopic content of the authentic biosynthetic precursor pool for any polymeric molecule—and therefore the isotopic content and composition of the newly synthesized population of molecules—is calculable without requiring any external measurements other than on the molecule itself. In the case of proteins, the true biosynthetic precursor pool is tRNA-bound amino acids at the site of ribosomal synthesis in a cell. No surrogate or extraneous measurements are needed, because the information about a molecule's biosynthetic precursor is contained in the internal labeling pattern of the molecule itself. This overcomes arguments and criticisms about possible biochemical pools and compartments (the "thing itself" is measured).
- The measurement comprises internal ratios in the molecules sampled. Accordingly, there is no effect of biochemical recovery, yield, concentration in the compartment analyzed, or differential sensitivity of instruments—all that matters is the relative abundances of mass isotopomers within the population of molecules analyzed.
- There is no radioactivity involved. Stable isotopes are safe for human use. Heavy water (^2H_2O) applications are ideal and almost universally applicable for combinatorial analysis of the major classes of biological polymers, including proteins (Busch et al., 2006; Claydon & Beynon, 2012; Claydon et al., 2012; Klionsky et al., 2012; Li et al., 2012; Price, Holmes, et al., 2012; Price, Khambatta, et al., 2012). Heavy water applications bring many practical advantages:

Labeling can be for very short periods (hours or days), because water extremely rapidly and freely equilibrates in all cells in living organisms, including humans; or labeling can be for very long periods (weeks or months). The half-life of body water in humans is 7–10 days (Shimamoto & Komiya, 2000) and subjects can take sips of heavy water every day as outpatients for periods as long as several months (Messmer et al., 2005; Price, Holmes, et al., 2012; Shankaran et al., 2014). Labeling does not need intravenous administration or medical supervision, which allows measurements to be performed under free-living, ambulatory real-life conditions.

- An independent metric of precursor pool labeling (body water enrichment) is easily measured and allows testing of the combinatorial model (see below), while permitting confirmation of compliance by the individual and being compatible with use under nonsteady-state experimental designs (e.g., ramp-up in labeling protocols; monitoring of the loss of labeled proteins after cessation of labeling—see below).
- Repeat labeling studies are not problematic, in principle, as the pattern of entry of new label can be modeled on top of preexisting label, and preexisting label can be subtracted (see below).
- In addition to measuring polymerization biosynthetic rates, other metabolic parameters can be calculated by use of combinatorial analysis. The calculation of n (number of labeling sites or monomeric units in the polymer) can be informative metabolically, when this varies rather than being a constant. If p can be independently determined, the calculated n may reveal relative contributions from the metabolic pathways contributing to a molecule. For example, the relative metabolic contributions from the glyceroneogenic versus glycolytic pathways to triose-phosphate pools in liver or adipose tissue are revealed by MIDA-calculated values of n (the number of exchanging H atoms) in the glycerol moiety of triacylglycerols (Chen et al., 2005). For proteins, if different values for n are present for certain AAs in different tissues (e.g., brain vs. peripheral tissues), this can provide evidence for the true biosynthetic site of proteins that may either be synthesized locally or synthesized elsewhere and imported (J. Price & K. Li, unpublished observations). This principle has powerful applications in flux metabolomics (see below).
- The timing of biosynthesis (date of birth) of a molecule can be established, when the precursor pool enrichment is time variant. For example, after cessation of ^2H_2O administration, the timing of biosynthesis of molecules sampled in a body fluid can be established by

comparing the labeling pattern in each molecule to the combinatorial patterns predicted from different points along the preceding body ^2H_2O time course. This technique has been key for CSF analyses of neuronal protein transport kinetics and has allowed investigators to establish the time for transport through axons to nerve terminals and into CSF (Fanara et al., 2012).

- One important feature of the principles underlying combinatorial probabilities in the equation for biosynthesis (MIDA) is that ratios of enriched mass isotopomer abundances for a molecule with a fixed n are a function only of p (independent of f), so that p can be inferred from isotopomer ratios regardless of the fraction of newly synthesized molecules that are present (Hellerstein & Neese, 1992).
- A corollary of this last point is that f can in principle be calculated from any isotopomer:

$$EM_3^*/EM_2^* = EM_3/EM_2 = \text{constant} \tag{1}$$

Rearranging Eq. 1:

$$EM_2/EM_2^0 = EM_3/EM_3^0 \tag{2}$$

This observation forms the basis for the filter criteria that ensure that the isotopomer pattern measured for a labeled peptide is what is expected for a given peptide (see below, filter criterion 8).

3. APPLICATION OF COMBINATORIAL ANALYSIS (MIDA) APPROACHES TO PROTEIN DYNAMICS

The focus of the remainder of this text will be on experimental considerations to ensure optimal use of combinatorial analysis (MIDA) approaches for the measurement of protein dynamics. We will first review the key factors that need to be taken into account, in terms of experimental design, analytic methods, and data interpretation. The operational details of the method will then be described.

3.1 General Protocol

The approach we developed for proteome-wide measurement of protein kinetics, as outlined in Fig. 1, begins with administration of heavy water to animal or human subjects, usually with an initial bolus to achieve rapid rise in body water enrichment followed by daily maintenance doses. Heavy water may be added to drinking water in animals or taken as daily sips (50–70 mL of water that is 70% enriched with ^{2}H) by human subjects, to maintain steady body water enrichment through time. We describe below the details of the labeling strategy, including the amount of label, the duration of label administration, and the timing of sample collection.

Heavy water is rapidly circulated throughout the body and is rapidly incorporated into free amino acids through biosynthetic reactions involving hydrogen exchange reactions with solvent water (Busch et al., 2006). Figure 3 illustrates the processes transferring deuterium from heavy water into covalent C–H bonds of an amino acid such as alanine. The number of C–H bonds incorporating deuterium varies among amino acids and ranges from 0 to 4 exchangeable hydrogen atoms in C–H bonds. Experimentally determined values for n of each amino acid are listed in Table 1.

In turn, proteins are constructed of amino acids and thereby carry their integrated isotope enrichment signature. Consequently, the isotope enrichment of each protein is encoded as a function of body water (precursor pool) enrichment, n of the total amino acid complement, and the FSR of the protein. We measure protein fractional synthesis by digesting the protein into tryptic peptides and analyzing isotopomer distributions of each peptide. Figure 4 illustrates how, as label is continually incorporated over a period of days, the isotopomer pattern of an example peptide will typically shift leading to decreased relative abundances of M_0 and M_1 isotopomers from

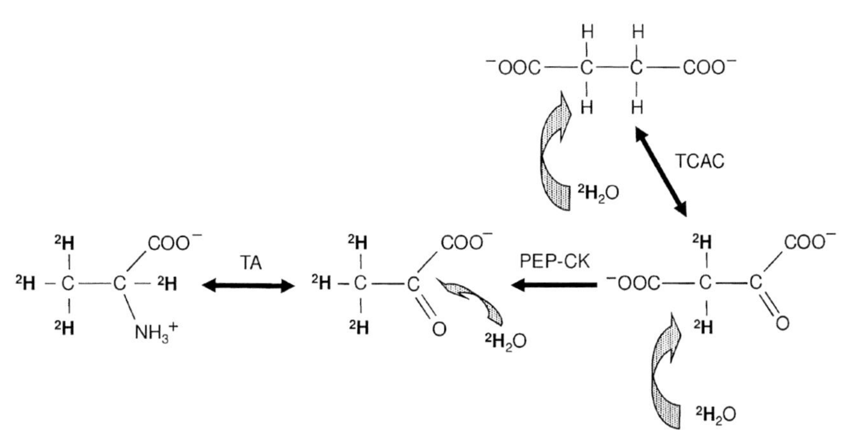

Figure 3 Example of ^{2}H incorporation into C–H bonds of alanine (n=4 C–H bonds labeled). TA, transaminase; PEP-CK, phosphoenolpyruvate carboxykinase; TCAC, tricarboxylic acid cycle. *Adapted from Busch et al. (2006).*

Table 1 Amino Acid *n* Values Used for Determining Parameters for Calculating EM_x^*.

Amino Acid	Abbreviation	Total H–C bonds	H_2O-derived H, *n*
Alanine	A	4	4
Arginine	R	7	3
Asparagine	N	3	2
Aspartate	D	3	2
Cysteine	C	3	2
Glutamate	E	5	4
Glutamine	Q	5	4
Glycine	G	2	2
Histidine	H	4	3
Isoleucine	I	10	1
Leucine	L	10	1
Lysine	K	9	1
Methionine	M	8	1
Phenylalanine	F	8	0
Proline	P	7	3
Serine	S	3	3
Threonine	T	5	0
Tyrosine	Y	7	0
Tryptophan	W	9	0
Valine	V	8	1

Values rounded to whole number.
Data adapted from a study with mice labeled with ^3H water (Commerford, Carsten, & Cronkite, 1983).

natural abundance values and increased relative abundances of M_2, M_3, and M_4 isotopomers. The degree of change in the relative abundance of each isotopomer from natural abundance values is a function of the molecular formula, the number of labeled hydrogen atoms (n), and *in vivo* metabolic label exposure (p), resulting in characteristic ratios of isotopomers in newly synthesized molecules that replace the preexisting natural abundance molecules (Eq. 1). This is illustrated in Fig. 5 which shows both positive and negative isotopomer excesses above or below baseline (nonisotopically enriched, or

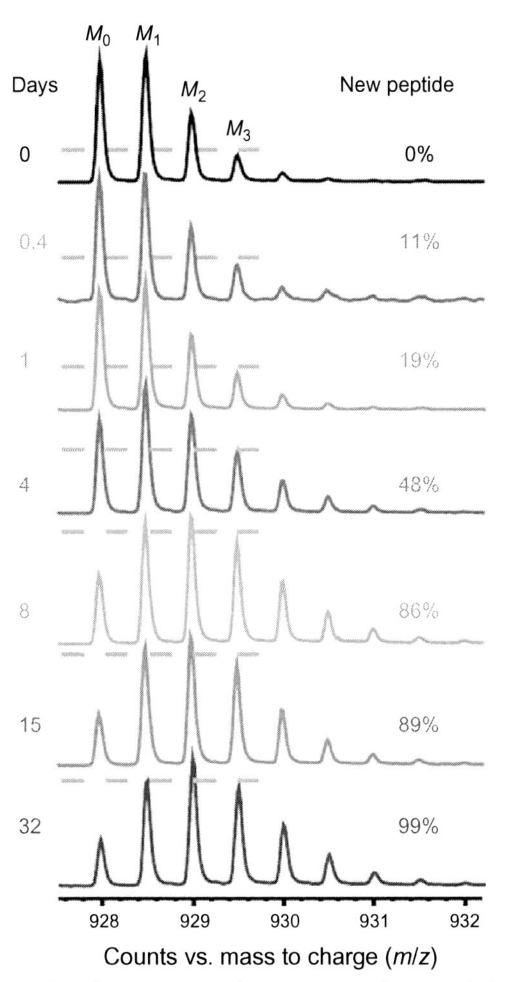

Figure 4 The relative abundances of peptide isotopomers (M_0 to M_4) change during metabolic labeling with ^2H_2O. The plot illustrates the change in isotope pattern of a tryptic peptide (FEDGDLTLYQSNAILR, $n = 29$) as it approaches 100% new during 32 days of continuous label. Isotope abundance relative to baseline (excess M_x or EM_x) changes as label is incorporated into this peptide; EM_0 and EM_1 become progressively more negative, while EM_2, EM_3, and EM_4 become more positive. *Adapted from Price, Khambatta, et al. (2012). © The American Society for Biochemistry and Molecular Biology.* (See the color plate.)

natural abundance) isotopomer abundances for 100% newly synthesized peptides over a range of precursor enrichments.

These combinatorial labeling assumptions are valid for biosynthesis at any moment in time, whether or not an isotopic steady state is present.

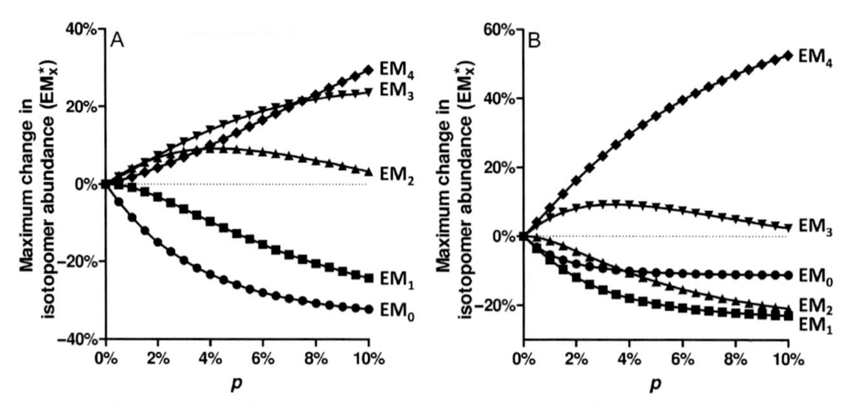

Figure 5 The maximum change in isotopomer abundance (EM_x^*) as a function of precursor pool enrichment (p) for two example peptides: (A) FEDGDLTLYQSNAILR (mass = 1853, n = 29) represents a typical tryptic peptide of average mass and n. (B) AAVAASGLNTMLEGNGQYTLLAPTNEAFEKIPSETLNR (mass = 3991, n = 79) represents a large peptide of high mass and n. Over the range of 1% p to 10% p, the preferred isotopomer for calculating f would be EM_0 for peptide A and EM_4 for peptide B.

Constant values of p do simplify calculations, but fractional synthesis can be measured accurately when p is not constant. Integrated biosynthesis rates over time can be calculated from average exposure to label in the precursor pool over the labeling period or, more formally, by computer modeling (Hellerstein & Neese, 1999; Messmer et al., 2005; Price, Khambatta, et al., 2012).

A key point to understand is that changes in relative abundances of isotopomers (EM_x) are unique to each peptide and are related to elemental composition and amino acid composition, specifically the total number of C–H bonds that can accommodate covalently bound deuterium derived from body water. For a tryptic peptide of intermediate mass and n (Fig. 5A), EM_0 typically displays the greatest divergence from zero over the range 0–10% p, and this is true of most tryptic peptides, except those with high masses (e.g., over ~3400). In contrast, for a large peptide with a mass ~4000 (Fig. 5B), EM_4 displays the greatest divergence from zero. For this reason, the choice of isotopomer for f calculations is often, but not always, EM_0. For simplicity, we will present our discussion of calculations based on EM_0.

Figure 6 shows the changes in isotopomer abundances of the example peptide (from Fig. 4) over a period of 32 days. Several observations are evident here. Fractional synthesis can be estimated from any isotopomer measured, but EM_1 has the least dynamic range and EM_0 has the greatest. Given adequate time, an isotopomer such as EM_0 will approach an asymptote and in this case

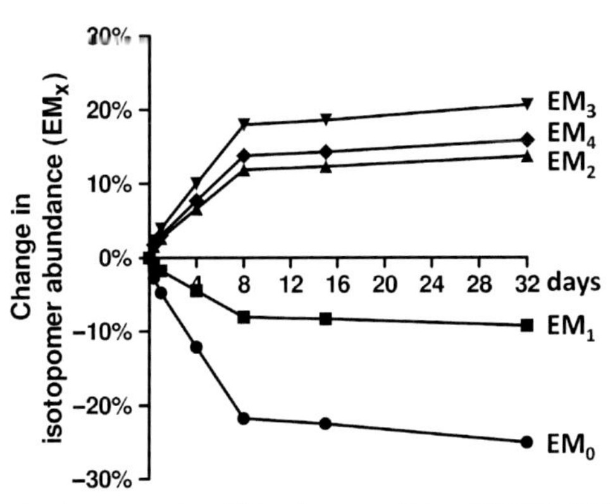

Figure 6 Measured changes in EM_x for peptide FEDGDLTLYQSNAILR ($n=29$) during 32 days of continuous label administration. The abundance of each mass isotopomer relative to baseline (EM_x) approaches an asymptote defined by the value at 100% replacement by newly synthesized proteins (EM_x^*). Prior to reaching the asymptote, the fraction of newly synthesized peptide may be calculated ($f = EM_x/EM_x^*$).

$EM_x^* = -0.253$. For this particular protein, the asymptote is almost reached by day 8 $\left(EM_0/EM_0^* = -0.217/-0.253 \text{ or } f = 86\% \text{ new}\right)$, meaning that reliable measurement of *f* will be optimal if samples are collected before ~8 days of continuous label. After that time, small changes in measured values of *f* will result in exaggerated changes in calculated rate constants (see below).

We are able to experimentally determine FSR of a protein by the following steps:

1. maintain relatively stable or measurable body water enrichments
2. sample at a specific time after initiation of heavy water administration
3. measure the precursor pool enrichment (p, ^2H_2O enrichment in body water) at that time
4. measure the mass isotopomer distributions of tryptic peptides derived from the protein
5. calculate the isotopic excess of the M_0 mass (expressed as EM_0) for each peptide (other targeted mass isotopomers can be monitored but EM_0 will be discussed here, for simplicity)
6. calculate the isotopic excess for 100% new peptide (expressed as EM_0^*) based on peptide molecular formula, peptide n, and p
7. divide measured EM_0 by the asymptotic EM_0^* to determine f (fraction new)
8. calculate FSR (k) from f and t (days of label), expressed as fraction new per day.

3.2 Analytical Requirements

1. Sufficient ion abundance is required for accurate quantitation of the relative abundances of mass isotopomers in molecules of interest. For screening (unbiased) approaches, accuracy must be sufficient in the "scan" mode or its equivalent (i.e., not targeted analyses). It should be recognized that quantitative accuracy and precision for mass isotopomer (or isotopologue) abundances has more stringent analytic demands than for simple detection of ions or their relative total abundances (Price, Khambatta, et al., 2012).
2. Signal to noise has to be adequate for reliable kinetic calculations. Isotopic signal is a function of the precursor pool enrichment achieved *in vivo* (p), the number of labeling sites (n in the selected peptide analyzed), and the biological turnover of the molecule targeted (f). Analytic performance in turn differs for different instruments, preparative methods, and biologic matrices. This is discussed in detail below.
3. Dynamic range for isotopomer signal is an important consideration especially in cases of low precursor pool enrichment. The isotopic enrichment or depletion of any mass isotopomer in a newly synthesized polymer (EM_x^*) represents mathematically the asymptotic value that is approached as newly synthesized molecules replace old ones. The lowest detectable threshold for EM_x^* should be set to $\sim 10\times$ the mass spectrometric instrument precision (SD) for the EM_x measurement. For example, if analytical precision is $\pm 0.5\%$ for an EM_x, then the minimum $|EM_x^*|$ should be 5%. Ideally, precursor pool enrichment should be great enough to generate $|EM_x^*|$ values well above the minimum. However, there are situations where $|EM_x^*|$ falls near or below the minimum, as may occur at early time points or in cases of slow FSRs. In these instances, it is useful to impose a filter (described below) to protein kinetic data sets that excludes measurements for peptides having insufficient $|EM_x^*|$ values for reliable quantitation.
4. Some knowledge about the biology and potential confounding variables can be helpful for protein kinetic analyses. For example, it is important to know *a priori* or to establish experimentally whether there is more than one biosynthetic pool contributing newly synthesized molecules to the compartment sampled; whether there is a quiescent, nondynamic pool (so that plateau labeling does not approach 100% new molecules); or whether there is a lag time for appearance of labeled molecules in the sampled compartment. Moreover, issues such as the molecular clearance

systems involved and the number of mixing compartments may have to be addressed in certain studies.

3.3 Other Factors to Take into Account for Biosynthesis Measurements

3.3.1 How Long to Label

The general time frame of turnover for the molecule(s) of interest determines the optimal labeling period. Too short label exposure period is problematic—not enough signal (label incorporation). Too long exposure is also a problem—once at plateau or close to it (e.g., if >~75% newly synthesized molecules are present in the population sampled), kinetic information from the transient can no longer be reliably calculated. Ideally, in a controlled experimental setting, molecular replacement (*f*) would be in the ~25–50% range, so that an increase or decrease in *f* can be reliably detected in the range of ~10–75% *f*. The duration of label and the timing of sample collection are particularly critical for measuring kinetics of intracellular or signaling molecules, which may turnover fully in seconds, minutes, or hours.

For *in vivo* measurement of protein kinetics, the length of the labeling period and the timing of sample collection are determined by the range of protein FSRs that can reasonably be measured (Fig. 7). There is, however,

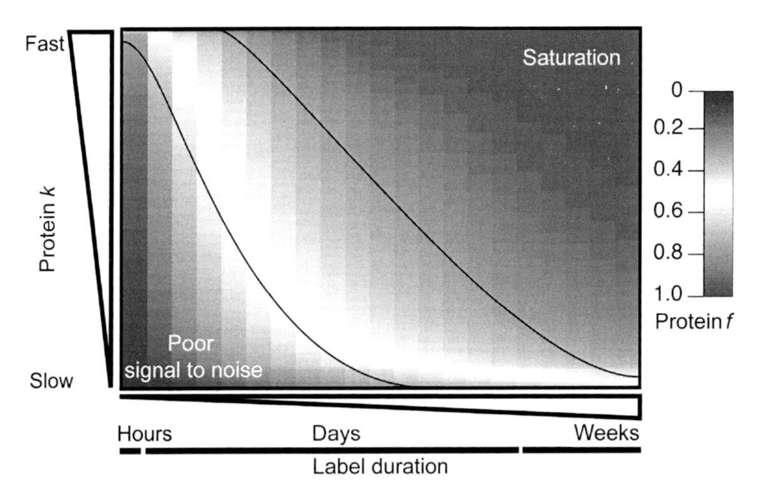

Figure 7 Depiction of the target range of *f* suitable for analysis by mass spectrometry. Slower turnover proteins (lower *k*) require more time to incorporate label to be measurable while faster turnover proteins (higher *k*) may become fully labeled after a certain duration of label exposure. The "sweet spot" is shown visually as the area between the two curves. (See the color plate.)

a wide range of FSR values in biological systems. For example, in plasma there are rapid turnover proteins such as VLDL apolipoprotein B (FSR ~6.6 pools d^{-1}, and slow turnover proteins such as serum albumin (FSR ~0.02 d^{-1} (Busch et al., 2006; Price, Holmes, et al., 2012)) and creatine kinase M-type (FSR ~0.01 d^{-1} Shankaran et al., 2014). A plasma sample collected after 7 days of continuous label would yield an interpretable FSR measurement for serum albumin, but the VLDL ApoB would be at or near 100% f after the first day of label and the creatine kinase-type M (CK-M) would have insufficient label incorporation to measure FSR with acceptable precision.

For first-time investigations of protein kinetics of a proteome that has not previously been studied, it is recommended to perform a pilot time course study, administering heavy water for various time periods, e.g., 7–14 days, and collecting samples at three or more time points. Collecting samples for a range of time points such as at 1, 3, 7, and 14 days can yield interpretable kinetics for proteins with FSRs (k values) ranging from ~0.01 to ~1.5 d^{-1}. Multiple time points are needed to determine whether proteins have reached a plateau in fractional synthesis during the time of label, which precludes kinetic inferences, as noted above. The observation of a plateau in f that is substantially less than 100% may indicate more than one biosynthetic source of the protein—i.e., release at different rates from more than one pool or from more than one tissue into the bloodstream.

Because, in our experience, the widest range of f values that can be measured with confidence in a single sample is ca. 10–75%; the lower bound for labeling is set by the analytical precision of the instrument and the upper bound is set by the maximum f value that can be interpreted as still approaching but not yet reaching a plateau (Table 2). Thus, the minimum number of days of label is set by the f of the slowest turnover (lowest FSR) proteins of interest. For example, a protein with a relatively slow FSR of 0.016 d^{-1} (see bottom row of Table 2) would require at least 6.4 days of label to reach 10% f (based on solving $f=1-e^{-kt}$ given $f=0.10$ and $k=0.016$ d^{-1}). However, at 6.4 days of label, the fastest turnover protein that could still be measured has an FSR of 0.25 d^{-1}. Samples would have to be collected at earlier time points to reliably measure FSR of proteins that turn over faster than 0.25 d^{-1}.

3.3.2 *"Virtual Biopsies"*

Because of the great sensitivity and dynamic range of mass spectrometers and their capacity to identify proteins in complex matrices, labeling kinetics

Table 2 Fractional Synthesis (f) in Relation to Fractional Synthesis Rate (FSR, or k) and Duration of Label Defined by the Equation $f = 1 - e^{-kt}$

	Duration of Continuous Label Administration (Days)										
FSR (d^{-1})	**0.1 (%)**	**0.2 (%)**	**0.4 (%)**	**0.8 (%)**	**1.6 (%)**	**3.2 (%)**	**6.4 (%)**	**13 (%)**	**26 (%)**	**51 (%)**	**102 (%)**
8	55	80	96	100	100	100	100	100	100	100	100
4	33	55	80	96	100	100	100	100	100	100	100
2	18	33	55	80	96	100	100	100	100	100	100
1	10	18	33	55	80	96	100	100	100	100	100
0.50	5	10	18	33	55	80	96	100	100	100	100
0.25	2	5	10	18	33	55	80	96	100	100	100
0.125	1	2	5	10	18	33	55	80	96	100	100
0.063	1	1	2	5	10	18	33	55	80	96	100
0.031	0	1	1	2	5	10	18	33	55	80	96
0.016	0	0	1	1	2	5	10	18	33	55	80

The minimum duration (days) of continuous label administration is determined by the slowest turnover protein of interest which yields an f above the threshold of detection of the instrument. The longest duration of label is determined by the fastest turnover protein of interest and by the upper limit of f which yields a reliable estimate of FSR. Values are f, expressed as percent new.

can often be measured in body fluids for proteins that escaped in trace amounts from tissues of interest. This approach has been termed a "virtual biopsy" of a tissue and is a powerful, minimally invasive strategy for development of clinical biomarkers (see below and Fanara et al., 2012). For example, we have shown that CK-M, a protein that is found predominantly (>95%) in skeletal muscle, can be identified in plasma in rodents and humans and that its labeling kinetics in plasma after heavy water administration are essentially identical to CK-M labeling from skeletal muscle samples (Shankaran et al., 2014). Moreover, the FSR of CK-M in plasma or muscle tissue correlates closely with FSRs of other cytosolic and structural proteins from muscle. Accordingly, measuring the labeling kinetics of plasma CK-M provides a minimally invasive "virtual biopsy" of skeletal muscle protein kinetics.

Similarly, two proteins that are part of the tissue fibrogenic pathway—lumican, which is involved in collagen fibril formation, and transforming growth factor-beta-induced protein, which is upregulated with collagen synthesis in fibrogenic states—can be detected and measured kinetically by LC/MS in plasma in human subjects after heavy water labeling (Emson, Decaris, Gatmaitan, Cattin, et al., 2014; Emson, Decaris, Gatmaitan, FlorCruz, et al., 2014). In patients with chronic liver disease due to hepatitis C infection or with fatty liver disease/nonalcoholic steatohepatitis who had concurrent liver biopsies, the FSRs of these two proteins in blood correlate closely with the FSR of liver collagens.

It is likely that other proteins in body fluids will also be identifiable that derive from and thereby reflect the protein dynamics in an inaccessible tissue of interest, as many proteins appear to leak from tissues into the circulation in trace amounts. It is worth noting that these minimally invasive markers of tissue protein kinetics are only possible because of the exquisite sensitivity of tandem MS for proteins in complex biologic matrices.

3.3.3 Repeat Labeling Protocols

It is important to be aware of, and be able to correct for, residual label in a molecule when repeated labeling protocols are used—as is often the case in longitudinal treatment studies (Fig. 8). The optimal approach is to obtain a repeat sample of tissue or body fluid sample prior to starting the second (repeat) label administration protocol. The residual label can then typically be subtracted from the new label incorporation pattern (see Fig. 8). Because a third tissue sample (biopsy) is often undesirable or impractical if a tissue biopsy is required, it is sometimes necessary to simulate the label present

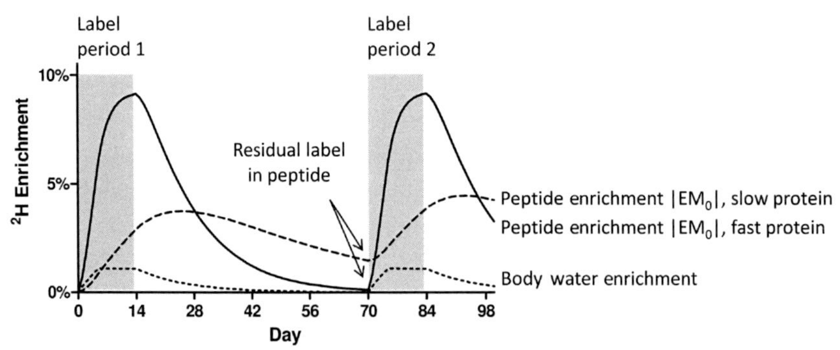

Figure 8 Time course of peptide enrichment for a repeated ^2H_2O labeling study design: fast and slow turnover proteins are shown for demonstration purposes. After cessation of the first heavy water labeling protocol, body water ^2H_2O enrichments and label retained in the fast-turnover protein die-away to baseline values by the start of the second labeling period 8 weeks later. Residual label can be corrected for by either modeling the protein delabeling curve or by taking a sample right before starting the second labeling protocol.

at the second baseline by modeling. The loss of labeled molecules since the end of the initial labeling period is calculated and the residual label must be subtracted from the label measured in the molecule after the second labeling protocol. To apply this modeling approach and avoid a third tissue sample, it is best to measure at least a subset of individuals *before* the second labeling period rather than after it, to validate the exponential decay model and calculations. These approaches are greatly helped by the availability of noninvasive techniques (e.g., see above, Section 3.3.2), as correction for any residual label can be carried out easily prior to repeat label administration protocols (see below).

An example is shown for a repeat labeling protocol, with measured or simulated second baselines (Fig. 8). The utility of a rapid turnover, preferably noninvasive protein kinetic marker, is apparent.

3.4 Calculations and Corrections to Ensure Optimal Kinetic Estimates

3.4.1 Establishing the Value for Expected Number of Monomeric Subunits (n, Number of Exchanging H–C Sites in Peptides)

Proteolytic peptides exhibit a value of n (the number of isotope incorporation sites) that is the sum of the incorporation sites in individual amino acids that comprise the peptide. The contribution to n of individual amino acids was determined in mouse by Commerford et al. (1983), who administered tritiated water to mice *in utero* and determined the extent of radioactivity incorporated into the individual protein-bound amino acids. We have confirmed the amino acid n values that Commerford et al. published for mouse proteins in two ways, both in humans and in rodents: by direct GC/MS measurement of deuterium enrichment in free amino acids and by MIDA calculation of total n in peptides following heavy water labeling experiments (Price, Guan, Burlingame, Prusiner, & Ghaemmaghami, 2010; Price, Holmes, et al., 2012). The findings in humans and in rodents by both approaches correlate well with the published values, supporting the use of the literature-derived component values for the number of labeled hydrogen atoms (n) incorporated into free amino acids (Table 1). In practice, investigators can use these published values or, if there is concern that an experimental system may have different underlying amino acid metabolism, use their own experimentally measured values of n (e.g., by best fit of labeling patterns in large numbers of peptides) for calculating the mass isotopomer pattern of proteolytic peptides from proteins synthesized during heavy water labeling experiments.

It is also possible, in principle, to calculate the value of n by MIDA based on the combinatorial pattern of labeling and the measured p (body water enrichment) for any peptide, when enrichments in more than one mass isotopomer are sufficient to meet analytic criteria for accuracy. Put differently, the mass isotopomer patterns measured from peptides after labeling can reveal their own n empirically. Lee et al. (1994) demonstrated elegantly that n can be determined experimentally by using combinatorial probability analysis as it is used for determining f. Instead of a reference table for p versus mass isotopomer pattern at a known value of n, one can generate a reference table for n versus mass isotopomer pattern at a known value of p. The true value of n can then be inferred from the experimental data. This technique is possible only when there exists an independent method for determining p; the measurement of body water ^2H enrichments during ^2H_2O incorporation experiments represents a situation that permits this application (Lee et al., 1994). Once the label enrichment p is measured for an experiment and the number of sites available for deuterium incorporation n is able to be estimated from experimental data on the mass isotopomer pattern of labeled tryptic peptides, the fraction of newly synthesized protein (f) can be computed. Experimental determination of n by MIDA can provide critical metabolic information when there are different pathways that lead to different values of ^2H incorporation from ^2H_2O in targeted metabolites (see below).

3.4.2 Analytic Accuracy and Precision for Measurement of Mass Isotopomer Abundances

The single most difficult problem facing the use of MIDA at present, both in theory and in practice, relates to quantitative accuracy of isotope ratio measurements, i.e., the analytic performance of mass spectrometers for relative abundance quantitation of mass isotopomers in molecules. MIDA is based on analysis of quantitative distributions in the context of a model of combinatorial probabilities. If the instrument generates inaccurate numbers, measured distributions will no longer reflect the actual isotopomeric patterns present. The true effect on kinetic estimates (p, f) will depend on the nature and extent of this experimental inaccuracy. Operational strategies to evaluate accuracy and precision of mass isotopomer abundance ratio measurements are discussed in sections below.

3.4.3 Constraints on ^2H_2O Administration and Factors to Consider in Designing Labeling Studies

The typical body water enrichment attained in biosynthetic measurements in humans is ~0.25–2.0% (Emson, Decaris, Gatmaitan, Cattin, et al., 2014;

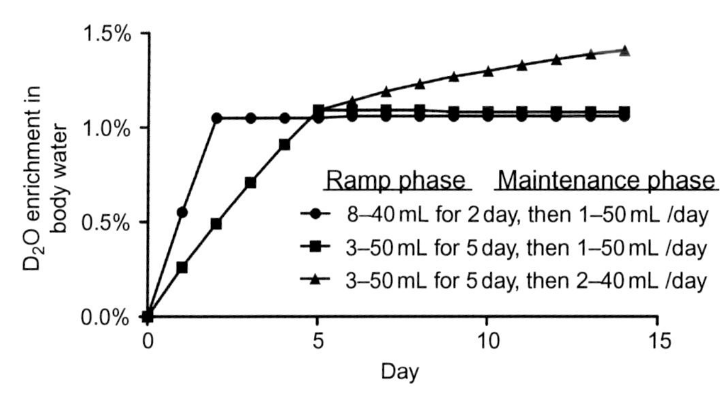

Figure 9 Time course of body water enrichments predicted in different inpatient (•) and outpatient (■,▲) protocols for administering heavy water. All loading protocols shown have a ramp phase and a maintenance phase, and differ in the number of doses of 70% ^2H_2O ingested per day, the volume of each dose, and the duration of each phase. The model assumes a typical male subject weighing 150 lb with a body water turnover rate of 8% a day.

Emson, Decaris, Gatmaitan, FlorCruz, et al., 2014; Hellerstein & Neese, 1999; Price, Holmes, et al., 2012; Shankaran et al., 2014). In animal models, body ^2H_2O enrichments can be 3.0–8.0% (Claydon & Beynon, 2012; Claydon et al., 2012; Klionsky et al., 2012; Li et al., 2012; Price, Holmes, et al., 2012; Price, Khambatta, et al., 2012). Examples of body water labeling protocols in humans are described in Fig. 9, which displays the predicted resulting body water enrichment trends over time for each protocol.

Biologic adverse effects in living systems are not apparent until body ^2H_2O levels are >15–20% (Jones & Leatherdale, 1991). The only toxicity limitation of heavy water administration at the levels that are typically achieved in people is the possibility of transient dizziness during the initial doses of ^2H_2O in Jones and Leatherdale (1991). This is believed to occur due to transient, small differences in water bulk flow properties between vestibuli in the inner ear during periods of changing body ^2H_2O levels, which is interpreted neurologically as motion in susceptible individuals. An estimated 1/30 people so exposed to rapid changes in body ^2H_2O enrichments can experience temporary dizziness or vertigo (Jones & Leatherdale, 1991), which resolves quickly. This side effect can be avoided by limiting the rate of ^2H_2O intake to avoid rapid changes in body ^2H_2O levels. In our experience in over 1500 subjects who have received ^2H_2O for these labeling protocols, there have been very few individuals who report side effects when each dose of ^2H_2O is less than 40 mL (e.g., 60 mL of 70% ^2H_2O) and when doses are taken at least 2–3 h apart.

This is an important experimental design feature for investigators using ^2H_2O labeling protocols in humans. When a rapid rise to plateau is desirable, we typically observe subjects in a metabolic ward or other medical settings and administer 40 mL ^2H_2O every 3 h.

3.4.4 *Impact of ^2H_2O Enrichment (p) Achieved* In Vivo *on Optimal EM_x^* to Measure in Peptides*

An important but subtle point about ^2H_2O labeling for polymers is worth noting. The relationship between mass isotopomer abundances (EM_x^*) and ^2H_2O is nonlinear and biphasic (Fig. 5; Hellerstein & Neese, 1999). For lower levels of isotope enrichment (p) or lower numbers of isotope incorporation sites (n), the largest change is isotopomer abundance occurs in EM_0. As enrichment increases or the n becomes large, however, the magnitude in change of EM_1^*, EM_2^*, EM_3^*, or EM_4^* can be larger than that of EM_0^* (Fig. 5).

The optimal isotopomer to select for FSR calculations depends on peptide molecular composition, summed n of amino acids, and the range of p used in an experiment. For human studies with 1–2% p and animal studies with 3–8% p, EM_0^* exhibits the largest change and is therefore optimal for most peptides, although other isotopomers may exhibit similar or greater changes for larger peptides with higher n. It is best to select the isotopomer having the widest dynamic range (e.g., most responsive to isotopic enrichment). This can be done using MIDA calculations to model each EM_x^* as a function of p, for any peptide, which our software program routinely calculates. Examples in Fig. 5 show that EM_0^* is optimal for an average tryptic peptide (mass = 1853, n = 29, Fig. 5A), whereas EM_4^* is the most responsive isotopomer for a large peptide (mass = 3991, n = 79, Fig. 5B). A larger EM_x^* as the denominator will result in smaller variations in the calculated f.

3.5 Optimal Design of ^2H_2O Labeling Experiments Based on Kinetic and Analytic Principles

3.5.1 *Minimal Duration and Level of Label Exposure (Also see Section 3.3.1)*

As noted above, instrument sensitivity and signal:noise considerations make higher enrichments of label in protein from ^2H_2O labeling generally an advantage.

Some simple calculations can help investigators in estimating analytic requirements relative to the precision and accuracy of their instrumental

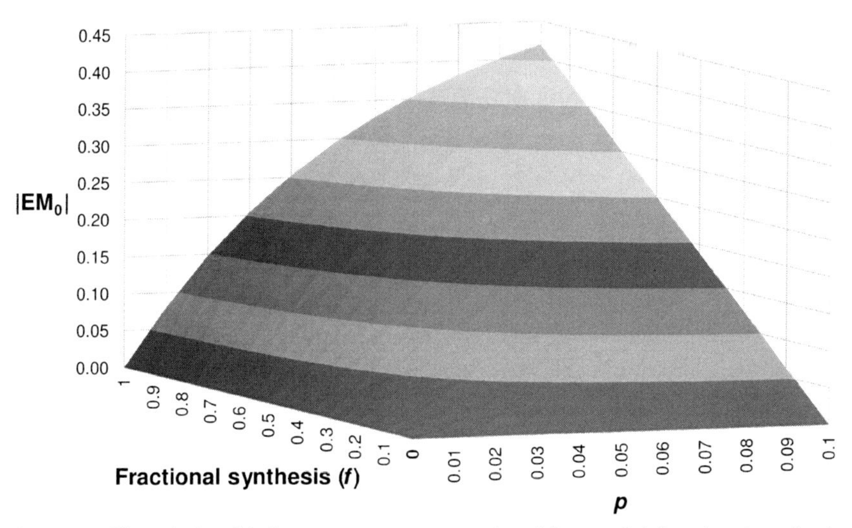

Figure 10 The relationship between precursor pool enrichment (p), fractional synthesis (f), and EM_0^* for peptide VLEDLRSGLF having mass = 1147, $n=17$. (See the color plate.)

analyses. An example of the relationship among p, f, and measured EM_0 is shown in Fig. 10 for a peptide with the sequence VLEDLRSGLF, molecular weight 1147.624, $n=17$. At a typical body water enrichment of ~1.0% in humans (Price, Holmes, et al., 2012), with an n of 30 for a peptide of mass 1500–2000, the optimal EM_x is typically EM_0 and the asymptotic enrichment achievable (EM_0^*) for a fully labeled peptide is ~10% (Price, Khambatta, et al., 2012). If analytic error is 0.2–0.5%, then it is desirable to achieve experimental values of EM_0 at least ~1.0%. Accordingly, the experimental goal for fractional synthesis values should be $f>10\%$ (i.e., $10\% f\times EM_x^*$ of $10\%=1\%$ measured EM_x). At $p=2\%$, EM_0^* is ~15%, so the goal can be $f>6\%$. These calculations can help investigators design duration of labeling protocols and desired body water ^2H_2O enrichments to achieve.

3.5.2 Maximal Duration of Label Exposure

Labeling for too long a period of time may be problematic. As a dynamic system that undergoes turnover approaches the state of being fully replaced with newly synthesized molecules, an asymptotic isotopic value is approached (Fig. 6), typically according to a rise-to-plateau exponential curve. As the plateau is approached, the change in label incorporation over any time period decreases and small differences between experimental observable values will cause large differences in kinetic estimates. The difference between 87.5%

labeled and 93.75% labeled, for example, reflects a full half-life of labeling—i.e., the same kinetic difference as between 0% and 50% labeled.

For this reason, f >75–80% are not optimal for kinetic experiments, and the experimental design of labeling studies should aim to avoid such high values by fitting the estimated values of k to the duration of label administration (Fig. 7, Table 2).

3.5.3 Calculation Model for ^2H_2O Exposure in Real-World Labeling Settings (When There Is Nonconstant ^2H_2O Exposure)

A labeling scenario in which a constant body water ^2H_2O enrichment is instantaneously achieved and maintained for the duration of the experiment is straightforward to interpret (Fig. 11A). While it is advisable to "front-load" heavy water intake in human subjects to approximate a bolus, a ramp in body water enrichment of heavy water often occurs over time that is not always negligible, and therefore constant body water enrichment cannot be assumed in most clinical scenarios. In such cases, the precise contribution from newly synthesized proteins to the perturbation of isotopomers in the biosynthetic protein pool is time varying—how much isotopic label a new protein contributes to the pool is dependent on the body water enrichment present at the time of protein synthesis. Conceptually, it is clear that the body water enrichment relevant to each protein is determined by the half-life of the protein. At the extreme cases, a protein pool that is extremely slow to turn over will contain proteins that were made at all labeling time points of the experiment, whereas for a rapidly turning over protein, recently synthesized proteins (synthesized at more recent body water enrichment levels) will dominate numerical representation within the pool. Thus, in the former case of slowly turning over proteins, a time-averaged heavy water enrichment can be used to approximate the peptide enrichment

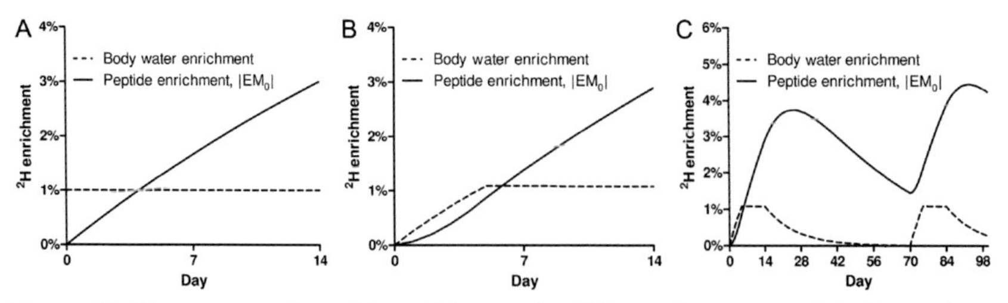

Figure 11 Time course of peptide enrichment for different heavy water labeling scenarios, for a representative tryptic peptide derived from a parent protein with FSR of 3% per day. Details described in text.

(EM_x^*) of the average newly synthesized peptide for the purpose of calculating f from experimental data. Conversely, very rapidly turned over pools have a reduced "memory" for p, because only recently synthesized proteins are retained within the pool at any given time. Therefore, the relevant precursor enrichment would be the body water enrichment values close to the day of sample collection, and for these proteins, the body water enrichment levels that had occurred previously can safely be ignored.

A globally applicable approach is to consider the time-varying nature of total protein enrichment within a simple two-part kinetic model. First, experimental measurements of body water enrichments are used to generate a continuous body water enrichment curve that varies over time, either by interpolation between measured time points or by evaluation in its own kinetic model of body water enrichments. For a given body water enrichment, the isotopic enrichment in a newly synthesized peptide ($|EM_x^*|$) can be calculated by the use of combinatorial probabilities (Hellerstein & Neese, 1999) based on n in the peptide (as described above). Next, the protein product pool is assumed to be homogeneous and at a quasi-steady state (explicitly meaning that the parameter's pool size and FSR do not vary in time over the time increment considered in the model). For a pool with pool size V, labeled peptides are introduced into the pool at a rate of $\text{FSR} \times V \times |EM_x^*|$, where $|EM_x^*|$ represents the enrichment of the newly synthesized peptide entering the population at each point in time (which, as a function of p, can be time varying). Concurrently, labeled peptides are degraded from the pool at a rate of $\text{FSR} \times V \times |EM_x^*|_t$, where $|EM_x^*|_t$ represents the average enrichment of the peptide population at any point in time. Then, the following mass balance of labeled peptides prevails for each time increment (delta_time).

$$\text{Delta}|EM_x|_t = \text{FSR} \times \text{delta\_time} \times \left(|EM_x^*| - |EM_x|_t\right) \qquad (3)$$

This kinetic model can be implemented in either SAAMII or Excel. The effect of a clinically common ramp in body water enrichment on the predicted peptide enrichment of a protein that turns over at 3% per day (CK-M, in this case) was simulated using this modeling approach (Fig. 11B).

The kinetic model for peptide enrichment described above can also be used to simulate the response to longitudinal study designs, in which two labeling periods of heavy water exposure are administered (Fig. 11C). In this design, the residual label in the peptide of interest dies away at the rate of turnover of the protein, and this exponential decay rate is distinct from the rate of body water turnover. In many cases, there will be residual label in proteins that needs to be considered and corrected for (either measured and subtracted or modeled and subtracted), if a repeat labeling protocol is to be performed (see Section 3.3.3).

3.6 Analytic Inaccuracy for Mass Isotopomer Abundances: Causes and Solutions

As noted above, a key requirement for applying combinatorial analysis methods to measurement of protein synthesis rates is that the mass isotopomer abundance ratios be not only precise (reproducible) but accurate—i.e., fit with theoretical distributions expected in natural abundance and isotopically perturbed states. Inaccuracy of peptide-level measurements can be caused by instrument bias. Some types of instruments are inherently more accurate than others, and it is important to test peptide isotope measurement accuracy using baseline unenriched as well as isotopically enriched standards. If an instrument yields poor accuracy but good precision, it may be possible to correct results using isotopically labeled standards, but this can be very complex, as there is no single labeled standard that can typically be synthesized to make a standard curve by combining with unlabeled material in known proportions. Each biosynthetic precursor pool enrichment achieved in a labeling study (p) will result in a different distribution of relative mass isotopomer abundances in target peptides. If relative abundances influence quantitation of ion signals in an isotopic envelope (Fagerquist, Hellerstein, Faubert, & Bertrand, 2001; Hellerstein & Neese, 1999), then a single labeled standard will not be adequate for correcting mass isotopomer distributions.

Inaccuracy of protein-level measurements can also result if there is high variability among the peptide-level kinetic measurements that are "rolled up" into protein-level kinetics. High standard deviations among peptide-level measurements could indicate unreliable protein measurements, if there are no biologic explanations for different kinetic behavior related to proteolytic processing, protein isoforms, or posttranslational modifications. High variability and poor accuracy may result from instrument factors relating to ionization as well as detection, particularly in complex matrices. Ion suppression or incomplete ionization may occur in the source and, importantly, there may be coeluting masses that overlap with the isotope envelope of some peptides. This may be indicated by much better reproducibility on replicate analyses of the same peptide and charge state (i.e., the same isotope envelope) than is observed for different peptides or charge states. In such cases, steps may be taken to reduce sample complexity, either through changes in peptide isolation (clean-up) or separation (chromatography).

A potentially interesting biologic issue should be noted in context of comparing or "rolling up" peptide-level measurements within a protein. Our general assumption has been that peptides within the same protein should turn over at the same rates, in the absence of known partial proteolytic processing (e.g., as occurs for certain neuropeptides or proteins that are involved in amplification cascades), since partial biosynthesis, repair and remodeling of intact proteins is not part of the canonical ribosomal protein biosynthesis pathway. For this reason, we typically use different peptides attributed to the same protein as replicate measurements for our kinetic calculations, and variations are treated as analytic noise. There could be exceptions to this assumption, however, if isoforms or partial proteolytic products of a parent protein exist with different turnover kinetics. It may prove interesting to explore this in detail and ask whether different proteolytic intermediates exhibit systematically different turnover rates for certain parent proteins. For some large "multimeric proteins" that we have observed (such as histones in nucleosomes), the turnover rates are clearly different. For other multimeric complexes, such as alpha subunits of collagen triple helices, the different subunits show essentially identical turnover rates *in vivo* and can be used as replicate values for collagen turnover (e.g., the alpha 1 or alpha 2 chains of type I collagen; Decaris et al., 2015; Decaris et al., 2014).

4. SAMPLE PROCESSING

There are no special requirements following heavy water labeling for collection, storage, or preservation of plasma, serum, urine, saliva, CSF, skin surface, or biopsies so long as samples are sufficiently handled to provide adequate peptide signal for detection. Importantly, there is no concern that hydrogen exchange will interfere with the deuterium label incorporated metabolically in peptides. There is no deuterium exchange between solvent water and C–H bonds in peptides *in vitro*. Unlike N–H and O–H bonds, C–H bonds are generally nonlabile under normal sample handling conditions. When incubated in physiologic solvents *ex vivo*, no exchange into or out of peptides in proteins is observed with solvent hydrogen atoms.

4.1 Fractionation Methods

Fractionation methods are particularly important for proteome dynamics because the depth of the proteome coverage for kinetic analysis is largely determined by sample handling prior to mass spectrometric analysis. Various options can be selected for, including gel electrophoresis, size exclusion, solubility (e.g., guanidine), centrifugation (e.g., lipoprotein fractions), SCX, high pH fractionation of peptides, immunoprecipitation, immunodepletion, and protein or peptide antibodies.

A typical workflow combining bottom-up proteomics methodologies with stable isotope labeling is shown in Figs. 1 and 12. A complex mixture of proteins is isolated from a biofluid such as blood, urine, or cerebrospinal fluid or a tissue biopsy. The measurement of proteome-wide FSRs requires deep analytical characterization of the biological sample under investigation. Thus, an important consideration when selecting sample preparation methods is the complexity and range of protein abundances within a sample. Biological samples that have a narrow range of protein abundances such as liver or brain benefit from the application of peptide-level fractionation methods that are orthogonal to reversed-phase liquid chromatography such as high pH and strong cation exchange methods (Schutzer et al., 2011; Wang et al., 2011). For biological sample types with asymmetric proteome abundance profiles such as blood or cerebrospinal fluid, undersampling during

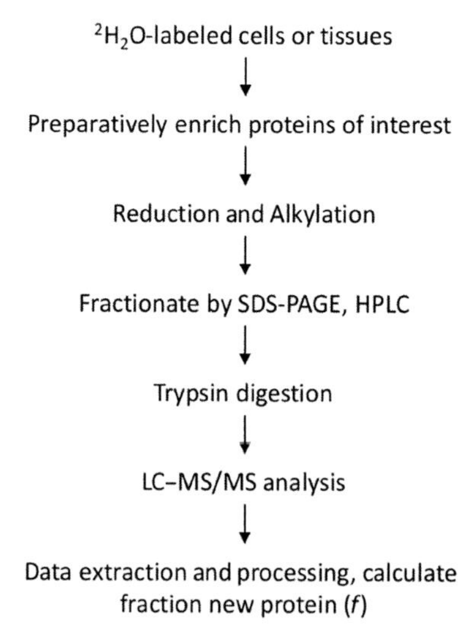

Figure 12 General sample preparation and analysis workflow for measuring protein kinetics.

LC–MS/MS analysis often results in great reduction in proteome coverage, which can effectively be mitigated by applying protein-level fractionation methods such as immunodepletion (Schutzer et al., 2011, 2013) and/or SDS-PAGE fractionation (Price, Holmes, et al., 2012).

In-solution digestion of isolated proteins is typically performed as follows: isolated protein is denatured and sulfhydryls are reduced following the addition of 50% TFE (99 + %; Sigma, cat# T6300-2) and DTT (>99 + %; Sigma, cat# D-5545). Cysteines are commonly alkylated by the addition of iodoacetamide (IAM 97%; Sigma, cat# I-670-9). Samples are then incubated at room temperature for 1 h in the dark before addition of 10 μL of 200 m*M* DTT stock solution to quench the remaining iodoacetamide. Proteins are digested following the addition of trypsin to the protein solution at a 1:50 ratio and incubated overnight at 37 °C. The digestion reaction is stopped by addition of 2–3 μL neat formic acid and dried by Speedvac followed by solubilization in 3% acetonitrile:97% water 0.1% TFA prior to LC–MS analysis.

5. ANALYTICAL METHODS

The approach that we have developed to analyze proteome-wide protein kinetics by LC–MS follows routine shotgun proteomics workflows, including protein fractionation and trypsin digestion or, alternatively, trypsin digestion followed by peptide fractionation. In either case, orthogonal methods are employed to improve breadth and depth of proteome coverage such that the number of fractions and the length of LC gradients do not exceed time and resource constraints of a laboratory or project.

5.1 MS Instrumentation

We have tested various MS detection modalities including quadrupole (by transitions of selected isotopomers), ion trap (by selective ion monitoring), and time of flight (TOF) (by MS-only) for peptide isotopomer measurements. Quadrupole instruments have sufficient isotope precision to detect isotope shifts of 1 Da or greater, such as within peptides incorporating D_3-leucine, for example, and isotope accuracy most likely will require standards. High-resolution ion trap instruments such as FT-ICR are capable of providing adequate isotope precision and accuracy for measuring peptide enrichments by selective ion monitoring, but have limited precision and accuracy in data-dependent mode (Erve, Gu, Wang, DeMaio, & Talaat, 2009; Mathur & O'Connor, 2009). There are potential applications of isotope analyses of daughter ions, such as immonium ions from amino acids, in MS/MS mode using ion traps (T. Angel, unpublished observations). We have found that TOF instruments operated in MS-only mode provide sufficient isotope accuracy and precision required to detect subtle changes in peptide isotopomer distributions achieved through body water enrichments of 1–5% (Claydon & Beynon, 2012; Claydon et al., 2012; Decaris et al., 2014; Klionsky et al., 2012; Li et al., 2012; Price, Holmes, et al., 2012; Price, Khambatta, et al., 2012).

5.2 Accuracy and Precision

Although TOF instruments can yield isotopomer precision of $\pm 0.2\%$ EM_0 under optimal conditions where peptide signal to noise is high and sample complexity is low (few coeluting species), in our experience, EM_0 precision decreases with increasing complexity of the matrix. Precision in complex matrices, where peptide signal to noise is lower, is approximately $\pm 0.5\%$. Accordingly, the analytical performance specifications we have determined for typical proteomics samples, using currently available commercial TOF instruments, are as follows:

1. Reproducibility of peptide EM_x: within-run and between-run SD from $\sim 0.3\%$ to 5%
2. Fractional synthesis precision from $\pm 1.5\%$ f at 5% p to $\pm 5\%$ f at 1% p

5.3 Limitations

Whatever the instrument, there are limits to the number of protein kinetic measurements that can be derived from MS analysis, and it may be substantially fewer than the number of proteins identified in a complex sample. In order to measure proteome-wide protein kinetics, we have developed data filters (details below) to remove peptide isotope clusters such as those with signals too low to overcome the effect of background noise or those with interference from coeluting species with overlapping masses. Data filtering of this type results in usable peptide isotope measurements for roughly 60% of the peptides identified for any set of sample fractions. Thus, some proteins which are identified may have few or no peptide measurements that pass filtering. This is especially true of the lowest abundance proteins in any sample fraction.

While an untargeted LC–MS approach is a useful research tool for screening potential protein biomarkers involved in disease pathways, it is ultimately a low-throughput approach and places a substantial burden on instrument time. Routine analysis of a small number of selected proteins (or peptides) of interest using a targeted approach requires more effort in sample preparation to isolate and purify targets, but makes more efficient use of instrument time as it requires fewer measurements per acquisition cycle. A high-throughput system suitable for clinical use might include multiplexed LCs for sample introduction at micro flow rates with minimal chromatography (see Section 6). Current limitations of TOF instrument sensitivity constrain the potential high-throughput applications to relatively high abundance proteins, but future MS developments could enable lower abundant targets in the future.

5.4 Overview of Method

The steps involved in measuring fractional synthesis (f) of proteins are outlined in Fig. 13. Precursor pool enrichment (body ^2H_2O enrichment, p) within each subject and at each time point is measured by GC–MS. Each protein/peptide fraction is analyzed by LC–MS, acquiring fragmentation spectra in MS/MS mode and precursor isotope spectra in MS mode. Depending on the instrument, MS/MS and MS scans may be performed in the same acquisition or in separate acquisitions requiring duplicate

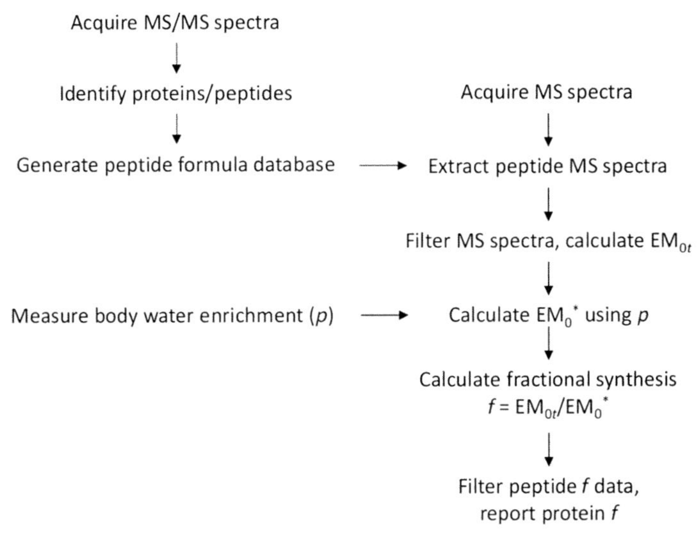

Figure 13 Sample analysis workflow for measuring protein kinetics. Measuring fractional synthesis (f) of proteins requires analysis of precursor pool enrichment (p), identification of proteins, and measurement of peptide isotopomer MS spectra.

injections. In either case, fragmentation spectra are searched to identify proteins. A nonredundant peptide formula and retention time database is generated and is used to extract peptide isotopomer MS spectra from the same MS/MS acquisition or a separate MS-only acquisition. Raw extracted MS spectra are filtered, and EM_x values are calculated for each peptide charge state observed. EM_x^* values are calculated for each peptide using precursor pool enrichment (p), as measured by GC–MS. Fractional synthesis (f) is calculated for each charge state of each peptide. Peptide f data pass through additional filters and f results are rolled up to the protein level. Details of these steps are described below.

5.5 GC–MS Analysis of Precursor Pool Isotope Enrichment

5.5.1 Body Water Enrichment

Body water precursor pool enrichment is measured in blood, serum, or saliva samples collected at each time point using the method described by McCabe et al. (2006) in which water is distilled from a fluid sample and transferred to acetone which is analyzed by GC–MS. Approximately 125 μL of serum, plasma, or saliva is pipetted into the center of a vial cap (E&K Scientific, 440011-N) held inverted and sealed with 2 mL polypropylene in a microcentrifuge tube (E&K Scientific, 640201). Tubes are incubated 12–24 h with caps submerged in a bed of glass beads at 80 °C with the ends of the tubes exposed to room temperature air circulating within a fumigation hood. Caps are replaced and tubes are spun briefly in a mini centrifuge. The distilled water samples are transferred to flip-cap microcentrifuge tubes (Corning Inc., 3620). A set of 10 water isotope standards ranging from 0 to 10 mol% is prepared by proportional mixing of pure H_2O with 99.9% D_2O (Sigma, 151882). To each sample and standard, 1 μL of 10 *M* NaOH and 5 μL HPLC grade acetone are added. After vortexing and 12–24-h incubation at room temperature, 300 μL HPLC grade hexane is added to each vial, which is immediately capped and vortexed for 10 s. A water droplet will appear below the solvent. 250 μL of the supernatant is transferred to a microcentrifuge tube containing ~100 μL sodium sulfate powder, capped immediately, and vortexed. After sulfate crystals settle, ~200 μL hexane is transferred to crimp top GC vial with fused glass insert (Sun-Sri, 500854) and capped (National Scientific, C4011-1A). Samples and standards are stored at 4 °C until GC–MS within a 1–2 days. GC–MS analysis is performed with an EI source, He carrier gas, and 30 m GC column (Restek, RTX-225, 0.32 mm ID × 0.1 μm film thickness) with acquisition parameters: split 20:1, inlet 250 °C, column flow 2.6 mL/min, column pressure ~8.75 psi, oven temp 70 °C (hold 2 min), ramp 60 °C/min to 220 °C (hold 1 min), runtime 5.5 min, source 230 °C, quad 150 °C, SIM ions m/z 58, 59, 60 (M_0, M_1, and M_2), dwell 15 ms.

5.6 LC–MS Analysis of Peptide Isotope Enrichment

5.6.1 MS Acquisition

Acquisition and data processing parameters will depend on the instrument used. Here, we describe details for using an Agilent 6550 Q-TOF with 1260 Chip Cube nano ESI source and Polaris HR chip. Fairly sharp peaks and consistent injection-to-injection retention times of ±12–18 s are achieved with Polaris chips using nano LC flow rates of 300–350 nL/min and LC gradients from 30 to 60 min. Longer gradients result in peak broadening and increased retention time variation which can negatively impact isotope precision. Typical acquisition parameters are used for data-dependent MS/MS analysis (6 Hz MS and 4 Hz MS/MS with up to 20 precursors per cycle). A slower acquisition rate (0.6 Hz MS) is used for a duplicate MS-only analysis to provide ~10× signal intensity relative to MS/MS analysis. An important consideration for data-dependent MS/MS analysis of isotopically enriched peptides is precursor selection. With the Agilent Acquisition software, we found that setting the isotope model to "unbiased" enabled proper selection of M_0 precursors, whereas setting it to "peptide" caused erroneous selection of M_1 precursors resulting in far fewer database hits, particularly for samples with precursor enrichments over 2%.

5.6.2 Peptide Identification

MS/MS spectra acquired with the Agilent Q-TOF are extracted and searched against the appropriate UniProt/SwissProt database using a proteomics search engine. Recommendations for Agilent Spectrum Mill Proteomics Workbench software are described:

1. Data file extraction parameters: scans with same precursor mass merged by spectral similarity within tolerances (retention time ±10 s, mass ±1.4 m/z), precursor minimum MS1 S/N = 10, ^{12}C precursor m/z assigned during extraction.
2. Search parameters: maximum number of missed cleavages = 2, minimum matched peak intensity = 30%, precursor mass tolerance = 10 ppm, product mass tolerance = 30 ppm, minimum number of detected peaks = 4, maximum precursor charge = 3.
3. Validation: search results are validated at the peptide and protein levels with a global false discovery rate of 1%.

4. Peptide list: for proteins with scores >11, a list of peptides with scores >6 and scored peak intensities >50% is exported from Spectrum Mill.

5.6.3 Extraction of Peptide MS Spectra

Peptide lists resulting from multiple MS/MS analyses are collated and condensed to produce one nonredundant peptide database per sample fraction. A script is used to generate peptide molecular formulas from identified sequences and to calculate peptide n and curve fit parameters for determining EM_0^* for each peptide from p (precursor enrichment). Nonspecific peptide IDs (e.g., multiple sequences matched to the same mass and retention time) and peptides having fewer than 8 amino acids or n below 15 are removed. The completed peptide database containing peptide molecular formulas, mass, and retention time is used to extract MS spectra from corresponding MS-only acquisition files using the Find-by-Formula algorithm in MassHunter Qualitative Analysis software. MS spectra are extracted with these parameters: EIC integration by Agile integrator, peak height >10,000 counts, unbiased isotope model, isotope peak spacing tolerance = 0.0025 m/z plus 12.0 ppm, mass and retention time matches required, mass match tolerance = ±12 ppm, retention time match tolerance = ±0.6 min, charge states $z = +2$ to +4, chromatogram extraction = ±12 ppm, EIC extraction limit around expected retention time = ±0.6 min. Compound reports containing raw peptide MS data and sample coding information are generated for each MS data file.

5.6.4 Calculations and Data Filtering

Compound reports generated by MassHunter Qualitative Analysis software are batch processed with a script which performs the following:

1. Collect isotope clusters (M_0–M_3 for peptide mass ≤2400 Da, M_0–M_4 for peptide mass >2400 Da).
2. Remove low abundance isotope clusters regardless of isotope enrichment. Remove if summed M_0–M_3 abundances <30,000.
3. Apply isotope pattern filter to remove isotope clusters with obvious interference in the form of erroneous high abundant masses in the isotope envelope. Remove if M_{diff} = Abs (M_x–M_{x+1}) >0.5 for mass <900 Da, >0.35 for mass 900–1100 Da, >0.25 for mass 1100–1800 Da, >0.2 for mass >1800 Da.

4. Calculate EM_x, EM_x^*, EM_x/EM_x^* for each mass isotopomer, for $x=0$, 1, 2, 3, 4.
5. Calculate protein-level f and peptide standard deviation of peptide-level f's within each protein.
6. Remove peptide-level f value if base peak abundance <20,000.
7. Remove peptide-level f value if $EM_0^* < -0.05$.
8. Remove peptide-level f value if variance among f's of usable isotopomers (EM_x/EM_x^* for $x=0$, 1, 2, 3, 4) >0.2.
9. Remove peptide-level f if value falls outside ±2 standard deviations of mean of peptide-level f's within the protein.
10. Report protein f data in a table collating results for all sample fractions of an experiment with proteins in rows and subjects, treatments, or time points organized into columns.

5.6.5 Data Review

After data filtering, protein f results should be inspected for the following:

1. Protein f estimates from few peptides. A lower limit of peptide-level observations may be set to two or more. Protein f measurements from few peptides may be viewed with less confidence than those from more peptides. This may exclude data for many of the lowest abundant proteins in sample fractions.
2. Agreement among peptides within each protein. An upper limit of SD (among usable peptide f measurements) may be set to ~20%.
3. Outliers. Some erroneous protein f measurements may pass data filtering but yield results far outside the expected range of values. Reviewing peptide-level results may reveal one or more data points skewing the mean high or low. Such values may result from false identifications or interfering masses overlapping the peptide isotope envelope.
4. Trends. Time course data and replicate injections, though costly, are useful for revealing noisy/unreliable peptide measurements that may pass data filtering.

5.6.6 Calculating FSR

FSR (or k) is calculated using Eq. (4).

$$k = \frac{\ln(1-f)}{t} \tag{4}$$

5.7 Data Analysis and Interpretation

5.7.1 Bioinformatics—Network, Pathway, and Ontology Analysis

Integration of information regarding the biological context of any given protein or set of proteins can greatly facilitate interpretation of complex, proteome-wide data sets, where many proteins are characterized and quantified in parallel. There are many publically available tools enabling the integration of metadata present in the public domain such as ontology terms (Gene Ontology, pathway, and subcellular localization terms), accessed from public repositories such as the Database for Annotation, Visualization and Integrated Discovery (DAVID) (Jiao et al., 2012), Kyoto Encyclopedia of Genes and Genomes (KEGG) (Kanehisa et al., 2014), the Universal Protein Resource (UniProt) (Jain et al., 2009), the Protein Analysis Through Evolutionary Relationships Classification System (PANTHER) (Mi, Muruganujan, & Thomas, 2013), and pathway and network tools such as STRING DB (Szklarczyk et al., 2011). Typically, ontological terms are retrieved from the selected database(s) based on an initial search of identified proteins, and numerical calculations (mean/median, standard deviation, and number of matching proteins) and statistical tests are performed per term, using corresponding protein fractional synthesis data.

5.8 Use of Other Stable Isotope Precursors

Using $^{2}H_2O$ for metabolic labeling in studies of protein kinetics has advantages but also limitations and may not be ideal in all circumstances. For example, measuring rapid turnover proteins with FSRs in the range of 4 d^{-1} or greater (i.e., half-lives <3 h) requires rapid increase in precursor pool enrichment for optimal calculations. This can be accomplished through administration of heavy water but other alternatives can also be useful. One alternative approach is to infuse a stable isotope-labeled amino acid (e.g., ^{13}C-, ^{15}N-, or ^{2}H-labeled) which can quickly distribute throughout the body and become incorporated into newly synthesized proteins. Infusion of ^{13}C- or ^{2}H-labeled leucine has been used in many studies of apolipoprotein synthesis rates in humans which tracked the incorporation of label into apolipoproteins, for example, hourly for periods of 12–24 h (e.g., Parks, Krauss, Christiansen, Neese, & Hellerstein, 1999; Welty, Lichtenstein, Barrett, Dolnikowski, & Schaefer, 2004). In these studies, kinetics of individual proteins can be measured by physical purification and digestion to amino acids followed by GC/MS analysis of isotope enrichment of leucine or other labeled amino acids, or measured by LC/MS analysis of intact peptides that contain the labeled amino acid (Fig. 14).

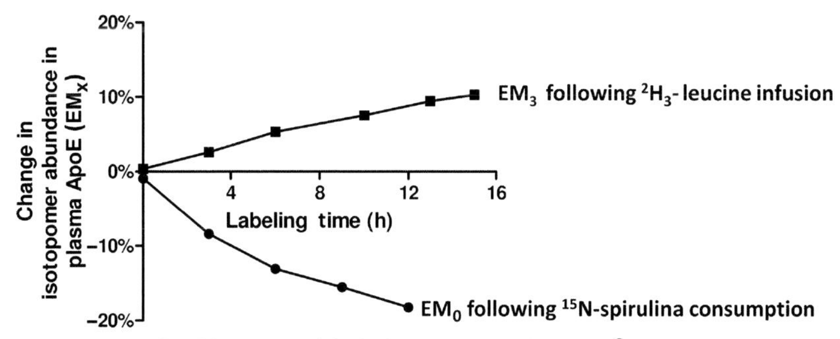

Figure 14 Use of stable isotope-labeled precursors other than ^2H_2O. Shown here is isotopic labeling of plasma apolipoprotein E in two human subjects. Following 16-h infusion with $^2H_3$-leucine (■) or 12-h oral consumption of ^{15}N spirulina (•), the changes in M_3 or M_0, respectively, in peptide SWFEPLVEDMQR were quantified by LC–MS.

Another alternative labeling approach is the use of ^{15}N-labeled amino acids. Algae such as spirulina can be grown on pure ^{15}N-labeled substrate and purified to produce a food-based isotope label with all amino acids fully (~99%) enriched in ^{15}N. Such material is now available through chemical supply companies. This approach has been applied for measuring proteome dynamics in the mouse (Price et al., 2010; Price, Holmes, et al., 2012). In humans, this approach may have the advantage of rapid label distribution throughout the body and rapid stabilization of precursor pool enrichment for short-term labeling studies. Initial testing of this approach in humans has shown that both the rise to plateau and label washout are rapid (J. Price et al., unpublished). Another advantage is that ^{15}N amino acids permit higher levels of precursor pool enrichment than is possible with ^2H_2O.

5.9 Calculation of *p* and *f* When the Complete Ion Spectrum Is Not Sampled

If the complete distribution of mass isotopomers is not sampled, some bias in calculation of *f* can be introduced (though values calculated for *p* remain valid Hellerstein & Neese, 1999; Price, Khambatta, et al., 2012). For any number of reasons, all of the ions in a mass isotopomer envelope (e.g., for convenience, to maximize dwell time on the most abundant ions or to avoid contaminating ions) may not be monitored. An important property of combinatorial probabilities is that the calculation of *p* is not affected by the ions selected for monitoring, as the internal pattern and relationships among excess mass isotopomers are fixed and characteristic of *p* and *n* regardless of which particular masses are monitored.

Surprisingly, this is not the case for calculation of *f*, which is affected by incomplete ion spectrum sampling, because of a mathematical feature of mixtures of numerical distributions (e.g., populations of mass isotopomers): dilution is not linear when the proportion of the total population monitored is different in the natural abundance and enriched populations (Hellerstein & Neese, 1999). Stated simply, when higher masses are not monitored, a mole of isotopically enriched molecules will contribute fewer ions to the total spectrum sampled than a mole of natural abundance molecules. The molecular mixture is thereby weighted in favor of the more completely sampled molecular population (the unlabeled population), and a correction has to be made to put equal weight on each molecular population in the mixture. The solutions to this confounding factor are straightforward: either monitor an essentially complete ion spectrum for the molecules under consideration or apply an algebraic correction for instances of significantly incomplete ion spectrum monitoring. This equation corrects for the proportion of ions monitored at the measured value of p present relative to the proportion monitored in unlabeled molecules (Hellerstein & Neese, 1999). By use of this correction factor, mixtures of labeled and unlabeled molecules are again reduced to linear combinations of mass isotopomers, from which dilution of molecules can be calculated simply. This correction is generally small and has no practical impact on most calculations because >98% of the ions within an isotopomeric envelope can typically be monitored for most labeled molecules.

5.10 Combining Quantitative and Kinetic Measurements

5.10.1 Calculation of Fractional Versus Absolute Flux Rates

1. *Label-free methods*: Quantitative measurements of protein abundance can be made with or without the use of protein or peptide labeling. Several *in vitro* and *in vivo* labeling techniques have been developed for MS-based protein quantification, typically building on bottom-up proteome profiling strategies (Angel et al., 2012). Determination of protein pool size by MS provides complementary information to protein fractional synthesis measurements. Label-free protein quantification, SILAC/SILAM, iTRAQ, and TMT labeling or enzymatic labeling with ^{18}O can be combined with and are compatible with metabolic labeling with heavy water and subsequent measurement of protein fractional synthesis. Briefly, label-free protein quantification can be accomplished following extraction of the total ion current of each isotope cluster/pattern for peptides identified in tandem MS/MS analyses from MS datasets. Extracted ion

abundances for all datasets are typically mean centered to one another to remove technical bias and relative protein abundances can then be inferred commonly from the median peptide abundance considering the top 50% most abundant peptides attributed to a protein.

2. *Isotopic labeling methods*: Alternatively, isotopic labeling of proteins for pool size quantification can be accomplished by the addition of a stable isotope precursor (such as ^{15}N-, ^{13}C-, ^{2}H-labeled amino acids) to cells during growth in culture (termed SILAC) or to living animals (termed SILAM), the latter typically by inclusion of isotope in food. Mixture of SILAC/SILAM-labeled reference samples with experimental samples is typically 1:1 and the relative abundances of peptides attributable to the exogenous labeled and the experimental samples are quantified by LC–MS analysis following enzymatic (typically tryptic) digestion. Methods employing global internal standards generated by SILAC labeling of different cell lines have enabled large-scale quantitative studies (Yates, Ruse, & Nakorchevsky, 2009).

A popular alternative quantitative labeling strategy is the use of isobaric tags for relative and absolute quantification (iTRAQ). In this approach, primary amino groups (the N-terminus and lysine side chains) of peptides are labeled. However, unlike SILAC, iTRAQ tags are only apparent following MS/MS analysis and detection of reporter ions in the mass spectrometer. iTRAQ can be multiplexed using many different labels (Zhang et al., 2013). However, it is important to note that some level of side reaction is unavoidable for most chemical derivatization procedures, and this may interfere with unbiased peptide and protein quantification.

It is also important to note in the context of biosynthetic measurements that these methods all measure relative pool sizes of individual proteins within a proteome. This parameter is not identical to the absolute mass of a protein in a tissue. The determination of absolute synthesis rates requires the measurement of the mass of the starting tissue and the total protein content of the tissue (mg protein/g tissue). Calculation of "absolute synthesis rates" (i.e., mass synthesized/time) has to take these factors into account. To calculate absolute synthesis rates for large datasets, the first step is to combine protein FSR and relative protein pool size. The normalized signal intensity quotient (Q) is calculated for each protein by correcting individual signal intensities against the treatment group mean. The within-proteome absolute synthesis rate for a given protein (i.e., the mass of a protein synthesized per unit time, normalized to other proteins in the proteome) can be determined by multiplying the fractional synthesis and the signal intensity quotient together ($k_{corr} \times Q$). Conversion to absolute synthesis rates then requires an estimate of the amount of total protein present in the proteome sampled, which may include measurement of the mass of the tissue.

6. FUTURE DIRECTIONS

6.1 Characterizing Global Proteome Kinetics Through Representative Proteins

A unique and potentially very informative feature of proteome dynamics is the capacity to identify coordinated changes or even global changes in synthesis or breakdown rates of proteins across the proteome. We have observed that certain interventions in human subjects, animal models or isolated cells increase or decrease turnover rates in the same direction for essentially all proteins detected across multiple gene ontologies. Long-term calorie restriction in mice, for example, results in markedly reduced synthesis and breakdown rates (turnover) of almost all proteins detected in both liver and muscle tissue (Fig. 15). Patients with advanced liver fibrosis due to hepatitis C viral infection exhibit generally elevated global turnover rates of liver proteins measured in biopsy tissue (Decaris et al., 2015; Emson, Decaris, Gatmaitan, Cattin, et al., 2014). In contrast, other physiologic alterations or disease conditions cause variable changes,

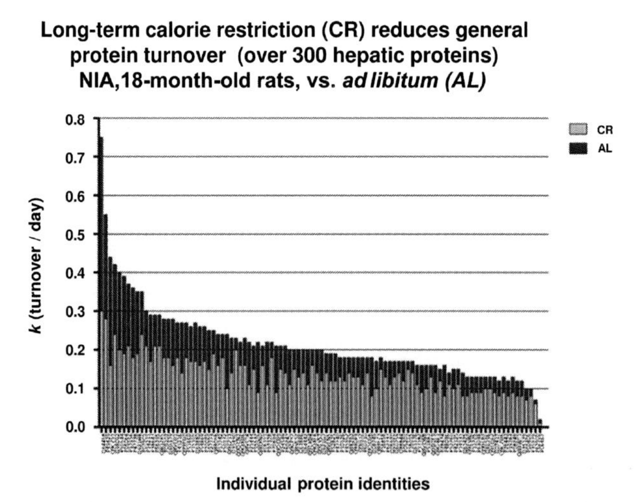

Figure 15 Global changes in protein turnover across the proteome can be detected by dynamic proteomics techniques. Long-term calorie restriction in 18-month-old rats reduces protein turnover across the proteome in liver, compared to *ad libitum*-fed rats. More than 300 proteins are shown. *Adapted from Price, Khambatta, et al. (2012).* (See the color plate.)

in both up and down directions, for the turnover of proteins individually or in different ontologies.

The basic observation that coordinated changes in turnover can occur across the proteome is of mechanistic interest in its own right, potentially reflecting global changes in translational efficiency or proteolytic systems, for example, and is uniquely capable of being evaluated by the proteome dynamics approach.

For characterization of global changes in dynamics across the proteome to become routinely applicable, particularly for clinical diagnostic testing, there would be obvious practical advantages if selected proteins could be measured to represent global changes. By identifying a subset of proteins that are measurable in every subject and can be targeted and concentrated, assays that are more reproducible and technically manageable may be devised, especially for auditing and reduction to routine clinical laboratory tests.

6.2 Characterizing Kinetics of a Parent Protein Through Selected, Targeted Peptides

One of the broad assumptions of peptide-based analyses of protein kinetics is that all the peptides that are unique to a protein should exhibit the same kinetic behavior, and that this should represent the behavior of the parent protein (see above). This simplifying assumption derives from the biochemistry of protein synthesis: polymerization synthesis occurs once, at the ribosome, as an ensemble for any protein; and there is not partial breakdown and resynthesis or other such core remodeling of the amino acid sequence once polymerized. The occurrence of posttranslational modifications or partial proteolytic cleavage does not alter this fundamental cell biologic principle. There can certainly be exceptions to this rule—if different isoforms, cleavage products, or posttranslationally modified species have different fates and functions, for example—but these kinetic variants in principle should be separable physically, as different species, and this phenomenon is not expected to apply to the majority of proteins in a cell.

Variations in measured kinetics among peptides are therefore generally taken to represent analytic variability rather than biologic differences. Each investigator should establish the reproducibility and accuracy of the peptides selected to represent the kinetics of proteins.

Applying the methods described above, multiple peptide-level kinetic measurements are rolled up to the protein level. If it is correct that individual peptides represent the kinetics of the parent protein, then the assay for protein kinetics might in principle be simplified by focusing analysis on selected peptides that represent the turnover of the parent protein. Indeed, a simplified, targeted approach, using representative peptides that are enriched or otherwise selected analytically, may provide a more sensitive and robust high-throughput approach. In particular, comparisons over time or among subjects are more defensible if the same peptides are compared, and making a GLP-compatible clinical assay will also require defined analytic targets such as specific peptides.

6.3 Applications of Combinatorial Analysis Approach with Heavy Water Labeling to Metabolic Fluxomics

6.3.1 Visualization of Molecular Fluxes Spatially in Histologic Tissue Specimens

The combinatorial analysis approach with heavy water labeling also has applications for nontargeted, global measurement of polymerization biosynthesis rates beyond the proteome. One area where this has recently been applied is in flux lipidomics in histologic tissue specimens, or "*in situ* kinetic histochemistry" (Louie et al., 2013).

This approach builds on the notion of MS imaging, which enables spatial mapping of molecular composition in histologic specimens. Previous images had only been static snapshots of molecules; however, kinetic information of flux through metabolic pathways had been lacking. We developed kinetic MS imaging (Louie et al., 2013), a novel technique integrating soft desorption/ionization MS with *in vivo* metabolic labeling of tissue with heavy water, to generate histologic images of the kinetics of biological processes. In brief, laser-induced desorption, rastered across a histologic tissue specimen, creates in a pixel-by-pixel manner ions with different labeling patterns, following heavy water labeling *in vivo*. Measurement of mass isotopomer distribution in these ions reveals their synthesis rates, characterized and visualized spatially and anatomically across the tissue—i.e., flux lipidomics represented anatomically.

Applied to a tumor, this approach revealed heterogeneous spatial distributions of newly synthesized versus preexisting lipids, with altered lipid synthesis patterns distinguishing region-specific intratumor subpopulations. Images also enabled identification and correlation of metabolic activity of specific lipids found in tumor regions of varying grade (Fig. 16). The spatial heterogeneity of biochemical dynamics may be a telling characteristic or fingerprint of a disease tissue—e.g., "hot spots," "cold spots," nodules, etc., in neoplastic conditions.

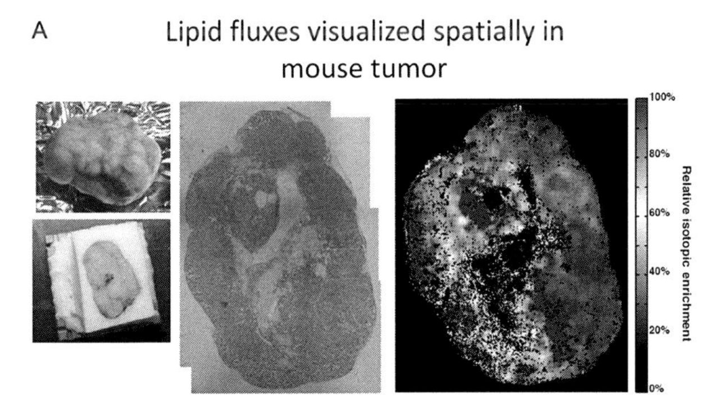

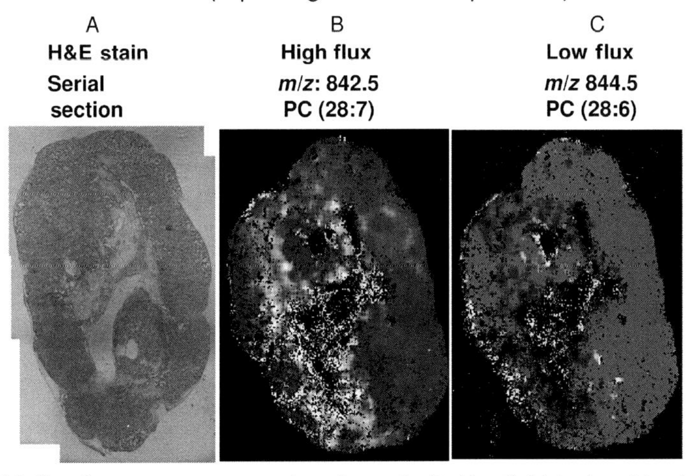

Figure 16 Kinetic mass spectrometry imaging or *in situ* kinetic histochemistry. (A). Lipid fluxes are visualized across tissue histologic specimens after *in vivo* heavy water labeling of a tumor-bearing mouse (triple-negative mammary cancer). Flux rates of the lipid molecule shown here reveal hot spots and cold spots within the anatomic bounds of the tumor. (B). Lipid turnover hot spots on histologic slide specimens correlate with more aggressive clinical behavior of triple-negative mammary tumors in mice. *Adapted from Louie et al. (2013).* (See the color plate.)

ACKNOWLEDGMENTS

We would like to note the critical contributions to this work by Richard Neese Ph.D., John C. Price Ph.D., Scott Turner Ph.D., Gregg Czerwieniec Ph.D., Mahalakshmi Shankaran Ph.D., Claire Emson Ph.D., Martin Decaris Ph.D., Carine Beysen Ph.D., Marc Colangelo Ph.D., Robert Busch Ph.D., Chancy Fessler, and Tim Riiff.

REFERENCES

Angel, T. E., Aryal, U. K., Hengel, S. M., Baker, E. S., Kelly, R. T., Robinson, E. W., et al. (2012). Mass spectrometry-based proteomics: Existing capabilities and future directions. *Chemical Society Reviews*, *41*(10), 3912–3928. http://dx.doi.org/10.1039/c2cs15331a.

Antoniewicz, M. R., Kraynie, D. F., Laffend, L. A., Gonzalez-Lergier, J., Kelleher, J. K., & Stephanopoulos, G. (2007). Metabolic flux analysis in a nonstationary system: Fed-batch fermentation of a high yielding strain of E. coli producing 1,3-propanediol. *Metabolic Engineering*, *9*(3), 277–292. http://dx.doi.org/10.1016/j.ymben.2007.01.003.

Baker, E. S., Liu, T., Petyuk, V. A., Burnum-Johnson, K. E., Ibrahim, Y. M., Anderson, G. A., et al. (2012). Mass spectrometry for translational proteomics: Progress and clinical implications. *Genome Medicine*, *4*(8), 63. http://dx.doi.org/10.1186/gm364.

Brown, M. S., & Goldstein, J. L. (1986). A receptor-mediated pathway for cholesterol homeostasis. *Science*, *232*(4746), 34–47.

Busch, R., Kim, Y. K., Neese, R. A., Schade-Serin, V., Collins, M., Awada, M., et al. (2006). Measurement of protein turnover rates by heavy water labeling of nonessential amino acids. *Biochimica et Biophysica Acta*, *1760*(5), 730–744. http://dx.doi.org/10.1016/j.bbagen.2005.12.023.

Chen, J. L., Peacock, E., Samady, W., Turner, S. M., Neese, R. A., Hellerstein, M. K., et al. (2005). Physiologic and pharmacologic factors influencing glyceroneogenic contribution to triacylglyceride glycerol measured by mass isotopomer distribution analysis. *The Journal of Biological Chemistry*, *280*(27), 25396–25402. http://dx.doi.org/10.1074/jbc.M413948200.

Choudhary, C., Weinert, B. T., Nishida, Y., Verdin, E., & Mann, M. (2014). The growing landscape of lysine acetylation links metabolism and cell signalling. *Nature Reviews. Molecular Cell Biology*, *15*(8), 536–550. http://dx.doi.org/10.1038/nrm3841.

Claydon, A. J., & Beynon, R. (2012). Proteome dynamics: Revisiting turnover with a global perspective. *Molecular & Cellular Proteomics*, *11*(12), 1551–1565. http://dx.doi.org/10.1074/mcp.O112.022186.

Claydon, A. J., Thom, M. D., Hurst, J. L., & Beynon, R. J. (2012). Protein turnover: Measurement of proteome dynamics by whole animal metabolic labelling with stable isotope labelled amino acids. *Proteomics*, *12*(8), 1194–1206. http://dx.doi.org/10.1002/pmic.201100556.

Commerford, S. L., Carsten, A. L., & Cronkite, E. P. (1983). The distribution of tritium among the amino acids of proteins obtained from mice exposed to tritiated water. *Radiation Research*, *94*(1), 151–155.

Decaris, M. L., Emson, C. L., Li, K., Gatmaitan, M., Luo, F., Cattin, J., et al. (2015). Turnover rates of hepatic collagen and circulating collagen-associated proteins in humans with chronic liver disease. *PLoS One*, *10*(4), e0123311. http://dx.doi.org/10.1371/journal.pone.0123311.

Decaris, M. L., Gatmaitan, M., Florcruz, S., Luo, F., Li, K., Holmes, W. E., et al. (2014). Proteomic analysis of altered extracellular matrix turnover in bleomycin-induced pulmonary fibrosis. *Molecular & Cellular Proteomics*, *13*(7), 1741–1752. http://dx.doi.org/10.1074/mcp.M113.037267.

Dietschy, J. M., & Brown, M. S. (1974). Effect of alterations of the specific activity of the intracellular acetyl CoA pool on apparent rates of hepatic cholesterogenesis. *The Journal of Lipid Research*, *15*(5), 508–516.

Emson, C. L., Decaris, M., Gatmaitan, L. F., Cattin, J., FlorCruz, S., Kelvin, L., et al. (2014). Identification of a putative serum protein biomarker for noninvasive measurement of fibrogenesis: Correlation with hepatic collagen synthesis and histological score in humans. In *Paper presented at the Keystone symposia, fibrosis: From bench to bedside.*

Emson, C. L., Decaris, M., Gatmaitan, L. F., FlorCruz, S., Holochwost, D., Angel, T. E., et al. (2014). Collagen synthesis rate distinguishes between early and late diffuse scleroderma subjects. In *Paper presented at the Keystone symposia, fibrosis: From bench to bedside, Keystone, CO*.

Erve, J. C., Gu, M., Wang, Y., DeMaio, W., & Talaat, R. E. (2009). Spectral accuracy of molecular ions in an LTQ/Orbitrap mass spectrometer and implications for elemental composition determination. *Journal of The American Society for Mass Spectrometry*, *20*(11), 2058–2069. http://dx.doi.org/10.1016/j.jasms.2009.07.014.

Evertts, A. G., Zee, B. M., Dimaggio, P. A., Gonzales-Cope, M., Coller, H. A., & Garcia, B. A. (2013). Quantitative dynamics of the link between cellular metabolism and histone acetylation. *The Journal of Biological Chemistry*, *288*(17), 12142–12151. http://dx.doi.org/10.1074/jbc.M112.428318.

Fagerquist, C. K., Hellerstein, M. K., Faubert, D., & Bertrand, M. J. (2001). Elimination of the concentration dependence in mass isotopomer abundance mass spectrometry of methyl palmitate using metastable atom bombardment. *Journal of the American Society for Mass Spectrometry*, *12*(6), 754–761. http://dx.doi.org/10.1016/S1044-0305(01)00227-6.

Fanara, P., Wong, P. Y., Husted, K. H., Liu, S., Liu, V. M., Kohlstaedt, L. A., et al. (2012). Cerebrospinal fluid-based kinetic biomarkers of axonal transport in monitoring neurodegeneration. *The Journal of Clinical Investigation*, *122*(9), 3159–3169. http://dx.doi.org/10.1172/JCI64575.

Gardner, J. L., Turner, S. M., Bautista, A., Lindwall, G., Awada, M., & Hellerstein, M. K. (2007). Measurement of liver collagen synthesis by heavy water labeling: Effects of profibrotic toxicants and antifibrotic interventions. *American Journal of Physiology. Gastrointestinal and Liver Physiology*, *292*(6), G1695–G1705. http://dx.doi.org/10.1152/ajpgi.00209.2006.

Hawkridge, A. M., & Muddiman, D. C. (2009). Mass spectrometry-based biomarker discovery: Toward a global proteome index of individuality. *Annual Review of Analytical Chemistry (Palo Alto, California)*, *2*, 265–277. http://dx.doi.org/10.1146/annurev.anchem.1.031207.112942.

Hellerstein, M. K. (2003). In vivo measurement of fluxes through metabolic pathways: The missing link in functional genomics and pharmaceutical research. *Annual Review of Nutrition*, *23*, 379–402. http://dx.doi.org/10.1146/annurev.nutr.23.011702.073045.

Hellerstein, M. K. (2004). New stable isotope-mass spectrometric techniques for measuring fluxes through intact metabolic pathways in mammalian systems: Introduction of moving pictures into functional genomics and biochemical phenotyping. *Metabolic Engineering*, *6*(1), 85–100.

Hellerstein, M. K., Christiansen, M., Kaempfer, S., Kletke, C., Wu, K., Reid, J. S., et al. (1991). Measurement of de novo hepatic lipogenesis in humans using stable isotopes. *The Journal of Clinical Investigation*, *87*(5), 1841–1852. http://dx.doi.org/10.1172/JCI115206.

Hellerstein, M. K., Kletke, C., Kaempfer, S., Wu, K., & Shackleton, C. H. (1991). Use of mass isotopomer distributions in secreted lipids to sample lipogenic acetyl-CoA pool in vivo in humans. *The American Journal of Physiology*, *261*(4 Pt. 1), E479–E486.

Hellerstein, M. K., & Neese, R. A. (1992). Mass isotopomer distribution analysis: A technique for measuring biosynthesis and turnover of polymers. *The American Journal of Physiology*, *263*(5 Pt. 1), E988–E1001.

Hellerstein, M. K., & Neese, R. A. (1999). Mass isotopomer distribution analysis at eight years: Theoretical, analytic, and experimental considerations. *The American Journal of Physiology*, *276*(6 Pt. 1), E1146–E1170.

Hentze, M. W. (1991). Determinants and regulation of cytoplasmic mRNA stability in eukaryotic cells. *Biochimica et Biophysica Acta*, *1090*(3), 281–292.

Hetenyi, G., Jr. (1982). Correction for the metabolic exchange of 14C for 12C atoms in the pathway of gluconeogenesis in vivo. *Federation Proceedings*, *41*(1), 104–109.

Hiller, K., Metallo, C., & Stephanopoulos, G. (2011). Elucidation of cellular metabolism via metabolomics and stable-isotope assisted metabolomics. *Current Pharmaceutical Biotechnology*, *12*(7), 1075–1086.

Jain, E., Bairoch, A., Duvaud, S., Phan, I., Redaschi, N., Suzek, B. E., et al. (2009). Infrastructure for the life sciences: Design and implementation of the UniProt website. *BMC Bioinformatics*, *10*, 136. http://dx.doi.org/10.1186/1471-2105-10-136.

Jiao, X., Sherman, B. T., Huang da, W., Stephens, R., Baseler, M. W., Lane, H. C., et al. (2012). DAVID-WS: A stateful web service to facilitate gene/protein list analysis. *Bioinformatics*, *28*(13), 1805–1806. http://dx.doi.org/10.1093/bioinformatics/bts251.

Jones, P. J., & Leatherdale, S. T. (1991). Stable isotopes in clinical research: Safety reaffirmed. *Clinical Science (London)*, *80*(4), 277–280.

Jucker, M., & Walker, L. C. (2013). Self-propagation of pathogenic protein aggregates in neurodegenerative diseases. *Nature*, *501*(7465), 45–51. http://dx.doi.org/10.1038/nature12481.

Kanehisa, M., Goto, S., Sato, Y., Kawashima, M., Furumichi, M., & Tanabe, M. (2014). Data, information, knowledge and principle: Back to metabolism in KEGG. *Nucleic Acids Research*, *42*(Database issue), D199–D205. http://dx.doi.org/10.1093/nar/gkt1076.

Kelleher, J. K., Kharroubi, A. T., Aldaghlas, T. A., Shambat, I. B., Kennedy, K. A., Holleran, A. L., et al. (1994). Isotopomer spectral analysis of cholesterol synthesis: Applications in human hepatoma cells. *The American Journal of Physiology*, *266*(3 Pt. 1), E384–E395.

Klionsky, D. J., Abdalla, F. C., Abeliovich, H., Abraham, R. T., Acevedo-Arozena, A., Adeli, K., et al. (2012). Guidelines for the use and interpretation of assays for monitoring autophagy. *Autophagy*, *8*(4), 445–544.

Lee, W. N., Bassilian, S., Ajie, H. O., Schoeller, D. A., Edmond, J., Bergner, E. A., et al. (1994). In vivo measurement of fatty acids and cholesterol synthesis using D2O and mass isotopomer analysis. *The American Journal of Physiology*, *266*(5 Pt. 1), E699–E708.

Lee, W. N., Bergner, E. A., & Guo, Z. K. (1992). Mass isotopomer pattern and precursor-product relationship. *Biological Mass Spectrometry*, *21*(2), 114–122. http://dx.doi.org/10.1002/bms.1200210210.

Li, L., Willard, B., Rachdaoui, N., Kirwan, J. P., Sadygov, R. G., Stanley, W. C., et al. (2012). Plasma proteome dynamics: Analysis of lipoproteins and acute phase response proteins with 2H2O metabolic labeling. *Molecular & Cellular Proteomics*. *11*(7). http://dx.doi.org/10.1074/mcp.M111.014209, M111 014209.

Louie, K. B., Bowen, B. P., McAlhany, S., Huang, Y., Price, J. C., Mao, J. H., et al. (2013). Mass spectrometry imaging for *in situ* kinetic histochemistry. *Scientific Reports*, *3*, 1656. http://dx.doi.org/10.1038/srep01656.

Mathur, R., & O'Connor, P. B. (2009). Artifacts in Fourier transform mass spectrometry. *Rapid Communications in Mass Spectrometry*, *23*(4), 523–529. http://dx.doi.org/10.1002/rcm.3904.

Mawuenyega, K. G., Kasten, T., Sigurdson, W., & Bateman, R. J. (2013). Amyloid-beta isoform metabolism quantitation by stable isotope-labeled kinetics. *Analytical Biochemistry*, *440*(1), 56–62. http://dx.doi.org/10.1016/j.ab.2013.04.031.

Mawuenyega, K. G., Sigurdson, W., Ovod, V., Munsell, L., Kasten, T., Morris, J. C., et al. (2010). Decreased clearance of CNS beta-amyloid in Alzheimer's disease. *Science*, *330*(6012), 1774. http://dx.doi.org/10.1126/science.1197623.

McCabe, B. J., Bederman, I. R., Croniger, C., Millward, C., Norment, C., & Previs, S. F. (2006). Reproducibility of gas chromatography-mass spectrometry measurements of 2H labeling of water: Application for measuring body composition in mice. *Analytical Biochemistry*, *350*(2), 171–176. http://dx.doi.org/10.1016/j.ab.2006.01.020.

Messmer, B. T., Messmer, D., Allen, S. L., Kolitz, J. E., Kudalkar, P., Cesar, D., et al. (2005). In vivo measurements document the dynamic cellular kinetics of chronic lymphocytic leukemia B cells. *The Journal of Clinical Investigation*, *115*(3), 755–764. http://dx.doi.org/10.1172/JCI23409.

Mi, H., Muruganujan, A., & Thomas, P. D. (2013). PANTHER in 2013: Modeling the evolution of gene function, and other gene attributes, in the context of phylogenetic trees. *Nucleic Acids Research*, *41*(Database issue), D377–D386. http://dx.doi.org/10.1093/nar/gks1118.

Neese, R. A., Schwarz, J. M., Faix, D., Turner, S., Letscher, A., Vu, D., et al. (1995). Gluconeogenesis and intrahepatic triose phosphate flux in response to fasting or substrate loads. Application of the mass isotopomer distribution analysis technique with testing of assumptions and potential problems. *The Journal of Biological Chemistry*, *270*(24), 14452–14466.

Ong, S. E., Blagoev, B., Kratchmarova, I., Kristensen, D. B., Steen, H., Pandey, A., et al. (2002). Stable isotope labeling by amino acids in cell culture, SILAC, as a simple and accurate approach to expression proteomics. *Molecular & Cellular Proteomics*, *1*(5), 376–386.

Papageorgopoulos, C., Caldwell, K., Shackleton, C., Schweingrubber, H., & Hellerstein, M. K. (1999). Measuring protein synthesis by mass isotopomer distribution analysis (MIDA). *Analytical Biochemistry*, *267*(1), 1–16. http://dx.doi.org/10.1006/abio.1998.2958.

Parks, E. J., Krauss, R. M., Christiansen, M. P., Neese, R. A., & Hellerstein, M. K. (1999). Effects of a low-fat, high-carbohydrate diet on VLDL-triglyceride assembly, production, and clearance. *The Journal of Clinical Investigation*, *104*(8), 1087–1096. http://dx.doi.org/10.1172/JCI6572.

Price, J. C., Guan, S., Burlingame, A., Prusiner, S. B., & Ghaemmaghami, S. (2010). Analysis of proteome dynamics in the mouse brain. *Proceedings of the National Academy of Sciences of the United States of America*, *107*(32), 14508–14513. http://dx.doi.org/10.1073/pnas.1006551107.

Price, J. C., Holmes, W. E., Li, K. W., Floreani, N. A., Neese, R. A., Turner, S. M., et al. (2012). Measurement of human plasma proteome dynamics with (2)H(2)O and liquid chromatography tandem mass spectrometry. *Analytical Biochemistry*, *420*(1), 73–83. http://dx.doi.org/10.1016/j.ab.2011.09.007.

Price, J. C., Khambatta, C. F., Li, K. W., Bruss, M. D., Shankaran, M., Dalidd, M., et al. (2012). The effect of long term calorie restriction on in vivo hepatic proteostasis: A novel combination of dynamic and quantitative proteomics. *Molecular & Cellular Proteomics*, *11*(12), 1801–1814. http://dx.doi.org/10.1074/mcp.M112.021204.

Schutzer, S. E., Angel, T. E., Liu, T., Schepmoes, A. A., Clauss, T. R., Adkins, J. N., et al. (2011). Distinct cerebrospinal fluid proteomes differentiate post-treatment Lyme disease from chronic fatigue syndrome. *PLoS One*, *6*(2), e17287. http://dx.doi.org/10.1371/journal.pone.0017287.

Schutzer, S. E., Angel, T. E., Liu, T., Schepmoes, A. A., Xie, F., Bergquist, J., et al. (2013). Gray matter is targeted in first-attack multiple sclerosis. *PLoS One*, *8*(9), e66117. http://dx.doi.org/10.1371/journal.pone.0066117.

Shankaran, M., Angel, T., Holmes, W., Turner, S., Miller, B., et al. (2014). Dynamic proteomics: A platform for proteome-wide interrogation of skeletal muscle anabolic response and non-invasive plasma biomarker discovery in humans and animal models. In *# 96. Paper presented at the 2014 new directions in biology and disease of skeletal muscle conference, Chicago, IL.*

Shimamoto, H., & Komiya, S. (2000). The turnover of body water as an indicator of health. *Journal of Physiological Anthropology and Applied Human Science*, *19*(5), 207–212.

Srere, P. (1994). Complexities of metabolic regulation. *Trends in Biochemical Sciences*, *19*(12), 519–520.

Szklarczyk, D., Franceschini, A., Kuhn, M., Simonovic, M., Roth, A., Minguez, P., et al. (2011). The STRING database in 2011: Functional interaction networks of proteins, globally integrated and scored. *Nucleic Acids Research*, *39*(Database issue), D561–D568. http://dx.doi.org/10.1093/nar/gkq973.

Turner, S. M., & Hellerstein, M. K. (2005). Emerging applications of kinetic biomarkers in preclinical and clinical drug development. *Current Opinion in Drug Discovery & Development*, *8*(1), 115–126.

Valastyan, J. S., & Lindquist, S. (2014). Mechanisms of protein-folding diseases at a glance. *Disease Models & Mechanisms*, 7(1), 9–14. http://dx.doi.org/10.1242/dmm.013474.

Wang, Y., Yang, F., Gritsenko, M. A., Wang, Y., Clauss, T., Liu, T., et al. (2011). Reversed-phase chromatography with multiple fraction concatenation strategy for proteome profiling of human MCF10A cells. *Proteomics*, *11*(10), 2019–2026. http://dx.doi.org/10.1002/pmic.201000722.

Waterlow, J. C., Garlick, P. J., & Millward, D. J. (1978). *Protein turnover in mammalian tissues and in the whole body*. Amsterdam/New York: North-Holland, sole distributors for the U.S.A. and Canada, Elsevier North-Holland.

Welty, F. K., Lichtenstein, A. H., Barrett, P. H., Dolnikowski, G. G., & Schaefer, E. J. (2004). Interrelationships between human apolipoprotein A-I and apolipoproteins B-48 and B-100 kinetics using stable isotopes. *Arteriosclerosis, Thrombosis, and Vascular Biology*, *24*(9), 1703–1707. http://dx.doi.org/10.1161/01.ATV.0000137975.14996.df.

Wolfe, R. R. (1984). Tracers in metabolic research: Radioisotope and stable isotope/mass spectrometry methods. *Laboratory and Research Methods in Biology and Medicine*, *9*, 1–287.

Yates, J. R., Ruse, C. I., & Nakorchevsky, A. (2009). Proteomics by mass spectrometry: Approaches, advances, and applications. *Annual Review of Biomedical Engineering*, *11*, 49–79. http://dx.doi.org/10.1146/annurev-bioeng-061008-124934.

Zhang, Y., Fonslow, B. R., Shan, B., Baek, M. C., & Yates, J. R., 3rd. (2013). Protein analysis by shotgun/bottom-up proteomics. *Chemical Reviews*, *113*(4), 2343–2394. http://dx.doi.org/10.1021/cr3003533.

Zilversmit, D. B. (1960). The design and analysis of isotope experiments. *The American Journal of Medicine*, *29*, 832–848.

CHAPTER EIGHT

Non-targeted Tracer Fate Detection

Daniel Weindl, André Wegner, Karsten Hiller[1]
Luxembourg Centre for Systems Biomedicine, University of Luxembourg, Esch-Belval, Luxembourg
[1]Corresponding author: e-mail address: karsten.hiller@uni.lu

Contents

Abstract

Stable isotopes have been used to trace atoms through metabolism and quantify metabolic fluxes for several decades. Only recently non-targeted stable isotope labeling approaches have emerged as a powerful tool to gain insights into metabolism. However, the manual detection of isotopic enrichment for a non-targeted analysis is tedious and time consuming. To overcome this limitation, the non-targeted tracer fate detection (NTFD) algorithm for the automated metabolome-wide detection of isotopic enrichment has been developed. NTFD detects and quantifies isotopic enrichment in the form of mass isotopomer distributions (MIDs) in an automated manner, providing

Methods in Enzymology, Volume 561
ISSN 0076-6879
http://dx.doi.org/10.1016/bs.mie.2015.04.003

the means to trace functional groups, determine MIDs for metabolic flux analysis, or detect tracer-derived molecules in general. Here, we describe the algorithmic background of NTFD, discuss practical considerations for the freely available NTFD software package, and present potential applications of non-targeted stable isotope labeling analysis.

1. INTRODUCTION

Since the recognition of cellular metabolism as a potential intervention point to treat complex diseases, stable isotope-assisted metabolomics techniques have emerged as a valuable tool to analyze intracellular fluxes (Chokkathukalam, Kim, Barrett, Breitling, & Creek, 2014). Stable isotope labeled tracers have been successfully applied to study metabolic fluxes in cell culture and *in vivo* in animals and humans (Bodamer & Halliday, 2001; Metallo et al., 2012; Missios et al., 2014; Sellers et al., 2015; Vallino & Stephanopoulos, 2000; Yoo, Stephanopoulos, & Kelleher, 2004).

In the past, such analyses have been rather targeted, mainly because isotopic enrichment from stable isotopes is not as readily detected as, for example, enrichment from radioactive isotopes. After a stable isotope labeling experiment, isotopic enrichment is typically determined by comparing the mass isotopomer distribution (MID) observed by mass spectrometric analysis with the MID to be expected from known natural isotope abundances (Pickup & McPherson, 1976). However, to determine the theoretical MID, knowledge of the elemental composition of the compound of interest is required (Wegner et al., 2014). For this reason, usually only a predefined list of compounds is analyzed for isotopic enrichment.

It is self-evident that such a targeted approach is suited to test a well-defined hypothesis, but fails to generate knowledge on unknown or unexpected parts of the metabolic network. Although many metabolic pathways are well described and known since many years, new reactions are continuously being identified (Dang et al., 2009; Michelucci et al., 2013), underlining the necessity for a global detection of isotopic enrichment. Recently, several tools for the non-targeted detection of isotopic enrichment in mass spectrometric data have been developed (Bueschl et al., 2012; Cho et al., 2014; Chokkathukalam et al., 2013; Creek et al., 2012; Hiller, Metallo, Kelleher, & Stephanopoulos, 2010; Hiller et al., 2013; Huang et al., 2014).

Alternatively, isotopic enrichment can be determined using isotope ratio mass spectrometry (IRMS) (Chace & Abramson, 1989) with high precision in a non-targeted manner. Unlike conventional mass spectrometric methods, IRMS cannot be used to determine MIDs. However, MIDs can provide interesting insights into metabolism and are the basis for ^{13}C-metabolic flux analysis (^{13}C-MFA) (Niedenführ, Wiechert, & Nöh, 2015; Sauer, 2006).

Here, we will present the non-targeted tracer fate detection (NTFD) algorithm (Hiller et al., 2010) which is able to detect and quantify isotopic enrichment in GC–EI-MS data in a non-targeted manner. We will introduce the theoretical background necessary to obtain meaningful results, discuss practical considerations, and present potential applications of non-targeted stable isotope labeling analysis.

Isotopic Isomers

Stable isotope labeling leads to isotopic isomerism. Usually, the following classes of isotopic isomers are distinguished:

Isotopologues

The widest class of isotopic isomers. Isomers that differ only in their isotopic composition with otherwise identical structure, including isotopic substitutions of different elements (IUPAC, 2012). For example, [U-^{13}C]acetate ([1,2-$^{13}C_2$]acetate), [1-^{13}C]acetate, and [2,2,2-D_3]acetate are isotopologic. In the context of carbon labeling experiments, the term is often implicitly used to only refer to ^{13}C-isotopologues.

Mass isotopomers

Mass isotopomers are groups of isotopologues of the same nominal mass or of isotopologues that cannot be separated by mass spectrometry (Hellerstein & Neese, 1999). Mass isotopomers are referred to as M_i, Mi, or $M+i$, where i indicates the increase in nominal mass as compared to the lightest isotopologue. For example, for carbon labeling of acetate, M_1 comprises [1-^{13}C]acetate and [2-^{13}C]acetate.

Isotopomers

Isotopomers are isotopologues with the same isotopic substituents but in varying positions (Murray et al., 2013). For example, [1-^{13}C]acetate and [2-^{13}C]acetate, but not [1,2-$^{13}C_2$]acetate are isotopomeric. Isotopomers cannot be separated by mass spectrometry.

MID

The mass isotopomer distribution (MID) is the relative abundance of the different mass isotopomers of a molecule. MIDs can be determined from mass spectrometric measurements and are often the basis for ^{13}C-MFA, a method for metabolic flux analysis based on stable isotope labeling and stoichiometric modeling, where a set of fluxes through a predefined metabolic network is determined that can best explain the experimentally observed isotopic labeling.

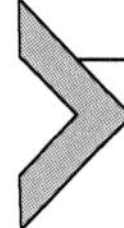

2. THEORETICAL BACKGROUND

2.1 Overview

Before discussing experimental and analytical considerations for the use of NTFD, we will provide the theoretical background for the non-targeted detection and quantification of isotopic enrichment that is necessary for the user to obtain meaningful results from NTFD. After the theoretical introduction, we will explain which parameters of the NTFD implementation (Hiller et al., 2013) are critical for each stage.

The NTFD algorithm operates on data from two MS measurements, one of an isotopically enriched (labeled) and one of a nonenriched (unlabeled) sample. After peak picking and mass spectral deconvolution, the mass spectrum of each compound from the labeled sample is paired with its counterpart from the unlabeled sample. Isotopically enriched compounds and mass spectrometric fragments are detected from the differences in these mass

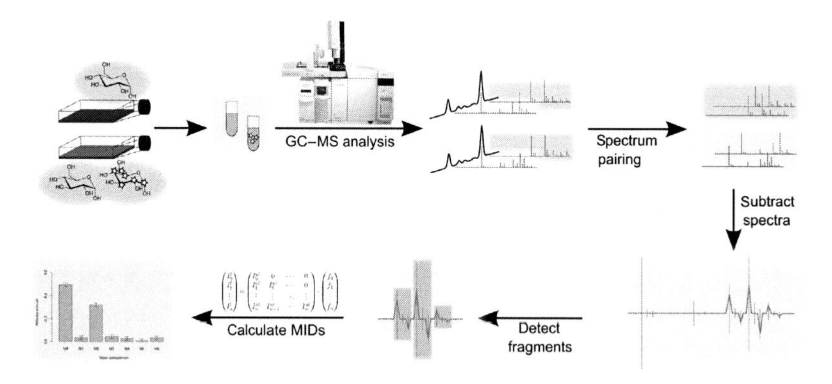

Figure 1 The NTFD algorithm operates on GC–EI-MS measurements of an isotopically enriched and nonenriched sample. For each compound, mass spectra from both measurements are paired. Isotopically enriched fragments are determined from characteristic patterns in the difference spectrum. For each of these enriched fragments the MID is determined. (See the color plate.)

spectra and the isotopic enrichment is quantified (Fig. 1). Along with the MID calculation, certain measures are determined to judge the quality of the obtained MID results.

The current NTFD software is designed to analyze low resolution GC–MS data. For high resolution LC–MS data, there are other tools available. These tools are beyond the scope of this chapter and we will only provide a short overview. Briefly, MetExtract (Bueschl et al., 2012) is a tool for the non-targeted detection of signals of unlabeled and uniformly labeled metabolites in LC–MS data. These information allow to differentiate between metabolites and analytical noise and the isotopologue ratios can be used for relative quantification of metabolite levels in two samples. However, MetExtract cannot be used to analyze complex mixtures of isotopologues. IsoMETLIN (Cho et al., 2014) is a web-based tool that identifies isotopologues of known compounds based on a computationally generated database and is thus rather targeted. It cannot identify unknown labeled metabolites. IsoMETLIN can make use of MS/MS spectra to resolve individual isotopomers. Such information can also be obtained from NTFD by analyzing specific mass spectrometric fragments (Wegner et al., 2014) or combinations thereof (Christensen & Nielsen, 1999). MzMatch-ISO (Chokkathukalam et al., 2013) and X^{13}CMS (Huang et al., 2014) are both R packages that allow for the non-targeted determination of MIDs and can be regarded as the LC–MS equivalents of NTFD.

We will now discuss the NTFD algorithm for the detection and quantification of isotopic enrichment in more detail.

2.2 Detection of Isotopically Enriched Fragments

When a compound is isotopically enriched, its mass spectrum changes. Assuming there are no isotope effects on ionization and fragmentation, the relative intensities of all fragments will be conserved. However, within each enriched isotopic peak cluster, the relative M_0 intensity will decrease and the intensity of the remaining peaks will increase as compared to the nonenriched spectrum (Fig. 2). The decrease in M_0 equals the absolute value of the summed increase in heavier mass isotopomer abundances, since we assume that the relative fragment intensities remain constant.

These changes become more obvious in the difference spectrum where a labeled fragment will lead to a positive peak followed by a negative peak with the same area under the peak (Fig. 2). These characteristic patterns are used to detect all labeled fragments.

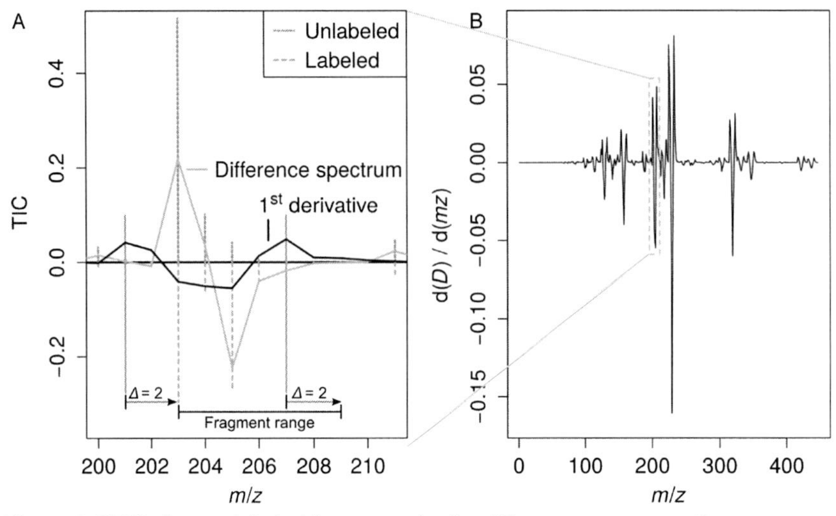

Figure 2 NTFD detects labeled fragments in the difference spectrum of mass spectra derived from a potentially isotopically enriched and nonenriched compound. (A) Mass spectrum of a fragment of the unlabeled compound and its labeled analogue, normalized to total ion current (TIC). Isotopic enrichment leads to characteristic peak patterns in the difference spectrum and its first derivative. Fragment boundaries are detected in the first derivative of the smoothed difference spectrum. The boundaries are shifted by 2 units due to the applied smoothing algorithm. (B) First derivative $\left(\frac{dD}{d(m/z)}\right)$ of the full difference spectrum of glutamine 4 TMS. Two positive peaks, separated by a negative peak, mark an isotopically enriched fragment. (See the color plate.)

For this detection of isotopic enrichment, the mass spectra of all compounds from the labeled and unlabeled sample need to be compared. Therefore, a spectrum and retention time (or retention index) matching (Stein, 1999) is performed between all mass spectra detected in the labeled and unlabeled samples during the previous step (Fig. 1). To ensure, only identical compounds are matched, this matching is highly stringent in terms of retention time and loose in terms of mass spectrum similarity. Therefore, there should not be any significant retention time shifts between the two measurements. The loose spectrum matching will ensure that the spectra of the labeled and unlabeled compound are also matched correctly in case of high isotopic enrichment, which can lead to very dissimilar mass spectra of the same compound.

In the case that replicate measurements have been provided, mass spectra from these chromatograms are matched first. This replicate mass spectra matching is highly stringent in terms of mass spectrum similarity, since there should not be any differences in isotopic enrichment and, thus, the mass spectra should be identical.

Once the mass spectra from the labeled ($I_{\mathrm{l}}^{\mathrm{raw}}$) and unlabeled sample ($I_{\mathrm{ul}}^{\mathrm{raw}}$) have been paired, they are analyzed for differences. Therefore, the peak intensities across the whole mass spectrum are normalized to a sum of 1:

$$I = \frac{I^{\mathrm{raw}}}{\sum_i I_i^{\mathrm{raw}}} \quad (1)$$

The normalized spectra are then subtracted to yield the difference spectrum D (Fig. 2):

$$D = I^{\mathrm{ul}} - I^{\mathrm{l}} \quad (2)$$

The detection of the characteristic patterns of labeled fragments is more robust in the first derivative of the difference spectrum (Fig. 2). Using a Savitzky–Golay filter with window size 5, a smoothed variant of the first derivative of the difference spectrum can be calculated very efficiently. In this first derivative, each isotopically enriched fragment will manifest as two positive peaks separated by a negative peak (Fig. 2). These two subsequent maxima indicate the boundaries of a labeled fragment shifted by 2 mass units due to the previous smoothing step. Therefore, by scanning the first derivative for such patterns, all isotopically enriched fragments are determined along with their m/z ranges. These ranges are required for the next step to quantify the isotopic enrichment.

2.3 Quantification of Isotopic Enrichment

The detection of isotopic enrichment in the previous step was only qualitative. This enrichment now needs to be quantified in terms of an MID for each labeled fragment. The raw MID observed in the mass spectrum of a compound after stable isotope labeling is a combination of natural isotope abundance and of artificial isotopic enrichment as derived from the tracer (Fig. 3).

When analyzing stable isotope labeling data, usually only the artificial enrichment derived from the tracer is of interest. Therefore, the raw mass isotopomer distribution as obtained from the mass spectrum needs to be corrected for natural isotope abundance. This is especially important in GC–MS analyses where often a chemical derivatization is part of the sample

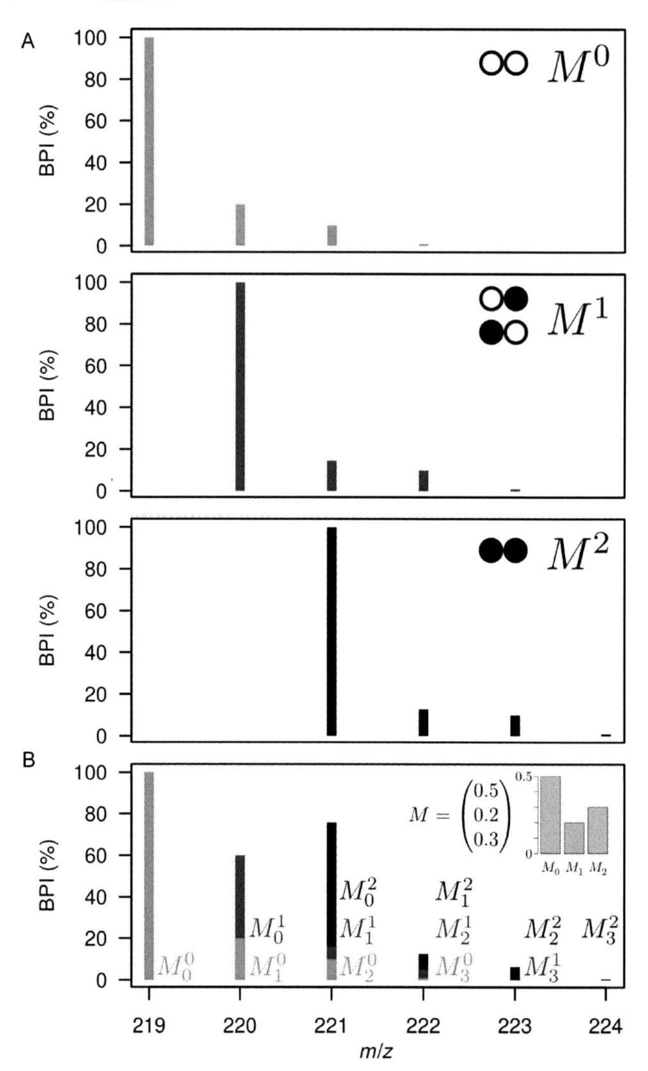

Figure 3 Artificial isotopic enrichment and natural isotope abundance determine isotopic peak patterns in a mass spectrum after stable isotope labeling experiments. Stable isotope labeling of a compound with two enrichable positions. (A) The mass spectra of all three mass isotopomers that can arise from stable isotope labeling show isotopic peaks due to the natural isotope abundance in the nonenriched positions. The natural isotope contribution, i.e., the relative intensity of the M+1 peak, decreases with increasing artificial enrichment. The relative M+0 intensity increases by the same value. Filled circle, heavy isotope; empty circle, light isotope. Signal intensity is scaled to base peak intensity (BPI). (B) After a stable isotope labeling experiment the mass spectrum of the given compound is a mixture of the natural MIDs of the three artificial mass isotopomers. This spectrum is the average of M^0, M^1, and M^2, weighted by the corrected MID M. (See the color plate.)

preparation, which introduces a large number of carbon or silicon atoms with high natural isotope abundances.

In targeted stable isotope labeling analyses, this correction for natural isotope abundances is usually performed based on the theoretical MID as expected from the elemental composition and the average natural isotope abundance of each element (Pickup & McPherson, 1976). However, this method is not applicable for non-targeted applications, since many compounds and fragments cannot be identified and, thus, their elemental composition is not available.

Jennings and Matthews (2005) presented an elegant method to correct raw MIDs for natural isotope abundance through comparison with the mass spectrum of the unlabeled compound. An adaptation of this approach is applied in NTFD.

In a ^{13}C-labeling experiment, a compound with n carbons can contain maximally up to $n\,^{13}$C isotopes from artificial tracer-derived enrichment. The remaining, non-enriched atoms are subject to natural isotope abundance (Fig. 3A). The relative peak intensities in the mass spectrum of such a compound represent the raw MID. The uncorrected MID I^1 of a fragment of such a labeled compound arises from natural MIDs of its "artificial" mass isotopomers as depicted in Fig. 3B. The relative abundances in I^1 are given as the summation of the natural MIDs of all $n + 1$ artificial mass isotopomers $M^0,\ldots,M^n$, weighted by their relative abundance or the corrected MID M:

$$I^1 = \begin{pmatrix} M_0^0 \cdot M_0 \\ M_1^0 \cdot M_0 + M_0^1 \cdot M_1 \\ \vdots \\ M_n^0 \cdot M_0 + M_{n-1}^1 \cdot M_1 + \cdots + M_0^n \cdot M_n \end{pmatrix} \tag{3}$$

$$= \begin{pmatrix} M_0^0 & 0 & \cdots & 0 \\ M_1^0 & M_0^1 & \cdots & 0 \\ \vdots & \vdots & \ddots & \vdots \\ M_n^0 & M_{n-1}^1 & \cdots & M_0^n \end{pmatrix} \cdot M = \mathbf{M}_{\text{corr}} \cdot M \tag{4}$$

with M_b^a as the relative abundance of the bth natural mass isotopomer of the compound containing a tracer isotopes. If the correction matrix $\mathbf{M}_{\text{corr}}$ is known, the corrected MID can be determined as

$$M = \mathbf{M}_{\text{corr}}^{-1} \cdot I^1 \tag{5}$$

To correct the raw MID for natural isotope abundance, we need to know the natural MIDs M^a of the artificial mass isotopomers. As mentioned above, they cannot be calculated from natural isotope abundances, since the elemental compositions are not available. However, we can obtain this information from the unlabeled mass spectrum I^{ul}. The first column of $\mathbf{M}_{corr}$, M^0, is the natural MID of the molecule, without any artificial enrichment. These values are exactly the relative peak intensities in the mass spectrum I^{ul} of the unlabeled compound, leading us with Eq. (3) to:

$$\mathbf{M}_{corr} = \begin{pmatrix} s_0^{ul} & 0 & \cdots & 0 \\ s_1^{ul} & M_0^1 & \cdots & 0 \\ \vdots & \vdots & \ddots & \vdots \\ s_n^{ul} & M_{n-1}^1 & \cdots & M_0^n \end{pmatrix} \tag{6}$$

The remaining M_b^a can be replaced in a similar way. The mass spectra of $M^1,\ldots,M^n$ look similar to the unlabeled spectrum, but are shifted to higher mass by $1,\ldots,n$ (Fig. 3). Furthermore, due to the fixed inclusion of stable isotopes in $M^1,\ldots,M^n$, the M_0 abundance will increase, and the other mass isotopomer abundances will decrease. Since the relative natural abundance of ^{13}C and the number of carbons in most metabolites are rather low, we assume that only the M_0 and M_1 abundances are affected by the artificial enrichment, and neglect the insignificant changes in the abundances of higher mass isotopomers.

When a carbon atom in a molecule is artificially substituted by a ^{13}C isotope, this atom is no longer subject to natural isotope abundance, so that the M_0 abundance of the molecule increases and the M_1 abundance decreases. When a ^{13}C isotopes are included, the ratio of M_1 and M_0 changes to (Jennings & Matthews, 2005):

$$\frac{M_1^a}{M_0^a} = \frac{M_1^0}{M_0^0} - a \cdot \frac{p_{^{13}C}}{p_{^{12}C}} \tag{7}$$

with $p_{^{12}C}$ and $p_{^{13}C}$ as the average natural isotope abundances of ^{12}C and ^{13}C, respectively. We define this correction term as:

$$c_a = a \cdot \frac{p_{^{13}C}}{p_{^{12}C}} \tag{8}$$

Since we assume that the relative abundances of the heavier mass isotopomers remain constant, the M_0 abundance increases by the same value as the M_1 abundance decreases. Therefore, their sum will not change:

$$M_1^a + M_0^a = M_1^0 + M_0^0 \tag{9}$$

Combining Eqs. (7), (8), and (9) provides a correction term $M1_{\text{corr}}$ to obtain M_0^a from M_0 as:

$$\frac{M_1^0}{M_0^0} \cdot M_0^a - c_a \cdot M_0^a = M_1^0 + M_0^0 - M_0^a \tag{10}$$

$$\frac{M_1^0}{M_0^0} \cdot M_0^a - c_a \cdot M_0^a + M_0^a = M_1^0 + M_0^0 + M_0^0 \cdot c_a - M_0^0 \cdot c_a \tag{11}$$

$$M_0^a = M_0^0 + \frac{M_0^0 \cdot c_a}{\frac{M_1^0}{M_0^0} + 1 - c_a} = M_0^0 + M1_{\text{corr}} \tag{12}$$

Since the sum of M_0 and M_1 need to be constant (Eq. 9), M_1 needs to decrease by the same amount as M_0 increases:

$$M_1^a = M_1^0 - \frac{M_0^0 \cdot c_a}{\frac{M_1^0}{M_0^0} + 1 - c_a} = M_1^0 - M1_{\text{corr}} \tag{13}$$

In summary, the natural mass isotopomer abundances M_b^a are given as:

$$M_b^a = \begin{cases} M_b^0, & \text{for } a = 0 \vee b > 1 \\ M_0^0 + M1_{\text{corr}}, & \text{for } a > 0 \wedge b = 0 \\ M_1^0 - M1_{\text{corr}}, & \text{for } a > 0 \wedge b = 1 \end{cases} \tag{14}$$

This enables us to populate the correction matrix $\mathbf{M}_{\text{corr}}$ accordingly and to determine the isotopic enrichment from Eq. (5) by least-squares regression without requiring any further knowledge on the given compound.

We explained this MID correction for the case of ^{13}C labeling, but it can be applied to any other isotopic element. Depending of the natural isotope abundance of the isotopic element, a M_2 correction may become necessary as described in Jennings and Matthews (2005). Combined labeling from different isotopic elements (e.g., ^{13}C and ^{15}N) cannot be corrected using this approach, since the contributions of the two isotopic species cannot be distinguished.

2.4 Quality Measures

To judge the quality of the calculated MIDs, NTFD provides three quality measures: the coefficient of determination, the sum of absolute values of the relative mass isotopomer abundances, and confidence intervals of the relative mass isotopomer abundances.

2.4.1 Sum of Absolute Values

Theoretically, all relative mass isotopomer abundances should be within the interval [0,1] and their sum should be equal to one:

$$\sum_{i=0}^{n} M_i = 1 \quad | \quad M_i \in [0, 1] \tag{15}$$

However, due to measurement errors, the regression can lead to negative mass isotopomer abundances. The sum of the absolute values

$$\sum_{i=0}^{n} |M_i| \tag{16}$$

can therefore be greater than one. This sum of absolute values provides a first, simple, yet effective measure for the accuracy of the calculated MIDs. The closer this value is to one, the more accurate are the MIDs.

2.4.2 Coefficient of Determination

If replicate measurements of the labeled and unlabeled sample are available, the coefficient of determination R^2 can be computed for every fragment. This value is defined as

$$R^2 = \frac{\sum_{i=1}^{n} (\hat{A}_i - \bar{A})^2}{\sum_{i=1}^{n} (A_i - \bar{A})^2} \quad | \quad R^2 \in [0, 1] \tag{17}$$

where A_i is the value of the given raw mass isotopomer abundance from the ith measurement, $\hat{A}_i$ the corresponding regression value from the best fitting MID, and $\bar{A}$ is the mean of the raw mass isotopomer abundances across all measurements.

To put it less formally: R^2 shows the fraction of the total variation in the measured values that can be explained by the linear regression model. Ideally, all variation is explained by the model ($R^2 = 1$). In the worst case, none of the variation can be explained by the regression model ($R^2 = 0$).

2.4.3 Confidence Intervals

The previous measures only provide an overall quality measure for the whole MID. To judge the precision of the individual mass isotopomer abundances, an alternative measure is required. When replicate measurements are provided, confidence intervals for the relative mass isotopomer abundances can be computed. The NTFD application reports 95% confidence intervals for the mass isotopomer abundances.

3. PRACTICAL CONSIDERATIONS

To obtain meaningful results from NTFD, certain critical points need to be taken into account. Starting with important experimental considerations, we will discuss how to avoid pitfalls in data analysis and highlight how common problems can be solved.

3.1 Experimental Considerations

A prerequisite for an NTFD analysis is a stable isotope labeling experiment and a parallel experiment with the unlabeled tracer. To ensure that the maximum number of labeled compounds is detected, both experiments must be performed under the same conditions using the same cell line and growth medium.

Although NTFD can work on single measurements, the use of replicate measurements is strongly recommended. Replicates can reduce the effect of analytical variations and will make the MID calculation more robust. If replicates are available, NTFD provides additional statistical quality measures for the calculated MIDs (Section 2.4). Usually, three measurements per experiment provide sufficiently good results.

3.1.1 Choice of Isotopic Tracer

For the design of a stable isotope labeling experiment, the proper choice and amount of tracer is crucial. This choice completely depends on the biological question asked (Metallo, Walther, & Stephanopoulos, 2009). For example, if the goal is to differentiate between metabolites and analytical background, all carbon substrates should be replaced by their fully labeled analogues. To completely label an organism, a defined minimal medium, if it is available for the organism of interest, is most conveniently used.

Generally, isotopic enrichment from any isotopic element can be analyzed with NTFD, as long as there is no significant effect on M_2 abundance, which would not be corrected for. However, special care has to be taken in case of deuterium. Deuteration is well known to affect chromatographic retention, and different isotopologues may be separated completely. This can turn out problematic for two reasons: (1) the isotopologue mixture can be spread out over multiple deconvoluted mass spectra and (2) the matching of the labeled and unlabeled spectrum may not succeed due to shifts in retention time.

A combination of multiple tracers is generally possible, but will further complicate the already difficult interpretation of the resulting MIDs. In most cases, it is advisable to distribute tracers among different experiments, except if the goal is to completely label an organism.

Once the tracer is chosen, the second important decision is the ratio at which it is introduced. On the one hand, a higher tracer abundance will lead to higher and therefore better detection of isotopic enrichment in downstream metabolites. On the other hand, complete isotopic enrichment may lead to very dissimilar mass spectra from the labeled and unlabeled sample, which cannot be matched properly. In such a case, labeled compounds may be overseen, since the detection of isotopic enrichment by NTFD depends on the correct matching of these mass spectra. However, in most cases, NTFD works well with high isotopic enrichment.

An important parameter for subsequent data analysis is the incubation time with the selected tracer. For many applications, it is desirable that the system reaches isotopic steady state, meaning that isotopic labeling does not change over time. The required incubation time can range from several minutes to several days and needs to determined from stable isotope labeling time-series of the specific system of interest. With *in vivo* labeling of complex organisms isotopic steady state may not be achievable at all. However, isotopic non-steady state does not affect the detection and quantification of isotopic enrichment *per se*, but needs to be accounted for during interpretation of the resulting MIDs.

3.1.2 GC–MS Measurements

Ideally, the labeled and the unlabeled samples are measured subsequently on the same instrument. Standard GC–MS temperature programs and settings for the analysis of trimethylsilyl (TMS) derivatives of metabolites can be found here (Sapcariu et al., 2014). In case measurements are from different runs or instruments, retention time stability must be ensured. To aid compound matching, a calibration mixture such as an *n*-alkane mix (usually C_{10} to C_{40}) should be measured to calculate the Kováts retention index (Kováts, 1958).

The usual metabolomics quality controls, like, e.g., blank samples, should be included in the measurement as a quality control measure. If a compound that is isotopically enriched is also found to be present in any blank measurements, the respective MIDs need to be interpreted with caution, since the relative M_0 abundance may be overestimated due to label dilution from the exogenous compound.

A similar problem is caused by detector saturation. If signal intensities lie outside the linear range of the detector, or if the detector is fully saturated, signal intensities are skewed or clipped. The relative mass isotopomer abundance determined by the respective ion may then be underestimated. This can, depending on the extent, strongly impact MID calculation for the respective fragments.

3.2 Data Analysis

As with most non-targeted methods, NTFD data analysis can be tedious and frustrating if the wrong parameters are applied. In the following section, the most important steps and the respective parameters controlling a successful analysis are explained. The NTFD implementation is freely available for Linux and Windows operating systems. The software and detailed documentation of the user interface, along with sample data and a step-by-step tutorial are available at http://ntfd.mit.edu. Briefly, there are four different dialogs the user can interact with (Fig. 4): (1) Selection of labeled measurements in netCDF or MetaboliteDetector (Hiller et al., 2009) format. (2) Selection of unlabeled measurements in netCDF or MetaboliteDetector format. (3) The settings tab containing the important parameters for compound and label detection. (4) The result tab showing the list of detected labeled compounds and their respective fragment MIDs.

3.2.1 Data Preprocessing

Before starting an NTFD analysis, it is important to check the quality of all chromatograms. The outcome of the analysis depends heavily on the correct preprocessing of the raw data. Before NTFD can start to detect isotopically enriched compounds, a peak picking and mass spectral deconvolution has to be performed. The raw data is scanned for peaks in the signal intensity which cross a certain threshold of absolute intensity and signal-to-noise ratio. Any signal below this threshold is discarded. All peaks detected at this stage are used for mass spectral deconvolution, that is, they are grouped to specific mass spectra according to their retention time. A narrower deconvolution window will ensure that closely eluting peaks are separated, and a broader deconvolution window is necessary for broad peaks or for deuterated compounds where different isotopologues can elute at slightly different retention times.

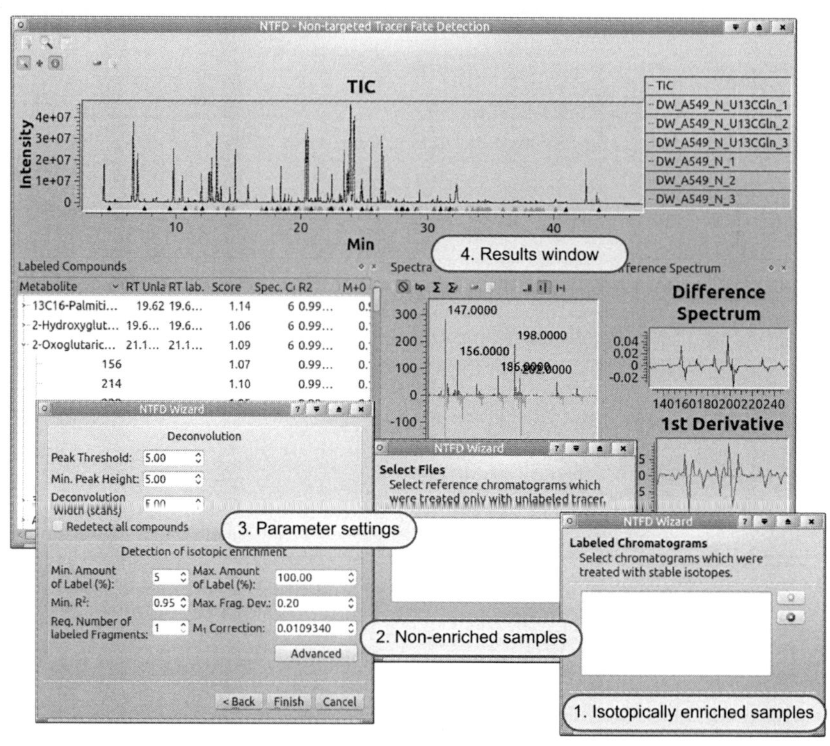

Figure 4 Graphical user interface of the NTFD implementation (Hiller et al., 2010). The program is freely available for Linux and Windows operating systems. After GC–MS measurement, (1) the files of the isotopically enriched and (2) the nonenriched samples need to be selected, (3) parameters for the detection and quantification of isotopic enrichment need to be adjusted, then (4) NTFD will determine all isotopically enriched compounds and fragments along with their MIDs. (See the color plate.)

This stage is very important, because it forms the basis on which all subsequent label detection will be performed. Therefore, the parameters should be well chosen. If the peak picking is too sensitive, it will increase processing time and may yield an increased number of false positives. On the other hand, this step needs to be sensitive enough to detect low-intensity isotopic peaks (Fig. 5). To avoid a high number of false positives during the detection of isotopic enrichment, low quality spectra are best filtered out already after deconvolution. For example, most EI-MS spectra with less than 20 peaks are usually of low quality. Although NTFD can directly process netCDF files, perform peak picking and spectrum deconvolution and the default parameters will often fit, it is advisable to check the settings used within a dedicated analysis software such as MetaboliteDetector (Hiller et al.,

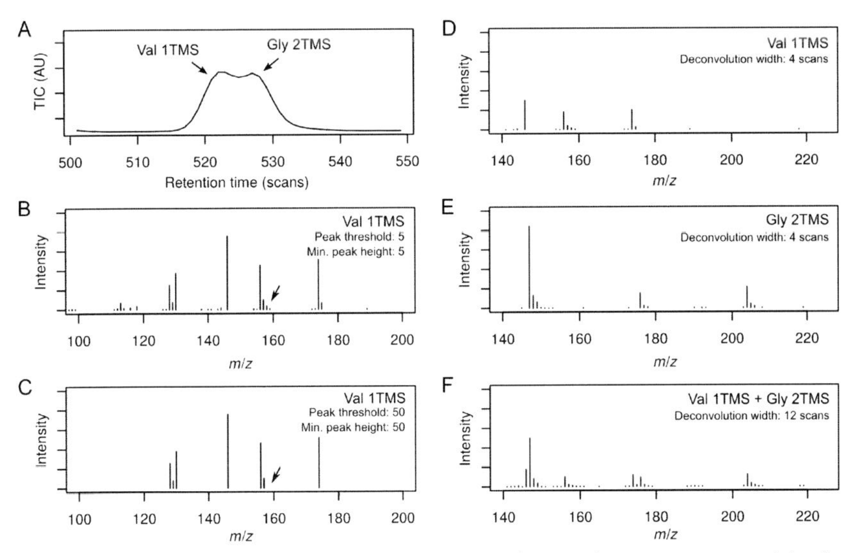

Figure 5 Proper peak picking and mass spectrum deconvolution is a prerequisite for meaningful NTFD results. (B and C) Insensitive peak detection will miss isotopic peaks or whole fragments and impair calculated MIDs. (D–F) Deconvolution settings need to be adjusted to properly extract pure mass spectra (D and E) of close-by chromatographic peaks of coeluting compounds (A). If deconvolution width is set too high, close-by chromatographic peaks will be assigned incorrectly to a single compound (F) and overlapping fragments will give rise to misleading MIDs.

2009). The advantage of MetaboliteDetector is that the results of these preprocessing steps can be visualized. Therefore, effects of different parameters can be evaluated more easily. Moreover, MetaboliteDetector can calculate the Kováts retention index (Kováts, 1958) and NTFD can directly use the generated files.

Critical parameters controlling this step (Fig. 4):

Peak threshold The minimum signal-to-noise ratio for every chromatographic peak (default value: 5).

Min Peak height The peak threshold in absolute signal intensity (default value: 5).

Deconvolution width The number of scans by which chromatographic peaks are allowed to differ to still be considered as part of the same mass spectrum (default value: 5).

Redetect all compounds The peak detection and spectrum deconvolution step only has to be performed once. However, if any of the settings are updated, it has to be repeated for all measurements.

3.2.2 Detection and Quantification of Isotopic Enrichment

While the detection of isotopically enriched fragments mainly depends on the correct peak picking and deconvolution parameters described above, the correct quantification additionally requires a correct M_1 correction (Section 2.3).

Critical parameters controlling this step (Fig. 4):

M_1 correction Needs to be adjusted for the isotopic element used in the tracer (Eq. 7). Only A +1 elements are currently supported. This value is the ratio of the natural relative abundance of the heavy p_h and light p_l isotope:

$$M1_{\mathrm{corr}} = \frac{p_h}{p_l}$$

Natural isotope abundances can be found in Berglund and Wieser (2011). Most common correction values are:

$$M1\mathrm{corr}_{^{13}\mathrm{C}} = \frac{p_{^{13}C}}{p_{^{12}C}} = \frac{0.0107}{0.9893} = 0.0108, \text{ for } ^{13}\mathrm{C} \text{ labeling}$$

$$M1\mathrm{corr}_{^{15}\mathrm{N}} = \frac{p_{^{15}N}}{p_{^{14}N}} = \frac{0.00364}{0.99636} = 0.00365, \text{ for } ^{15}\mathrm{N} \text{ labeling}$$

$$M1\mathrm{corr}_{^{2}\mathrm{H}} = \frac{p_{^{2}H}}{p_{^{1}H}} = \frac{0.000115}{0.999885} = 0.000115, \text{ for } ^{2}\mathrm{H} \text{ labeling}$$

3.2.3 Data Filtering and Quality Control

One important part of data analysis is to exclude false positives from the result set. Although NTFD offers several different parameters to exclude false positively identified fragments or fragments with incorrectly calculated MIDs, it is absolutely essential to go back and consult the original data before interpreting the results. For example, fragment overlap, which impairs MID calculation, can be easily identified by inspecting the unlabeled mass spectrum (Fig. 6). Furthermore, compounds which have not been detected as isotopically enriched need to be carefully checked before interpreting any negative results.

Critical parameters controlling this step (Fig. 4):

Tuning the following parameters will suffice for most applications. For additional *Advanced* parameters, Fig. 4, which are described in the NTFD documentation, the default values should be used.

Minimum R^2 Results can be filtered by their R^2. All fragments with a coefficient of determination below this threshold will be discarded.

Maximal fragment deviation All fragments for which the sum of absolute values of relative mass isotopomer abundances deviates from 1 by more than this value will be discarded.

Required number of labeled fragments Compounds with less labeled fragments will be discarded.

Minimal and maximal isotopic enrichment Compounds for which the detected enrichment is very low are often false positives; moreover isotopic enrichment cannot be higher than the original tracer ratio. Results can be filtered for $min \leq 1 - M_0 \leq max$.

3.2.4 Controlling Sensitivity and Specificity

The above-described parameters allow the user to adjust the sensitivity and specificity of the label detection to the specific purpose of the study. For example, if the aim of the study is to validate active metabolic pathways, the information on isotopic enrichment *per se* is of more interest than the accuracy of the quantitative enrichment. In that case, parameters should be optimized towards high sensitivity. Specifically the following settings could be applied: *Required number of labeled fragments* = 1, *Minimal isotopic enrichment* = 1, $R^2 = 0.9$, and *Maximal fragment deviation* = 0.2. As described above, it is of utterly high importance with these sensitive settings to carefully check the results for false positives. However, if the goal of the study is to quantitatively compare MIDs between two conditions, settings should be stricter. As a rule of thumb, the R^2 should be above 0.98 and the sum of absolute values should be less than 1.02 for MID interpretation. To get reliable MIDs, the following settings could be used: *Required number of labeled fragments* = 2, *Minimal isotopic enrichment* = 5, $R^2 = 0.98$, and *Maximal fragment deviation* = 0.02. Nevertheless, it is advisable to start with sensitive settings and optimize towards the best trade-off between specificity and sensitivity.

4. NTFD APPLICATIONS

Most tools for the non-targeted detection of stable isotope labeling have been developed only recently. Therefore, it is not surprising that there have not been many non-targeted stable isotope labeling analyses yet. We

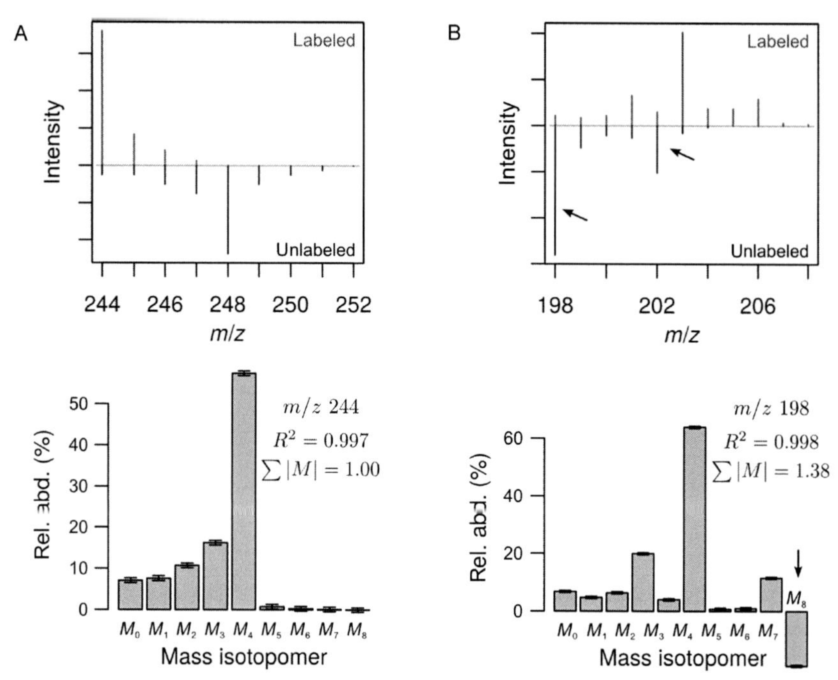

Figure 6 Quality measures of MID determination. MIDs of 2-oxoglutaric acid 2TMS 1MeOX determined after incubating mammalian cells with [U-^{13}C]glutamine. (A) Fragment *m/z* 244: High quality mass spectra without fragment overlap yield well-determined MIDs (small confidence intervals, $R^2 \approx 1.00$, $\sum|M| \approx 1.00$). (B) Overlapping fragments impair calculated MIDs. Fragments *m/z* 198 *m/z* 202 are overlapping causing a high negative M_9 abundance in the determined MID. Such MIDs are unusable, despite low confidence intervals and high R^2. (See the color plate.)

will present previously published as well as additional potential applications of non-targeted detection of isotope labeling.

4.1 Detecting Genuine Metabolites

In non-targeted metabolomics one is often confronted with a large number of chromatographic–mass spectrometric features where it is unclear whether these are artifacts from sample preparation, compounds from complex growth media, or native metabolites. After cultivating an organism on fully ^{13}C enriched substrates, one can be sure that any carbon-containing compound that is not isotopically enriched is not produced by the organism but is of exogenous origin. Therefore, non-targeted detection of stable isotope labeling has most commonly been applied to distinguish metabolites from analytical background and to determine mass spectrometric ions originating from the labeled and the unlabeled compound to use them for relative quantification of metabolite levels (Bueschl et al., 2013; de Jong & Beecher, 2012; Weindl, Wegner, Jäger, & Hiller, 2015; Yang, Hoggard, Lidstrom, & Synovec, 2013; Zhou, Tseng, Huan, & Li, 2014).

4.2 ^{15}N Tracing

Not only ^{13}C, but also ^{15}N labeling can provide interesting insights into metabolism. For mammalian cells, glutamine and glutamic acid are the main nitrogen sources. Therefore, [U-^{15}N]glutamine can be applied as a tracer to label most nitrogen containing metabolites in mammalian cells, which can in turn be detected with NTFD. In a more quantitative approach, Gaglio et al. (2011) analyzed MIDs after [α-^{15}N]glutamine labeling in a non-targeted manner to reveal altered glutamine utilization in K-Ras transformed cells.

4.3 Analysis of Mass Spectrometric Fragmentation

The non-targeted detection of stable isotope labeling in mass spectrometric fragments can be interesting for compounds with unknown fragmentation pathways. MIDs of individual fragments provide valuable information on elemental composition and origin of the respective atoms. Moreover, the MID can reveal overlapping fragments of the same mass (Wegner et al., 2014). For this purpose, we embedded the NTFD algorithm into the Fragment Formula Calculator (FFC) software (Wegner et al., 2014). FFC determines elemental composition and potential substructures of mass spectrometric fragment ions by a graph-based combinatorial approach. It can include information from stable isotope labeled spectra to rule out certain combinations.

4.4 Metabolism of Xenobiotics

In pharmaceutical research, absorption, distribution, metabolism, and excretion (ADME) studies are conducted to determine the fate of a drug or other xenobiotics within an organism. Such studies often use radiolabeled compounds or analyze stable isotope labeling in a targeted manner (Mutlib, 2008). NTFD can provide insights into metabolism of xenobiotics without the need of radiolabeled compounds. After applying a stable isotope labeled drug to an organism, NTFD can be used to detect all measured drug metabolites in a non-targeted manner.

4.5 Model Validation

Genome-scale metabolic networks (Thiele et al., 2013) are used for the stoichiometric modeling of metabolism to infer intracellular metabolic fluxes (Lewis et al., 2014). These metabolic networks are generated from transcriptomics or proteomics data. However, the presence of a certain mRNA or even protein does not tell anything about actual enzymatic activities. To this end, non-targeted tracer fate analyses can be used to determine metabolic pathways that are active under any given conditions. Furthermore, NTFD can be used to test the comprehensiveness of a metabolic network reconstruction. If a reconstruction is comprehensive, all isotopically enriched compounds should be identifiable. Otherwise, additional compounds need to be included.

4.6 Tracing Functional Groups

Dedicated tracers can be used in combination with NTFD to trace functional groups through metabolism or detect products of specific types of reactions. Methylation reactions and associated one-carbon metabolism are important for many anabolic processes as well as for detoxification of many xenobiotics and inactivation of endogenous substances such as catecholamines (Meiser, Weindl, & Hiller, 2013). The methyl groups are derived directly or indirectly from methyl-tetrahydrofolate. The main contributor to this pool is serine, or more specifically C-3 of serine. Therefore, [3-^{13}C]serine can be applied to label the one-carbon pool and all methylation products.

4.7 Cofactor Tracing

Lewis et al. (2014) and Fan et al. (2014) used deuterated glucose tracers to selectively label the cellular NADH or NADPH pool. [3-^{2}H]glucose specifically labels NADPH that is generated in the pentose phosphate pathway. Other glucose molecules that are undergoing glycolysis and are not entering the pentose phosphate pathway will lose the deuterium label to water. The deuterium from the labeled NADPH will be transferred as hydride ion by NADPH-dependent enzymes, mostly reductases and dehydrogenases, to their substrates. These products can then be detected by NTFD in a non-targeted manner. Analogously, [4-^{2}H]glucose can be used to label the cellular NADH pool and to trace NADH-dependent reductions (Lewis et al., 2014).

4.8 Non-targeted Metabolic Turnover Analysis

Non-targeted metabolic turnover analysis, was presented by Nakayama, Tamada, Tsugawa, Bamba, and Fukusaki (2014) to detect metabolic pathways affected by external perturbations. To this end, they performed principal component analysis (PCA) on a time-series of relative mass isotopomer abundances. Such analyses can reveal metabolic vicinity and group unidentified compounds into discrete pathways. More such novel workflows are needed to make full use of the information contained in metabolome-wide stable isotope labeling datasets.

5. SUMMARY AND OUTLOOK

The field of non-targeted stable isotope labeling analysis is still very young and algorithms and tools for the automated metabolome-wide detection of isotopic enrichment have only become available in recent years. Here, we presented the NTFD algorithm, explained how NTFD can be used to determine isotopic enrichment in a non-targeted manner, and finally highlighted potential applications. The advancement of current tools and the development of novel workflows will help to benefit from the high potential of non-targeted stable isotope labeling data to provide further insights into the structure and dynamics of metabolic networks.

ACKNOWLEDGMENT

This project was supported by the Fonds National de la Recherche (FNR) Luxembourg (ATTRACT A10/03).

REFERENCES

Berglund, M., & Wieser, M. E. (2011). Isotopic compositions of the elements 2009. *Pure and Applied Chemistry*, *83*(2), 397–410. http://dx.doi.org/10.1351/PAC-REP-10-06-02.

Bodamer, O. A., & Halliday, D. (2001). Uses of stable isotopes in clinical diagnosis and research in the paediatric population. *Archives of Disease in Childhood*, *84*(5), 444–448. http://dx.doi.org/10.1136/adc.84.5.444.

Bueschl, C., Kluger, B., Berthiller, F., Lirk, G., Winkler, S., Krska, R., et al. (2012). MetExtract: A new software tool for the automated comprehensive extraction of metabolite-derived LC/MS signals in metabolomics research. *Bioinformatics*, *28*, 736–738. http://dx.doi.org/10.1093/bioinformatics/bts012.

Bueschl, C., Kluger, B., Lemmens, M., Adam, G., Wiesenberger, G., Maschietto, V., et al. (2013). A novel stable isotope labelling assisted workflow for improved untargeted LC-ESI-HRMS based metabolomics research. *Metabolomics*, *10*, 1–16. http://dx.doi.org/10.1007/s11306-013-0611-0.

Chace, D. H., & Abramson, F. P. (1989). Selective detection of carbon-13, nitrogen-15, and deuterium labeled metabolites by capillary gas chromatography-chemical reaction

interface/mass spectrometry. *Analytical Chemistry, 61*(24), 2724–2730. http://dx.doi.org/10.1021/ac00199a009.
Cho, K., Mahieu, N., Ivanisevic, J., Uritboonthai, W., Chen, Y.-J., Siuzdak, G., et al. (2014). isoMETLIN: A database for isotope-based metabolomics. *Analytical Chemistry, 86*(19), 9358–9361. http://dx.doi.org/10.1021/ac5029177.
Chokkathukalam, A., Jankevics, A., Creek, D. J., Achcar, F., Barrett, M. P., & Breitling, R. (2013). mzMatch-ISO: An R tool for the annotation and relative quantification of isotope-labelled mass spectrometry data. *Bioinformatics, 29*(2), 281–283. http://dx.doi.org/10.1093/bioinformatics/bts674.
Chokkathukalam, A., Kim, D.-H., Barrett, M. P., Breitling, R., & Creek, D. J. (2014). Stable isotope-labeling studies in metabolomics: New insights into structure and dynamics of metabolic networks. *Bioanalysis, 6*(4), 511–524. http://dx.doi.org/10.4155/bio.13.348.
Christensen, B., & Nielsen, J. (1999). Isotopomer analysis using GC-MS. *Metabolic Engineering, 1*(4), 282–290. http://dx.doi.org/10.1006/mben.1999.0117.
Creek, D. J., Chokkathukalam, A., Jankevics, A., Burgess, K. E. V., Breitling, R., & Barrett, M. P. (2012). Stable isotope-assisted metabolomics for network-wide metabolic pathway elucidation. *Analytical Chemistry, 84*(20), 8442–8447. http://dx.doi.org/10.1021/ac3018795.
Dang, L., White, D. W., Gross, S., Bennett, B. D., Bittinger, M. A., Driggers, E. M., et al. (2009). Cancer-associated IDH1 mutations produce 2-hydroxyglutarate. *Nature, 462*(7274), 739–744. http://dx.doi.org/10.1038/nature08617.
de Jong, F. A., & Beecher, C. (2012). Addressing the current bottlenecks of metabolomics: Isotopic Ratio Outlier AnalysisTM, an isotopic-labeling technique for accurate biochemical profiling. *Bioanalysis, 4*(18), 2303–2314. http://dx.doi.org/10.4155/bio.12.202.
Fan, J., Ye, J., Kamphorst, J. J., Shlomi, T., Thompson, C. B., & Rabinowitz, J. D. (2014). Quantitative flux analysis reveals folate-dependent NADPH production. *Nature, 510*(7504), 298–302. http://dx.doi.org/10.1038/nature13236.
Gaglio, D., Metallo, C. M., Gameiro, P. A., Hiller, K., Danna, L. S., Balestrieri, C., et al. (2011). Oncogenic K-Ras decouples glucose and glutamine metabolism to support cancer cell growth. *Molecular Systems Biology*, 7, 523. http://dx.doi.org/10.1038/msb.2011.56.
Hellerstein, M. K., & Neese, R. A. (1999). Mass isotopomer distribution analysis at eight years: Theoretical, analytic, and experimental considerations. *The American Journal of Physiology, 276*(6 Pt 1), E1146–E1170.
Hiller, K., Hangebrauk, J., Jäger, C., Spura, J., Schreiber, K., & Schomburg, D. (2009). MetaboliteDetector: Comprehensive analysis tool for targeted and nontargeted GC/MS based metabolome analysis. *Analytical Chemistry, 81*(9), 3429–3439. http://dx.doi.org/10.1021/ac802689c.
Hiller, K., Metallo, C. M., Kelleher, J. K., & Stephanopoulos, G. (2010). Nontargeted elucidation of metabolic pathways using stable-isotope tracers and mass spectrometry. *Analytical Chemistry, 82*(15), 6621–6628. http://dx.doi.org/10.1021/ac1011574.
Hiller, K., Wegner, A., Weindl, D., Cordes, T., Metallo, C. M., Kelleher, J. K., et al. (2013). NTFD—A stand-alone application for the non-targeted detection of stable isotope-labeled compounds in GC/MS data. *Bioinformatics, 29*(9), 1226–1228. http://dx.doi.org/10.1093/bioinformatics/btt119.
Huang, X., Chen, Y.-J., Cho, K., Nikolskiy, I., Crawford, P. A., & Patti, G. J. (2014). X(13)CMS: Global tracking of isotopic labels in untargeted metabolomics. *Analytical Chemistry, 86*(3), 1632–1639. http://dx.doi.org/10.1021/ac403384n.
IUPAC. (2012). Gold book—compendium of chemical terminology. International Union of Pure and Applied Chemistry. (Version 2.3.2 2012-08-19).
Jennings, M. E., & Matthews, D. E. (2005). Determination of complex isotopomer patterns in isotopically labeled compounds by mass spectrometry. *Analytical Chemistry*, 77(19), 6435–6444. http://dx.doi.org/10.1021/ac0509354.

Kováts, E. (1958). Gas-chromatographische Charakterisierung organischer Verbindungen. Teil 1: Retentionsindices aliphatischer Halogenide, Alkohole, Aldehyde und Ketone. *Helvetica Chimica Acta*, *41*(7), 1915–1932. http://dx.doi.org/10.1002/hlca.19580410703.

Lewis, C. A., Parker, S. J., Fiske, B. P., McCloskey, D., Gui, D. Y., Green, C. R., et al. (2014). Tracing compartmentalized NADPH metabolism in the cytosol and mitochondria of mammalian cells. *Molecular Cell*, *55*(2), 253–263. http://dx.doi.org/10.1016/j.molcel.2014.05.008.

Meiser, J., Weindl, D., & Hiller, K. (2013). Complexity of dopamine metabolism. *Cell Communication and Signaling*, *11*(1), 34. http://dx.doi.org/10.1186/1478-811X-11-34.

Metallo, C. M., Gameiro, P. A., Bell, E. L., Mattaini, K. R., Yang, J., Hiller, K., et al. (2012). Reductive glutamine metabolism by IDH1 mediates lipogenesis under hypoxia. *Nature*, *481*(7381), 380–384. http://dx.doi.org/10.1038/nature10602.

Metallo, C. M., Walther, J. L., & Stephanopoulos, G. (2009). Evaluation of 13C isotopic tracers for metabolic flux analysis in mammalian cells. *Journal of Biotechnology*, *144*(3), 167–174. http://dx.doi.org/10.1016/j.jbiotec.2009.07.010.

Michelucci, A., Cordes, T., Ghelfi, J., Pailot, A., Reiling, N., Goldmann, O., et al. (2013). Immune-responsive gene 1 protein links metabolism to immunity by catalyzing itaconic acid production. *Proceedings of the National Academy of Sciences of the United States of America*, *110*(19), 7820–7825. http://dx.doi.org/10.1073/pnas.1218599110.

Missios, P., Zhou, Y., Guachalla, L. M., von Figura, G., Wegner, A., Chakkarappan, S. R., et al. (2014). Glucose substitution prolongs maintenance of energy homeostasis and lifespan of telomere dysfunctional mice. *Nature Communications*, *5*, 4924. http://dx.doi.org/10.1038/ncomms5924.

Murray, K. K., Boyd, R. K., Eberlin, M. N., Langley, G. J., Li, L., & Naito, Y. (2013). Definitions of terms relating to mass spectrometry (IUPAC Recommendations 2013). *Pure and Applied Chemistry*, *86*(7), 1515–1609. http://dx.doi.org/10.1351/PAC-REC-06-04-06.

Mutlib, A. E. (2008). Application of stable isotope-labeled compounds in metabolism and in metabolism-mediated toxicity studies. *Chemical Research in Toxicology*, *21*(9), 1672–1689. http://dx.doi.org/10.1021/tx800139z.

Nakayama, Y., Tamada, Y., Tsugawa, H., Bamba, T., & Fukusaki, E. (2014). Novel strategy for non-targeted isotope-assisted metabolomics by means of metabolic turnover and multivariate analysis. *Metabolites*, *4*(3), 722–739. http://dx.doi.org/10.3390/metabo4030722.

Niedenführ, S., Wiechert, W., & Nöh, K. (2015). How to measure metabolic fluxes: A taxonomic guide for 13C fluxomics. *Current Opinion in Biotechnology*, *34*, 82–90. http://dx.doi.org/10.1016/j.copbio.2014.12.003.

Pickup, J. F., & McPherson, K. (1976). Theoretical considerations in stable isotope dilution mass spectrometry for organic analysis. *Analytical Chemistry*, *48*(13), 1885–1890. http://dx.doi.org/10.1021/ac50007a019.

Sapcariu, S. C., Kanashova, T., Weindl, D., Ghelfi, J., Dittmar, G., & Hiller, K. (2014). Simultaneous extraction of proteins and metabolites from cells in culture. *MethodsX*, *1*, 74–80. http://dx.doi.org/10.1016/j.mex.2014.07.002.

Sauer, U. (2006). Metabolic networks in motion: 13C-based flux analysis. *Molecular Systems Biology*, *2*, 62. http://dx.doi.org/10.1038/msb4100109.

Sellers, K., Fox, M. P., Bousamra, M., 2nd., Slone, S. P., Higashi, R. M., Miller, D. M., et al. (2015). Pyruvate carboxylase is critical for non-small-cell lung cancer proliferation. *The Journal of Clinical Investigation*, *125*(2), 687–698. http://dx.doi.org/10.1172/JCI72873.

Stein, S. (1999). An integrated method for spectrum extraction and compound identification from gas chromatography/mass spectrometry data. *Journal of the American Society for Mass Spectrometry*, *10*(8), 770–781. http://dx.doi.org/10.1016/S1044-0305(99)00047-1.

Thiele, I., Swainston, N., Fleming, R. M. T., Hoppe, A., Sahoo, S., Aurich, M. K., et al. (2013). A community-driven global reconstruction of human metabolism. *Nature Biotechnology*, *31*(5), 419–425. http://dx.doi.org/10.1038/nbt.2488.

Vallino, J. J., & Stephanopoulos, G. (2000). Metabolic flux distributions in corynebacterium glutamicum during growth and lysine overproduction. *Biotechnology and Bioengineering*, *67*(6), 872–885. http://dx.doi.org/10.1002/bit.260410606, (Reprinted from Biotechnology and Bioengineering, (41), 633–646 (1993)).

Wegner, A., Weindl, D., Jäger, C., Sapcariu, S. C., Dong, X., Stephanopoulos, G., et al. (2014). Fragment Formula Calculator (FFC): Determination of chemical formulas for fragment ions in mass spectrometric data. *Analytical Chemistry*, *86*(4), 2221–2228. http://dx.doi.org/10.1021/ac403879d.

Weindl, D., Wegner, A., Jäger, C., & Hiller, K. (2015). Isotopologue ratio normalization for non-targeted metabolomics. *Journal of Chromatography. A*, *1389*, 112–119. http://dx.doi.org/10.1016/j.chroma.2015.02.025.

Yang, S., Hoggard, J. C., Lidstrom, M. E., & Synovec, R. E. (2013). Comprehensive discovery of 13C labeled metabolites in the bacterium Methylobacterium extorquens AM1 using gas chromatography-mass spectrometry. *Journal of Chromatography A*, *1317*, 175–185. http://dx.doi.org/10.1016/j.chroma.2013.08.059.

Yoo, H., Stephanopoulos, G., & Kelleher, J. K. (2004). Quantifying carbon sources for de novo lipogenesis in wild-type and IRS-1 knockout brown adipocytes. *Journal of Lipid Research*, *45*(7), 1324–1332. http://dx.doi.org/10.1194/jlr.M400031-JLR200.

Zhou, R., Tseng, C.-L., Huan, T., & Li, L. (2014). IsoMS: Automated processing of LC-MS data generated by a chemical isotope labeling metabolomics platform. *Analytical Chemistry*, *86*(10), 4675–4679. http://dx.doi.org/10.1021/ac5009089.

Isotopomer Spectral Analysis: Utilizing Nonlinear Models in Isotopic Flux Studies

Joanne K. Kelleher*^,†,1^, Gary B. Nickol*

*Department of Pharmacology and Physiology, George Washington University Medical School, Washington, USA

†Department of Chemical Engineering, MIT Cambridge, Cambridge, Massachusetts, USA

[1]Corresponding author: e-mail address: jkk@mit.edu

Contents

Abstract

We present the principles underlying the isotopomer spectral analysis (ISA) method for evaluating biosynthesis using stable isotopes. ISA addresses a classic conundrum encountered in the use of radioisotopes to estimate biosynthesis rates whereby the information available is insufficient to estimate biosynthesis. ISA overcomes this difficulty capitalizing on the additional information available from the mass isotopomer

Methods in Enzymology, Volume 561
ISSN 0076-6879
http://dx.doi.org/10.1016/bs.mie.2015.06.039

labeling profile of a polymer. ISA utilizes nonlinear regression to estimate the two unknown parameters of the model. A key parameter estimated by ISA represents the fractional contribution of the tracer to the precursor pool for the biosynthesis, *D*. By estimating *D* in cells synthesizing lipids, ISA quantifies the relative importance of two distinct pathways for flux of glutamine to lipid, reductive carboxylation, and glutaminolysis. ISA can also evaluate the competition between different metabolites, such as glucose and acetoacetate, as precursors for lipogenesis and thereby reveal regulatory properties of the biosynthesis pathway. The model is flexible and may be expanded to quantify sterol biosynthesis allowing tracer to enter the pathway at three different positions, acetyl CoA, acetoacetyl CoA, and mevalonate. The nonlinear properties of ISA provide a method of testing for the presence of gradients of precursor enrichment illustrated by *in vivo* sterol synthesis. A second ISA parameter provides the fraction of the polymer that is newly synthesized over the time course of the experiment. In summary, ISA is a flexible framework for developing models of polymerization biosynthesis providing insight into pools and pathway that are not easily quantified by other techniques.

1. INTRODUCTION

1.1 Historical Review

Isotopic tracers have a distinguished history in biomedical research going back to the 1930s (Shoenheimer & Rittenberg, 1935). These early studies utilized stable isotopes, chiefly, ^{2}H and ^{13}C. However, from the late 1940s, for over 40 years, radioisotopes were the primary form of isotopes used in biomedical investigations. Radiolabeled tracers for many metabolites were readily available and labeling could be quantified with high sensitivity using the relatively high-throughput method ofliquid scintillation counting. A chief disadvantage of radioisotopes in these protocols was that a separate method was required to identify and quantify the chemical amount of the labeled metabolite. The advent of lower cost mass spectrometers and personal computers in the 1980s, combined with the needs of clinical investigators, turned the tide again toward stable isotopes. These switches between the types of isotopes may seem to be a simply an historical note. However, as we discuss below, important consequences flow from the choice of radioisotope versus stable isotope for metabolic studies. Specifically, it was the appreciation of an important advantage of stable isotopes that led to the development in the early 1990s of the methods isotopomer spectral analysis (ISA) (Kelleher & Masterson, 1992; Kharroubi, Masterson, Aldaghlas, Kennedy, & Kelleher, 1992) and mass isotopomer distribution analysis (MIDA) (Hellerstein & Neese, 1992; Hellerstein, Wu, Kaempfer,

Kletke, & Shackleton, 1991). These methods utilize stable isotopes to overcome a key limitation of radioisotopic methodology of the day. This chapter consists of an overview of the ISA method, its development, and applications. Before delving into ISA itself, we first review a historical conundrum in the use of isotopes for estimating biosynthesis which was driving force behind the development of ISA and MIDA.

1.2 Isotope Incorporation Conundrum

In the peak period of radioisotope use, investigators became aware of a major difficulty, which thwarted their efforts to estimate biosynthesis and appeared to be an inherent property of the design of metabolic isotope incorporation studies. Consider a radioisotope incorporation study with the goal of measuring the rate of palmitate biosynthesis as diagramed in Fig. 1.

Here, the starting tracer is a labeled metabolite, perhaps [U-^{14}C]glucose or [U-^{14}C]acetate, which is metabolized to labeled acetyl CoA at rate, *a*. Because the pathways from glucose to acetyl CoA have several inputs, the starting tracer, injected into an animal, or added to cell culture media, will inevitable be diluted by the influx of endogenous compounds as it travels to the lipogenic precursor pool, the last pool for dilution prior to polymerization. Flux "*b*" represents the rate of this dilution. Biosynthesis of palmitate, measured in acetate units, occurs at rate *c*, and it is *c* that the investigator would like to measure. However, it is clear that the isotopic enrichment of the starting tracer will be decreased as it travels to the lipogenic precursor pool, and this is the source of the conundrum.

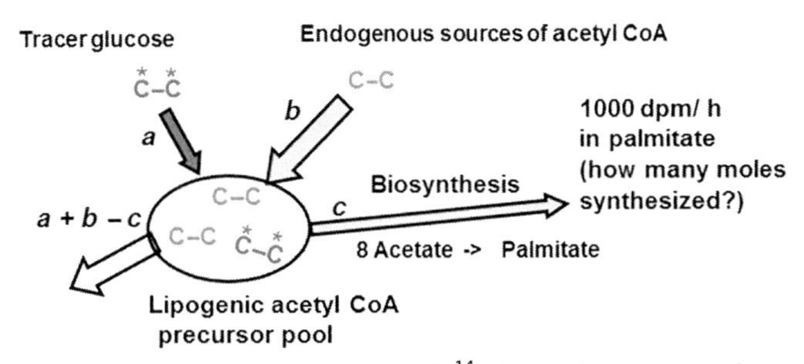

Figure 1 Isotopic biosynthesis conundrum. A ^{14}C-labeled precursor of acetyl CoA is added to a biological system synthesizing palmitate. Tracer enters at rate "*a*," endogenous sources of acetyl CoA enter at rate "*b*," and synthesis occurs at rate "*c*." The amount of radioactivity in the product, palmitate, is measured after a timed incubation. Estimating biosynthesis rate, moles/time requires the value of *a* relative to *b*, to estimate the dilution.

For radioisotopes, the term specific activity represents isotopic enrichment or the amount of radioactivity per mole of compound. Specific activity is reported as Curies/mole or Becquerels/mole, which are measured experimentally as disintegrations/time observed in a liquid scintillation counter. A Curie, for example, being 2.22×10^{12} disintegrations per minute (dpm). To estimate the true rate of biosynthesis of a product, the specific activity of the immediate precursor pool is required (i.e., the circled compounds in Fig. 1). The general radioisotope model equation for biosynthesis of a product from labeled precursor is:

$$\text{product synthesis rate} = \frac{\text{(product labeling rate)}}{\text{(immediate precursor specific activity)}} \quad (1)$$

In Fig. 1, the product labeling rate was determined to be 1000 dpm/h. The investigator knows the specific activity of the tracer when it was added to the system, but measuring the specific activity of the diluted lipogenic precursor is fraught with difficulties. In the example of acetyl CoA, multiple intracellular pools have been detected (Zhang et al., 1994), and homogenizing the cell will mix them. Therefore, it is not possible to isolate the appropriate precursor pool and measure its specific activity.

This situation is not unique to lipid biosynthesis. Perhaps, the most widely studied example of tracer incorporation is *in vivo* protein synthesis, which is estimated by the flux of labeled amino acids, usually leucine into proteins. Despite the fact that leucine is an essential amino acid, it undergoes significant isotopic dilution exchanging with endogenous leucine pools prior to incorporation into proteins. Approaches for estimating the specific activity of the leucine immediate precursor pool for biosynthesis include isolating leucine tRNA, overwhelming the endogenous pool with a flooding dose of label, and the reciprocal pool method. This last approach uses ketoisocaproic acid specific activity as a surrogate measurement for the leucine precursor pool. Despite the many years studying these approaches, debate continues regarding the optimal approach and possible size of errors involved in the various approaches to estimating protein synthesis (Wolfe & Chinkes, 2005). In summary, radioisotope studies often fail to yield valid estimates of biosynthesis due to the unknown dilution of the precursor.

This conundrum was clearly understood by metabolic investigators working in the 1960s and 1970s (Dietschy & Brown, 1974; Hetenyi, 1979; Krebs, Hems, Weidemann, & Speake, 1966). It appeared to be an inherent property of isotopes that limited the information gained from isotope incorporation studies. The resolution to this conundrum came with the development of MIDA and ISA, stable isotope methods for tracer incorporation that first took advantage of differences in the properties of radioisotopes and stable isotopes. To begin, we review a few key details of metabolic biosynthesis and the use of stable and radioisotopes.

1.3 Biosynthesis as Polymerization

A general property of all living organisms is the synthesis of relatively large molecules as polymers of metabolic subunits. Nucleic acids, proteins, lipids, ketone bodies, glucose, and glycogen are all polymers of smaller subunits. The stoichiometry of these biosyntheses is well understood: two lactate molecules form one glucose, eight acetate molecules form one palmitate. Slightly more complex relationships may be constructed for steroids such as cholesterol. Thus, many of the key molecules of interest in metabolic studies fall into the classification of polymers, and polymers are essential for the ISA and MIDA methods. A second key factor in the development of ISA and MIDA was an appreciation of the differences between radioisotopes and stable isotopes as commonly used for metabolic studies. The techniques we refer to are liquid scintillation counting for radioisotopes and chromatography coupled to mass spectrometry exemplified by gas chromatography/mass spectrometry (GC/MS). To illustrate these differences, consider that both radioisotopes and stable isotopes may be purchased at near 100% labeling but the signal to noise is far greater for radioisotopes. By this, we mean that radioisotope methods are superior for quantifying the amount of labeled material when a very small number of labeled molecules are present in a large pool of unlabeled molecules. As an example of the differences, we compare ^{14}C- and ^{13}C-labeled glucose as metabolic tracers for a typical cell culture biosynthesis experiment. Assume that the culture dish contains 10 ml of 10 m*M* glucose and will be incubated with labeled glucose for 2 days in an effort to measure the rates of fatty acid synthesis as depicted in Fig. 1. We now compare the protocols for estimating biosynthesis using either a radioisotope, ^{14}C or a stable isotope, ^{13}C.

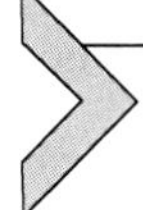

2. COMPARING RADIOISOTOPES AND STABLE ISOTOPES

2.1 ^{14}C Radioisotope Cell Incorporation Example

We first consider a radioisotope study. Ten micro-Curie of [U-^{14}C]glucose is usually quite adequate for a biosynthesis study in a cell culture plate. Ten micro-Curie corresponds to 22×10^6 dpm. This is a powerful signal as a liquid scintillation counter can accurately quantify 100 dpm above background. An additional fact is that 10 μCi of [U-^{14}C]glucose is only about 30 nmol of glucose. If the culture dish contains 10 ml of 10 m*M* glucose, adding 30 nmol of labeled glucose does not alter the glucose concentration in the media; ^{14}C-labeled molecules will represent only 3 in every 10,000 glucose molecules. For this reason, radiolabeled tracers are considered truly massless. When experiments are performed under these circumstances labeling the cells for a few days, the investigator may find 10^4–10^6 dpm in a labeled product such as palmitic acid. However, as noted above, true biosynthesis rate is not easily obtained due to the classic conundrum described above. On the positive side, liquid scintillation counting will detect all of the ^{14}C atoms incorporated into palmitate and report the total dpm in the product. Radioisotopes are a good choice of tracer if the goal is to estimate the total labeling of a class of molecules such as all fatty acids, all sterols, or all proteins. Radioisotopes are also a good choice if the fraction of total molecules that will become labeled is very small. Liquid scintillation counting can detect a few labeled molecules in a large pool of unlabeled molecules because the unlabeled molecules are invisible to the counter. But this feature is also a major disadvantage because the chemical amount of the molecule of interest must be measured by a separate method. In summary, radioisotopes are ideal for relatively few applications if the alternative is a stable isotope of the same element. However, an understanding of the differences between stable and radioisotopes allows an investigator to optimally design isotopic studies.

2.2 ^{13}C Stable Isotope Cell Incorporation Example

Compared to radioisotopes, stable isotopes are characterized by a much lower signal to noise due to the natural abundance of ^{13}C. It can be difficult to accurately estimate biosynthesis in a lipid if less than 1% of the molecules are labeled due to natural abundance of ^{13}C. Consider again the typical cell culture experiment described above, but now using a stable isotope labeling. In this case, incubation media may be prepared with 99% [U-^{13}C]glucose. Almost every glucose carbon atom in the cell culture medium will be ^{13}C labeled; and glucose-derived, labeled, 2-carbon units will be incorporated into multiple sites along the palmitic acid molecule. ^{12}C will enter the pathway only to the extent that palmitate is synthesized from molecules other than media glucose. For this type of study, GC/MS is commonly used to quantify each of the isotopomers of palmitate from $M0$ through $M16$. Figure 2 summarizes the key differences between radiolabeled isotopes and stable isotopes.

Radioisotopes provide information only in one form, total dpm incorporated. Thus, it is possible to write only one equation to describe the incorporation data resulting from an experiment. Stable isotopes provide the fractional amount of labeling for each mass isotopomer observed. In other words, one can measure both overall enrichment as well as the number of isotopes present in each molecule. As depicted in Fig. 2, a unique equation can be written representing the relative amount of each mass isotopomer observed. Therefore, the information content for the ^{13}C study exceeds that of the ^{14}C study in this example. Both ^{14}C and ^{13}C tracers at 10 mM in the media will undergo equal dilution in the pathway due to endogenous sources of acetyl CoA as shown in Fig. 1. However, only the ^{13}C tracer experiment will carry information about the intracellular dilution at the immediate precursor pool into the polymer product, palmitate. It was the appreciation of this fact that led to the development and applications of ISA and MIDA.

Radioisotope study using liquid scintillation counting

Stable isotope study using organic GC/MS

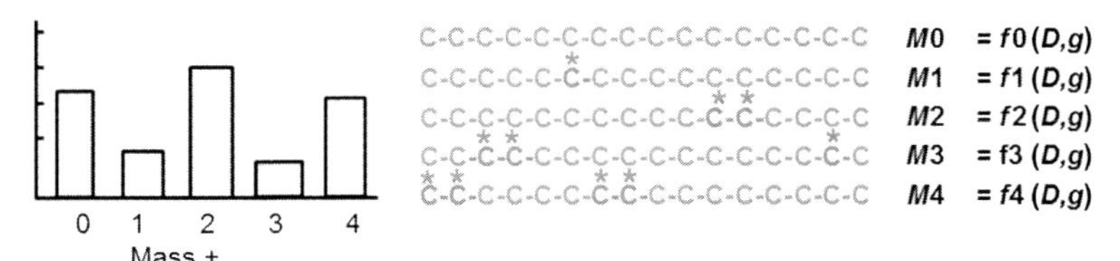

Figure 2 Information in radiolabeled versus stable isotope GC/MS experiments. Liquid scintillation counting of radiolabeled compounds typically yields only one element of information, the amount of label in the product. Unlabeled product is not detected. Organic mass spectrometry of stable isotope labeled product yields the amount of each mass isotopomer of the product, *M*0, *M*1, etc., corresponding to the number of heavy atoms in the product. A mathematical model can be used to write a function corresponding to each value in terms of the unknown parameters *D* and *g*(*t*).

2.3 Defining Stable Isotope Enrichment

A variety of terms have been used to express the degree of enrichment of stable isotopes. Unfortunately, the meaning of some of them is not clear or has evolved over time. A first requirement is that the formula for expressing isotopic enrichment must be compatible with the model to be used. Here, we investigate models based on probabilities and models that predict fractions of labeling. Accordingly, the terminology for compounds labeled with ^{13}C must represent the amounts of each mass isotopomer of a compound or ion from $M0$ (all ^{12}C atoms) to Mn (all ^{13}C atoms). We have used two different terms for indicating the relative amounts of each isotopomer of a molecule or ion, fractional abundance (FA) and moles percent excess (MPE). FA and MPE as used by our group are identical and represent the ratio of amount of a specific mass isotopomer, Mx to the sum of all isotopomers of the relevant compound or ion. Using MPE or FA, each isotopomer is represented as the fraction it contributes to the total. The sum of all isotopomers is 1. The FA of each isotopomer is less than or equal to 1 and is identical to the probability of that isotopomer in the population.

Unfortunately, a widely used term in biomedical isotope research, tracer-tracee ratio (Cobelli, Toffolo, & Foster, 1992; Wolfe & Chinkes, 2005) is not suitable for ISA and fractional synthesis techniques described here. The problem with the tracer-tracee ratio is that the value of $M0$ is set equal to 1. There is no requirement that all isotopomers sum to 1 or to any constant value. If a labeled isotopomer is more abundant than $M0$, it will have a tracer-tracee ratio greater than 1. This method of expressing isotopic abundance cannot be used for models based on probability or where the result is a fraction such as fractional synthesis. Moreover, in developing advanced techniques for predicting isotopomer distribution in pathway using elementary metabolic units, the MPE or FA approach is required (Antoniewicz, Kelleher, & Stephanopoulos, 2007; Jazmin et al., 2014).

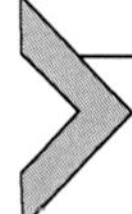

3. ISA METHODOLOGY

3.1 Preliminary Studies

Our laboratory was active in the period when the transition from radioisotopes to stable isotopes was well underway. While using radioisotopes, we investigated the role of glutamine in cancer cells and the possibility glutamine was a net source of carbon for lipid synthesis. This would occur via reverse or reductive flux through isocitrate dehydrogenase. We were aware of earlier work with radioisotopes (D'Adamo & Haft, 1965; Sabine, Kopelovich, Abraham, & Morris, 1973) indicating that this pathway was feasible. We investigated the pathway in a microorganism, *Dictyostelium,* where the pathway was not detected (Kelleher, Kelly, & Wright, 1979) and in rat AS30D hepatoma cells, where the reductive carboxylation pathway was active and led us to conclude that glutamine was a carbon source for lipogenesis via reductive carboxylation (Holleran, Briscoe, Fiskum, & Kelleher, 1995). However, because of the classic conundrum described above, we were unable to quantify the total flux of glutamine to lipid because we did not know the fractional contribution of glutamine carbon to the acetyl CoA pool that was the precursor for lipogenesis.

In exploring the move from radioisotopes to stable isotopes, we discovered that stable isotope methodology offered a new opportunity to resolve the longstanding conundrum of isotopic biosynthesis because a model could be built with a unique equation for the fraction abundance of each isotopomer. Independently, Hellerstein and colleagues had also made this discovery (Hellerstein & Neese, 1992; Hellerstein et al., 1991). Paul Lee's lab also made contributions to the development of these techniques (Lee, Bergner, & Guo, 1992; Lee, Byerley, Bergner, & Edmond, 1991). All of the approaches cited here are based on the same theory. The most important differences lie in the methods used to evaluate the model. In developing ISA, we focused on the nonlinear nature of the model equations and the requirement for nonlinear regression to estimate the parameter values. The discussion here will emphasize this aspect of the ISA model.

3.2 Classic ISA Model

It has been said that a model has a twofold function: it should provide a simplified description of the physiological process and it should provide a framework for quantitative analysis of those processes. The first step in developing a metabolic model is to diagram the pathway and identify the unknown parameters to be estimated as depicted in Fig. 3.

Shown here is the classic ISA model to quantify the biosynthesis of a fatty acid. A labeled metabolite is introduced into a biological system and metabolized along the pathway leading to *de novo* lipogenesis. Immediately prior to lipogenesis, the labeled metabolite enters the lipogenic precursor acetyl CoA pool. An important unknown parameter is the fractional contribution of the labeled metabolite to this pool, D. D is often described simply as the precursor dilution. The model assumes that D is a constant; it does not vary during

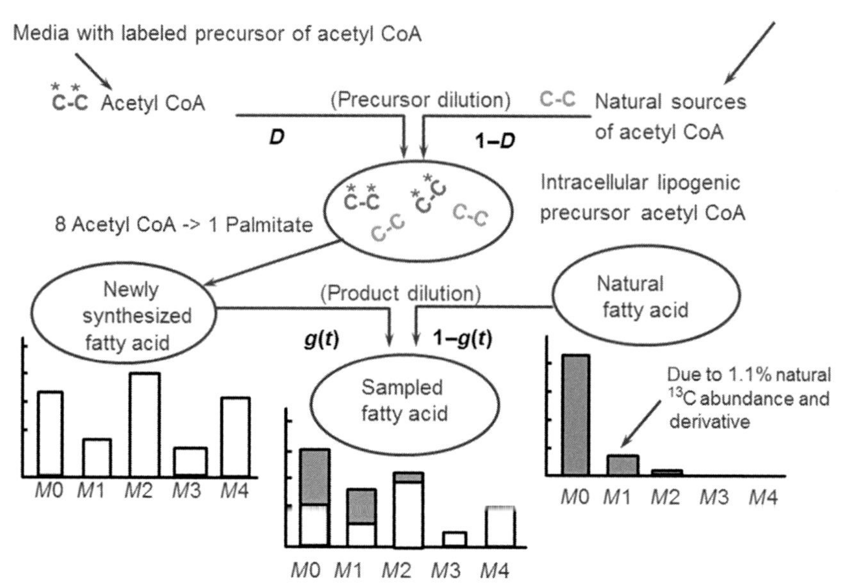

Figure 3 Classic ISA model for biosynthesis of a polymer. Label enters at acetyl CoA and mixes with endogenous sources of cytosolic acetyl CoA to form the lipogenic precursor pool. D and $g(t)$ are the two unknown parameters. Following biosynthesis, the newly synthesized palmitate is diluted with preexisting pools of palmitate of natural abundance. The goal of ISA is to estimate the values of D and $g(t)$ from the sampled palmitate pool. (See the color plate.)

the course of the experiment. Biosynthesis entails polymerization of acetyl CoA units from the lipogenic precursor pool into a fatty acid. Acetyl units are selected randomly from the mixture comprising the lipogenic precursor pool. Following biosynthesis, the newly synthesized fatty acid with its incorporated label will mix with preexisting fatty acids characterized by a profile consistent with the natural abundance of ^{13}C. A second unknown parameter is the fraction of newly synthesized fatty acid in the sampled pool. The parameter, $g(t)$, represents this product dilution. The labeling of the sampled pool is expected to increase over time. Hence, $g(t)$ is a time-dependent parameter with "t" representing the elapsed time of the incubation. Note that D and $g(t)$ are both fractions without units. They simply represent the flow into common pools from the relevant biological sources. Experimenters have access only to the sampled fatty acid pool and may determine its mass isotopomer distribution. From Fig. 3, it is apparent that there are two unknown parameters in this classic ISA model. Thus, at least two independent equations are required to estimate these parameters. Fortunately, ^{13}C-labeling studies in cell culture will often provide an overdetermined set of equations so that the goodness of fit of the model to the data may be assessed. Additionally, because of the surplus of equations, the model may be expanded to situations requiring additional parameters. The MIDA model developed by Hellerstein and colleagues describes the same two parameters using different symbols, p instead of D and f instead of $g(t)$.

With the model diagram in hand, the next step is to write the equations that predict the FA of each isotopomer of product. The equations, based on combinatorial probability theory, depend on the stoichiometry of the product formation. The first two equations for the synthesis of the 16 carbon fatty acid palmitate from acetate are presented in Fig. 4.

The complete ISA model for palmitate will contain 17 equations for product palmitate $M0$ through $M16$. The subsequent equations contain a large number of terms but are easily assembled from simple probability theory. Although each isotopomer equation is unique, they share a common pattern. Each equation is assembled as a linear sum of two possible forms of palmitate, new synthesis, and preexisting as shown in Fig. 4.

An important feature of the model is that the FA of each two carbon isotopomer of acetate derived from the tracer, Tx, and natural abundance acetate, Nx, are inputs to the model. Tracer acetate labeling may be a mixture of any combination of $M0$, $M1$, and $M2$ isotopomers as long as it differs from the distribution for natural abundance acetate. These must be measured or

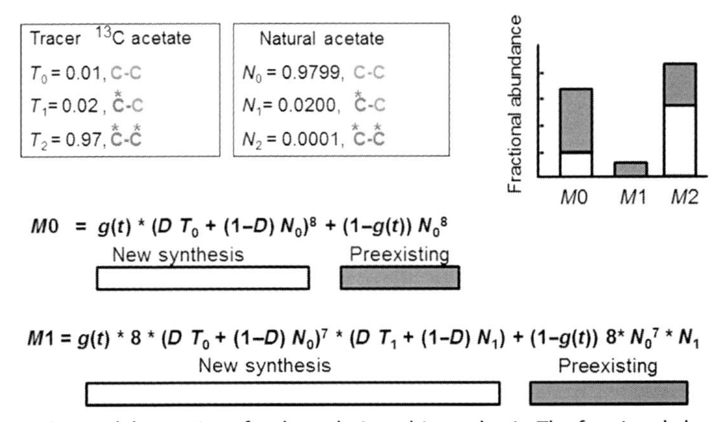

Figure 4 ISA model equations for the palmitate biosynthesis. The fractional abundance of the three isotopomers of C–C subunits from the tracer and from the natural abundance acetate present in the cell is inputs to the model. The first 2 of the 17 model equations are shown with the sections of the equations representing new synthesis and preexisting palmitate marked in yellow and green, respectively. (See the color plate.)

estimated by the investigator. It is required that the sum of the amounts of isotopomers of acetate generated by the tracer will equal 1 as will the sum of natural abundance acetate isotopomers. The model also contains the isotopomer abundance profile for any derivatization agent attached to the ion to be monitored and the natural abundance of all noncarbon elements in the fragment ion of interest. Including these values in the model does away with the necessity of performing corrections of the raw data to account for natural abundance. In our first studies, we performed corrections on the raw data before modeling following the lead of many investigations at the time. However, we settled on methods where the raw data was simply converted to FAs and the model was designed to simulate these data including all natural abundance and derivatization effects.

The ISA equations predict the FA of each mass isotopomer of the product. In addition to the values for the tracer and natural acetate FA, these equations contain the two unknowns, D and $g(t)$ (Fig. 4), which we want to estimate. The parameter, D, represents the fractional contribution of the tracer to the intracellular lipogenic acetyl CoA pool. The parameter, $g(t)$, represents the fraction newly synthesized but it should not be confused with the fractional synthesis rate. By this we mean that the value of $g(t)$ is the actual fractional amount of newly synthesized product in the sampled pool. However, because the pool may turn over and its labeling rate is not linear, the fraction of newly synthesized divided by the time, t, is not equivalent to the fractional synthesis rate. It is possible to estimate fractional synthesis with a series of estimates of $g(t)$ but simply dividing by time should be avoided because of the nonlinearity of fractional synthesis. Also $g(t)$ is sometimes abbreviated simply as "g" but it is important to note that it is time dependent.

3.3 Dynamic Steady State

The ISA model is based on the concept of "dynamic steady state" where the metabolic rates are constant and the isotopic labeling enrichment of all intermediates in the pathway quickly reaches a constant value throughout the system under study. However, the product of the biosynthesis would only be partially labeled over the time course of the experiment. Dynamic steady state is exemplified by a cell culture experiment where cells are transferred to media containing a labeled precursor such as glucose or acetate and exposed to this media for several days. The experiment is quenched and the end product of biosynthesis such a palmitate would be analyzed. We assume the pathway intermediates are small pools with rapid turnover times that become fully labeled relatively fast compared to biomass. However, the end product lipids, nucleic acids, and proteins turnover much slower and only a portion of the product would be newly synthesized during the course of the experiment. The remainder of the product would be either preexisting product molecules or product molecules taken up from the media. In addition, we assume that the cell population is relatively homogeneous with respect to the biosynthesis of the product. These conditions are required for the validity of the classic ISA model.

Dynamic steady state also applies to *in vivo* human studies where tracer is infused for several hours and end products such as plasma sterols are assayed as products (Clarenbach et al., 2005; Lindenthal et al., 2002). Although the values for the two parameters are usually much smaller *in vivo* and the number of isotopomers of product is fewer, the assumptions of dynamic steady state are required to utilize the classic ISA model. We have encountered situations *in vivo* where the assumptions of dynamic steady state do not apply and will discuss a specific example below. However, before we can estimate the parameters D and $g(t)$, it is important to review one additional feature of the model, its nonlinearity.

4. LINEAR VERSUS NONLINEAR MODELS

In addition to the choice of stable versus radiolabeled isotopes, isotopic metabolic studies may be differentiated based on the type of model used to analyze the data. Often tutorials on tracer methodology divide tracer studies into two broad classes "tracer dilution" and "tracer incorporation." Tracer dilution is characterized by studies where a labeled compound is injected into the blood stream at a constant rate and the rate of endogenous appearance of that compound is estimated by isotope dilution at steady state. Tracer incorporation is exemplified by studies estimating the rate of biosynthesis by injecting a precursor molecule and measuring the rate of incorporation of a precursor molecule into a larger product molecule. Protein and DNA synthesis, measured by the incorporation of labeled leucine or thymidine, are examples of tracer incorporation as are ISA and MIDA. While the division between tracer dilution and tracer incorporation has merit, we propose that a fundamental division in isotopic studies is between those specified by linear versus nonlinear models. Linearity is an important property of a model because it affects the methods used to estimate the parameters.

Models consist of one or more equations and these equations containing terms, which may be linear or nonlinear. The classic definition is that linear terms are of the first degree in the dependent variables and their derivatives. Linear equations are simply sums of linear terms. All other equations are nonlinear. See the recent text by DiStefano for more details about this important distinction (DiStefano, 2014). In examining our ISA model, we noticed that the equations were not simply linear (Fig. 4). The $g(t)$ is a linear term, but D, because it is raised to powers greater than one, is a nonlinear term. Furthermore, because the model is nonlinear in D, the proper way to estimate the unknown parameters utilizes nonlinear regression or related techniques.

Although biomedical research includes a number of prominent nonlinear models, this fact has not been emphasized. For example, a classic biomedical nonlinear equation is the Michaelis–Menten equation for enzyme kinetics. To solve for the parameters, V_{Max} and K_m, biomedical investigators commonly transform and linearize the equation so that linear regression may be used. Thus, the nonlinear properties of the model are obfuscated. Double reciprocal plots and other transformations, developed before computers were widely available, continued to be used to estimate parameters in enzyme kinetics and receptor binding despite the fact that these practices do not properly consider the error in the measurements (Martin, 1997). Thus, familiarity with nonlinear models and the best methods for estimating parameters of nonlinear models was not common in biomedical research at the time ISA was developed. We were not able to locate software packages for nonlinear regression to estimate parameters for the ISA type of model and wrote our own nonlinear regression software using the Levenberg–Marquardt algorithm (Draper & Smith, 1981) and the programming technique developed by our colleagues (Nelson & Homer, 1983). Today, suitable software is available in many packages such as Matlab and even Excel.

5. ESTIMATING ISA PARAMETER VALUES USING NONLINEAR REGRESSION

The approach to solving for the parameters of ISA begins with initial values as input to the software, or best guesses for the parameters. Starting with these initial values the ISA model simulates the FA values for each isotopomer (Fig. 5).

Panel A depicts the GC/MS data for each isotopomer plotted alongside the guess values used to initiate the nonlinear regression. The sum of squared error (SSE) is a function of these values as plotted in Panel C.

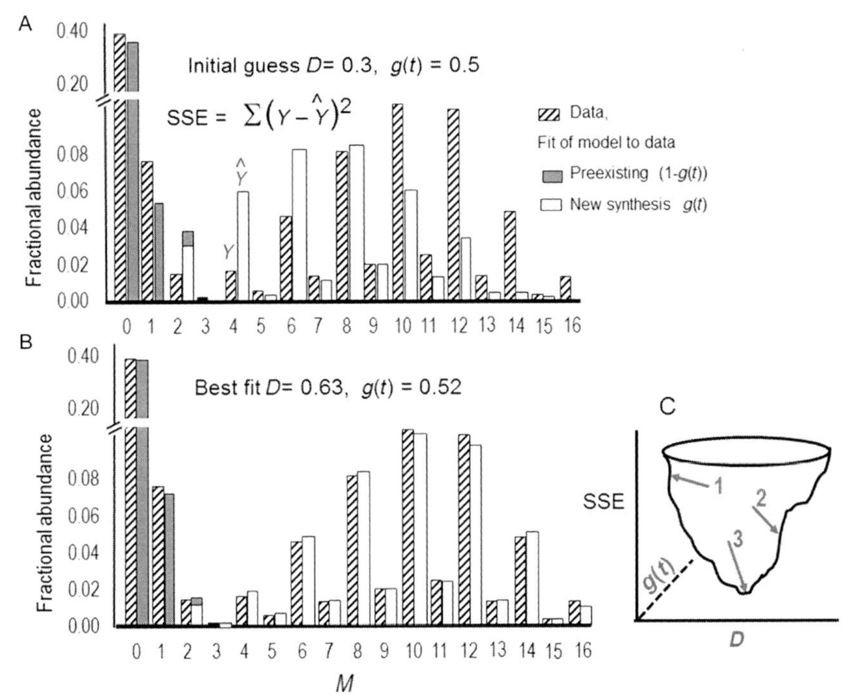

Figure 5 Solving for ISA parameters using nonlinear regression. Experimental data and model fits for hepatoma cells in medium 2 m*M* containing [1,2-^{13}C]acetate (2 m*M*). GC/MS data converted to fractional abundance values are plotted as stripped bars. ISA output simulations for each mass isotopomer plotted to the right of data. (A) Initial simulation with best guess starting values. (B) Best fit results found with ISA. (C) Plot of error *Z*-axis versus the two parameters, *D* and $g(t)$. (See the color plate.)

Arrow 1 represents the error observed with these initial values. After each iteration, the program calculates the SSE for the fit of the model to the data. SSE is the objective function determining the fit of the model to the data. The role of the Levenberg–Marquardt algorithm is to choose the next set of D and $g(t)$ values searching for the best fit with the lowest SSE. Iterations continue until the best fit is found. Panel B depicts the best fit solution with D and $g(t)$ values as shown. Arrow 3 in panel C depicts the error at the best fit.

As depicted in Fig. 5, the models used for cell culture labeling of lipids are usually overdetermined. A large number of isotopomers are detected. An equation predicting the FA of each isotopomer is contained within the ISA model. Hence, the number of independent equations is greater than the number of unknowns (two). With overdetermined models the issue of weighting the data becomes relevant. For cell culture studies, we have not found any advantage using weighting other than assigning all equations a weight of 1. However, we do advocate using the data for all of the detected isotopomers as inputs to model. Approaches using just the minimum number of equations, so that the model is not overdetermined, are essentially giving a weight of 0 to the data not used. If the model is not overdetermined, it will not be possible to estimate the goodness of fit of model to data. Additionally, the model must be overdetermined to reveal situations where the classic ISA model (Fig. 3) is not adequate to describe the data.

6. EXAMPLES OF ISA MODELS

6.1 Reassessing Reductive Carboxylation with ISA

With the ISA modeling in hand, we sought to quantify the fluxes of glutamine to lipid via reductive carboxylation that we had observed earlier with radioisotopes (Holleran et al., 1995). The experimental design reflected that of the radioisotope studies except that we switched from AS30D rat hepatoma cells to HepG2 human hepatoma cells. Replicate dishes were incubated with either [U-^{13}C]glutamine or [5-^{13}C]glutamine (4 m*M*) in cell culture medium. We sought to quantify the contribution of the reductive carboxylation pathway to lipogenesis by comparing it to the glutaminolysis pathway, which was the best described pathway for glutamine metabolism in rapidly dividing cells at the time (Board & Newsholme, 1996; Newsholme, Crabtree, & Ardawi, 1985; Fig. 6). Although a number of publications argued in favor of the glutaminolysis pathway, quantitative isotopic studies had not been performed. The rationale for the choice of isotopes to quantify glutamine fluxes is that flux of glutamine to lipid via glutaminolysis results in loss of label at position 5 of glutamine while reductive carboxylation retains it resulting in incorporation of ^{13}C–^{12}C units (Fig. 6). For comparison, [U-^{13}C]glutamine labels lipid with ^{13}C–^{13}C units via either pathway.

To estimate the relative fluxes, we evaluated triglyceride palmitate synthesis in replicate dishes with each glutamine tracer and used ISA to estimate the *D* value for each tracer. For this scheme, the *D* value for [U-^{13}C]glutamine estimates the total fraction of acetyl CoA derived by glutamine using either pathway. The ratio of *D* for [5-^{13}C]glutamine over *D* for [U-^{13}C]glutamine indicates the fraction of the glutamine in lipogenic acetyl CoA that used the reductive carboxylation pathway. Since, the two glutamine tracers yielded different tracer acetate fragments, it was not surprising that the

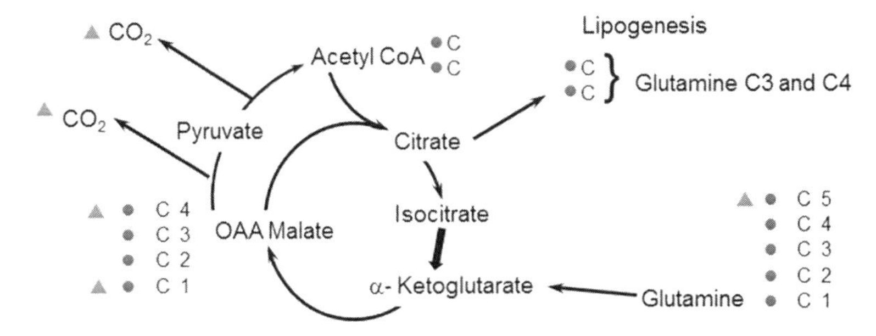

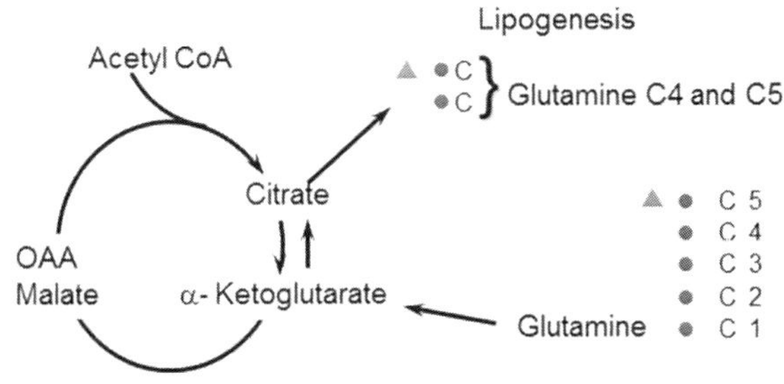

Figure 6 Comparison of pathways for fatty acid biosynthesis from glutamine. (A) Glutaminolysis pathways for fatty acid biosynthesis. (B) Reductive carboxylation pathway. The fate of [U-^{13}C]glutamine carbon atoms (circles) versus [5-^{13}C]glutamine (triangles) reveals the difference in lipid labeling following each path.

experimental FA profile of the resulting palmitate isotopomers were very different as seen by comparing the stripped bars for the two tracers (Fig. 7). However, when ISA solved for the two parameters there was a good fit to the model for both tracers and the D and $g(t)$ values were statistically identical for both tracers. Recall that the ISA D value is the fractional contribution of tracer acetate in the precursor pool regardless of whether the acetate coming from the tracer is singly or doubly labeled. Since there was no difference in the D value for the two tracers, we conclude that all of the glutamine flux to lipid was via reductive carboxylation. Furthermore, the D value of 0.4 indicates that 40% of newly synthesized palmitate is derived from glutamine. Thus, glutamine carbon makes a major contribution to lipogenesis.

We have measured this flux in a number of additional cell lines. For a brown fat cell line, we found similar results that glutamine contributed

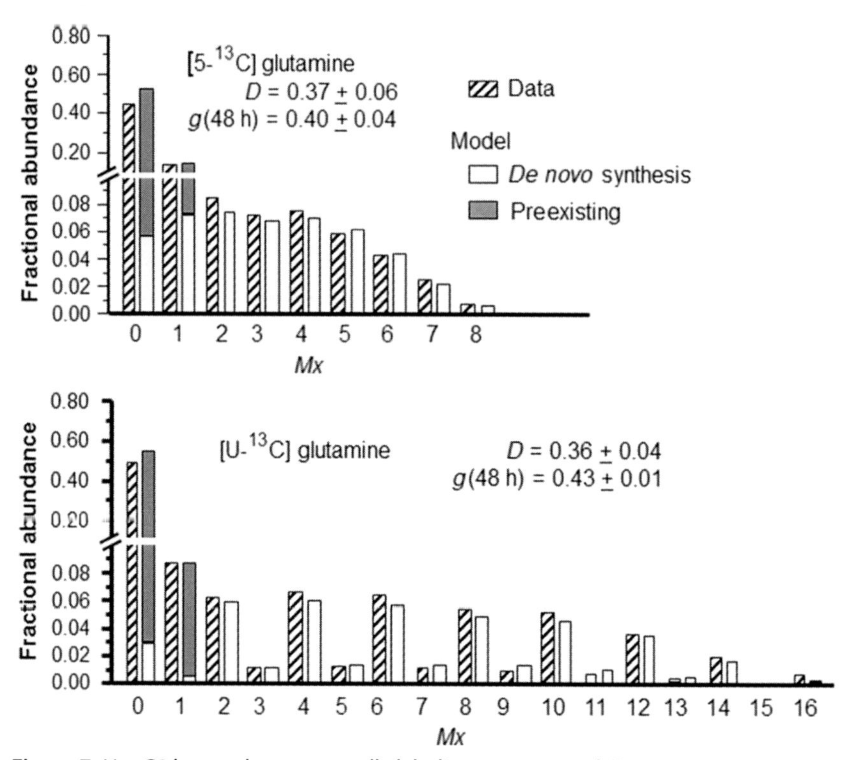

Figure 7 HepG2 human hepatoma cells-labeling patterns and ISA parameter estimates. Cells incubated with either [5-^{13}C]glutamine or [U-^{13}C]glutamine. Striped bars represent actual data plotted alongside model results indicating relative amounts of newly synthesized palmitate and preexisting at 48 h for best fit parameter values. (See the color plate.)

30–40% of the acetyl units for *de novo* lipogenesis and 85–95% of the flux to citrate was via the reductive carboxylation pathway (Yoo, Antoniewicz, Stephanopoulos, & Kelleher, 2008). Reductive carboxylation changes dramatically with hypoxia in cancer cells, nontransformed proliferating cells, and activated T cells rising from less than 20% contribution to acetyl CoA to nearly 80% (Metallo et al., 2012). Each tumor cell line represents an independent series of genetic transformation events leading to uncontrolled growth. Thus, we expect that reductive carboxylation will vary across cell lines. The challenge going forward is to utilize the quantitative information about this pathway available through ISA to better understand metabolic regulation in cancer cells. In summary application of requirements, these finding clearly demonstrate the power of ISA to detect the path of lipogenic of glutamine and to quantify its fractional contribution.

6.2 Modifying ISA to Quantify Fluxes from Multiple Precursors

ISA has the potential to evaluate the role of multiple precursors measuring their contribution to the lipogenic acetyl CoA pool. This is illustrated by a study focusing the role of ketone bodies versus glucose as an oxidative substrate and a precursor for the lipogenic acetyl CoA pool in hepatoma cells (Holleran, Fiskum, & Kelleher, 1997). A first step was expanding the classic ISA model (Fig. 3) to allow multiple precursors, glucose and acetoacetate in this example (Fig. 8). The interest in ketone body metabolism in hepatoma cells was motivated by previous studies. Although adult liver cells do not metabolize ketone bodies, we found that AS30D hepatoma cells do metabolize ^{14}C-acetoacetate, primarily to lipid and CO_2. Additionally, we found high activity of succinyl CoA acetoacetyl CoA transferase (SCOT) in AS30D hepatoma cells (Briscoe, Fiskum, Holleran, & Kelleher, 1994). The expression of this enzyme, which catalyzes the first step in the cytosolic metabolism of acetoacetate, is very low in adult hepatocytes. Similar observations had been reported in other hepatoma cells (Fenselau, Wallis, & Morris, 1975; Hildebrandt, Spennetta, Elson, & Shrago, 1995). However, the flux through this enzyme was previously unknown.

ISA provided the ideal tool to test the relative preference for acetoacetate over glucose as a lipogenic precursor. All of the previous flux studies had utilized radioisotopes, so it was previously not possible to confirm the role of acetoacetate in hepatoma cells. The ISA analysis followed the fate of

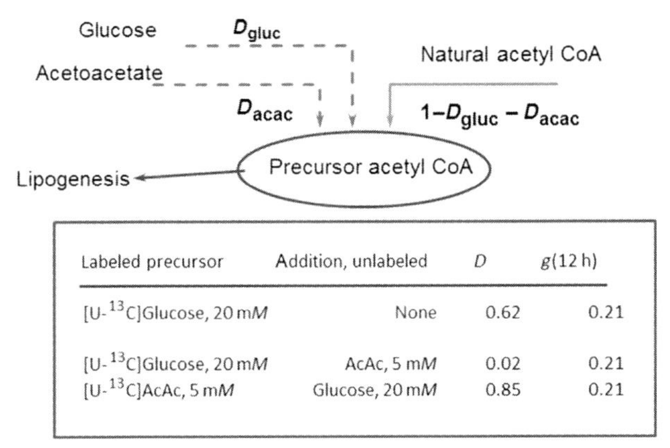

Labeled precursor	Addition, unlabeled	D	g(12 h)
[U-^{13}C]Glucose, 20 mM	None	0.62	0.21
[U-^{13}C]Glucose, 20 mM	AcAc, 5 mM	0.02	0.21
[U-^{13}C]AcAc, 5 mM	Glucose, 20 mM	0.85	0.21

Figure 8 ISA model and results for competition between glucose and acetoacetate in As-30D hepatoma cells. Precursor contributions of distinct precursors for lipogenesis are depicted with the balance from natural labeled sources of acetyl CoA. ISA estimates of tracer fluxes are shown indicating preference for acetoacetate.

[U-^{13}C]glucose and [U-^{13}C]acetoacetate in a variety of media combinations. As before the values of precursor dilution, *D*, and fractional new synthesis, *g*(*t*), were estimated (Fig. 8). This analysis revealed that, in the absence of acetoacetate, glucose is the major source of lipogenic acetyl CoA, supplying 62% of the acetate units for lipogenesis. However, the addition of acetoacetate to the media almost abolishes the flux of glucose to acetyl CoA, as acetoacetate carbon diminishes the glucose contribution and accounts for 85% of the acetate units in the lipogenic precursor pool. This study demonstrates the use of ISA to provide insights into the competition for fluxes in various pathways when duplicate experiments are performed in the same chemical medium altering only the ^{13}C-labeled substrate. The last two rows of the table are an example of this experimental design. In summary, these studies highlight the potential of ISA to uncover regulatory relationships in pathways leading to precursor pools.

To further explore the contributions of glucose and acetoacetate in hepatoma cells, substrate oxidation studies were designed using ^{14}C-labeled tracers. ^{14}C-labeled tracers are suitable for estimating the oxidation of a labeled substrate to CO_2. Radiolabeled CO_2 atoms will constitute only a tiny fraction of total CO_2, especially if bicarbonate CO_2 buffers are used, yet the radiolabeled signal is readily detected when metabolic $^{14}CO_2$ is released by cells and trapped in a basic solution. The use of ^{13}C-labeled tracers to measure $^{13}CO_2$ is feasible but requires isotope ratio mass spectrometry or nondispersive infrared spectrometry and is chiefly used for clinical studies (Schadewaldt et al., 1997). We had previously developed a method for assessing key parameters of TCA cycle fluxes using measurements of the labeled CO_2 released from different tracers and call this method the "$^{14}CO_2$ ratios technique" (Kelleher & Bryan, 1985). The technique requires identical culture dishes labeled with tracers of acetate and pyruvate, [2-^{14}C] or [3-^{14}C]pyruvate and [1-^{14}C] or [2-^{14}C]acetate and measuring the $^{14}CO_2$ production rate from each tracer. When pyruvate $^{14}CO_2$ ratios are compared to acetate ratios, an estimate of fraction of pyruvate metabolizes through pyruvate carboxylation versus pyruvate dehydrogenase results. Applying the $^{14}CO_2$ ratios technique to assess the role of acetoacetate versus glucose in hepatoma cells revealed that addition of acetoacetate to the media dramatically shifted the flux of glucose-derived pyruvate into the TCA cycle via the pyruvate carboxylase route rather than the through pyruvate dehydrogenase (Holleran et al., 1997). These experiments illustrate the combination of multiple types of isotopic techniques to explore metabolic pathways.

6.3 ISA Model of Sterol Synthesis, a Pathway with Multiple Inputs

Fatty acids provide a clear but simple model of polymerization starting with acetyl CoA. Cholesterol synthesis is more complex because of loss of some of the carbon atoms along the biosynthesis route. In addition cholesterol synthesis may have a greater variety of precursors. In choosing a model for cholesterol synthesis, an investigator could use a simple model based on acetyl CoA as the only relevant precursor and using the classic stoichiometry:

$$15\,\text{Acetate C2} + 12\,\text{Acetate C1} \rightarrow 1\,\text{Cholesterol}. \tag{2}$$

This relationship assumes independent probabilities for the incorporation at each position on the cholesterol molecule and may be used for labeled precursors metabolized to acetyl CoA which will be labeled on either C1 or C2 but not both. To allow for precursors with multiple labeled carbons on a single molecule and to allow for input into the biosynthesis at multiple additional points, a more complete model is required (Fig. 9).

The major change in this model is that there are now multiple precursor dilution parameters. As with the classic model, parameter D represents the labeled precursor entering at acetyl CoA. Parameter, E, allows flux from acetoacetyl CoA directly to the cholesterol acetoacetyl CoA precursor pool

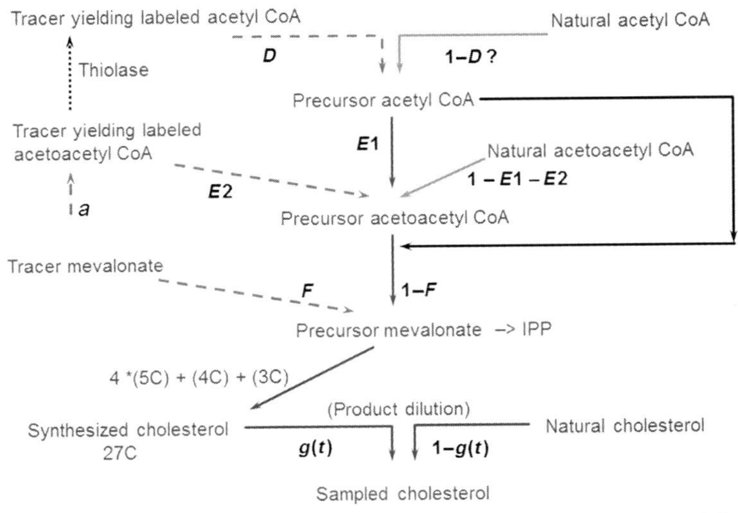

Figure 9 ISA model for mammalian cholesterol synthesis. Sterol synthesis modeled as a series of condensation reactions. Labeled precursor may be derived from tracers yielding acetyl CoA, acetoacetyl CoA or mevalonate, dashed lines. Tracer acetoacetate enters at "*a*".

avoiding the conversion of acetoacetyl CoA to acetyl CoA. Parameter, *F*, allows for including tracer mevalonate in the medium. We analyzed cholesterol synthesis in HepG2 cells using this more complete model and a variety of labeled substrates (Kelleher et al., 1994). Although mevalonate is not normally found in extracellular media, we exposed cells in culture media to labeled mevalonate to determine if it could replace typical endogenous substrates and found that $F=0.94$ indicating a preference for mevalonate from the media over endogenous mevalonate synthesis. The model was also useful in evaluating the path of acetoacetate to cholesterol. Consider labeled acetoacetate from the media entering the pathway at point "*a*" in Fig. 9. Acetoacetate may be converted directly to acetoacetyl CoA if the SCOT enzyme is active. Once converted, the molecule may now travel by either of two paths. It may go directly to acetoacetyl CoA for mevalonate synthesis (path *E*2) or it may be hydrolyzed to acetyl CoA (path *D*). Using a model allowing both possibilities and ISA we estimated that the *E*2 pathway was detectable but contributed only 8% to the acetoacetyl CoA pool for mevalonate precursor pool. Allowing both pathways produced a far better fit of the data to the model than analysis of models allowing only one of the two paths. In summary, this work emphasizes the concept that the basic ISA model is a framework that may be modified for a variety of more complex polymer biosynthesis reactions.

6.4 ISA Model Investigating Varying Precursor Contribution

As described above, ISA is a nonlinear model because the precursor dilution parameter, *D*, is raised to various powers in the equations predicting the FA of each mass isotopomer (Fig. 4). A consequence of the nonlinear terms is that the model is dependent on the value of *D* remaining constant over the course of the experiment. The model assumes that the tracer incubation or infusion is conducted such that the time for the initial flux of tracer into the precursor pools is negligible relative to the total time course of the experiment. After this period *D* must remain constant. This is not required for the linear term, $g(t)$ fraction newly synthesized. Consider first variation in the linear term, $g(t)$ when *D*, the nonlinear term, is a constant. If an experiment is conducted labeling for 48 h with variations in the rate of synthesis of the polymer, the resulting isotopomer labeling could be investigated with ISA. The value of g(48h) estimated by ISA will be the average value of g over the time course. *D* would be estimated correctly. It is a fundamental property of ISA that it is insensitive to variations in the rate of synthesis. Now consider variation in the value of the precursor dilution. If *D* varies the model will not perform as expected. Variations in nonlinear parameter values do no result in simple averages. The data resulting from this situation will not produce correct estimates of either *D* or *g*(48 h), using the classic form of ISA (Fig. 3) which is dependent on the assumption that *D* is constant. In actuality of course, there will be an initial time period before the tracer equilibrates in the precursor pools. ISA assumes that product synthesized in this initial period is negligible over the entire time course of the experiment. Although the model dependency on a constant value for *D* may appear as a weakness, this weakness can be exploited to gain new information. In summary, it is important to understand which terms in a model are linear or nonlinear and to consider the consequences of linearity when interpreting the results of a model.

We took advantage of the nonlinear term for precursor contribution in studies of sterol synthesis in dog liver (Bederman, Kasumov, et al., 2004; Bederman, Reszko, et al., 2004). The immediate objective was to test the hypothesis that the precursor dilution varied over the length of the hepatic sinusoid. If the precursor enrichment was not constant, estimating rates of synthesis from tracer incorporation data would be problematic. The experiment was designed to yield high levels of labeling of sterols so that the model would be overdetermined and the hypothesis could be fully explored. In contrast, most clinical studies using ^{13}C isotopes are necessarily conducted in situations where the total enrichment of the product sterol is not large, often only *M*0, *M*1, and *M*2 sterols will be detected. Under these conditions, the data are insufficient to test the hypothesis of a gradient even though the existence of a gradient could invalidate the approach. Thus, the significance of this study investigating gradients in the precursor enrichment was to determine whether reliable estimates of sterol synthesis were feasible with the techniques in use at the time.

For the experimental protocol, conscious dogs were prefitted with transhepatic catheters and were infused with [1,2-^{13}C]acetate directly into a branch of the portal vein to expose the liver cells to acetate at high enrichment. After a 6 h infusion, we isolated sterols from plasma and tissues. We focus our analysis on lathosterol, a cholesterol precursor which was heavily labeled by the infusion protocol. We observed masses *M*0 through *M*24 for lathosterol. Thus, we had a model with 25 equations to test the hypothesis of varying precursor enrichment using ISA (Fig. 10). We first tested the classic ISA model (Fig. 3), with a constant value for *D*. This model returned a very poor fit to the actual data (Fig. 10A). On further analysis, it became clear that ISA could not determine a single value for *D*, which would yield a good fit of model to data.

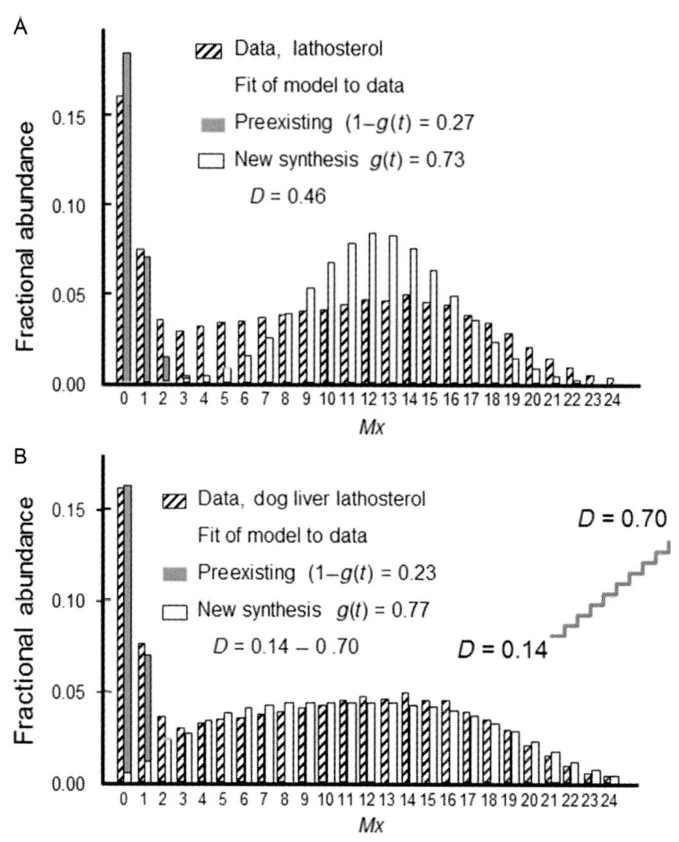

Figure 10 ISA analysis of lathosterol biosynthesis *in vivo*. (A) Lathosterol isotopomer profile from *in vivo* labeling using [1,2-^{13}C]acetate, where stripped bars representing GC/MS data are plotted alongside the best fit from the standard model using a constant value for *D*. (B) Data and best fit results for modified ISA model varying *D* between two values selected by the model in linear steps. (See the color plate.)

We then developed a modification of ISA allowing variation in the value of *D*. This model would have three parameters. These were two values for *D*, *D*(low), and *D*(high) and the value for fraction new synthesis *g*(*t*). The model allowed *D* to vary in 10 equal steps between *D*(low) and *D*(high) and combined these results to simulate a gradient. Once this modified version of ISA was prepared, we again analyzed the lathosterol data. As with other ISA protocols, we supplied initial guesses for the unknown parameters *D*(low), *D*(high), and *g*(*t*). ISA changed the values of the parameters and found the best fit. The best fit with this new, three-parameter model was a dramatic improvement (Fig. 10B). The sum of squared error dropped significantly, and this improvement justified adding the additional parameters to the model. ISA estimated that the contribution of acetate to the acetyl CoA pool for sterol synthesis varied between 14% and 70% in the liver. Thus, the nonlinearity of parameter D was used to gain new insight into the metabolic function of hepatic cells.

The requirement for a gradient three-parameter model to fit the data may be due to any of several physiological conditions. It may results from gradients in acetate concentration or enrichment across the liver. Alternatively, it may result from metabolic heterogeneity in acetate utilization in lipogenic cells. This heterogeneity may be one that occurs in space across the liver; or it may be heterogeneity that results from changes in the liver over the time of the infusion. To investigate this further, we measured the acetate concentration and enrichment at the portal vein and the hepatic artery and found that both the acetate concentration and enrichment dropped significantly across the liver. These measurements supported a role for a gradient in acetate as a major factor responsible for the observed data. However, heterogeneity of cells could not be ruled out in this study and should be considered as a possible source of variation in ISA studies where the data do not fit the classic model.

The *in vivo* ISA study of hepatic lipogenesis described here illustrates several important properties of ISA. First, ISA models using all isotopomers have the advantage of providing information which tests the underlying assumptions. If only two independent equations and data for only two isotopomers had been used to estimate the parameters, the gradient would not have been apparent and biosynthesis would have been estimated incorrectly. Second, when an ISA model does not fit the data, the source of the difficulty is likely to be with the nonlinear, D, parameter. A linear parameter, $g(t)$ for example, may vary over the liver space or the time course of the experiment, but these variations are invisible to the ISA model, as linear parameters report averages. Thus, an appreciation of the behavior of linear and nonlinear terms in models allows investigators to identify difficulties and modify models as required to estimate parameters and gain insights into the underlying physiology of a given biological system.

7. CONCLUDING THOUGHTS

The presentation here has focused on works that illustrate both applications of the classic ISA model and the flexibility of this model, which allow it to be adapted to address other interesting questions pertaining to the quantification of metabolic fluxes. Many other investigators have contributed to this field, expanding our understanding of isotopic flux methodology. We see a rich future for investigators building rigorous models and testing them with metabolic data.

REFERENCES

Antoniewicz, M. R., Kelleher, J. K., & Stephanopoulos, G. (2007). Elementary metabolite units (EMU): A novel framework for modeling isotopic distributions. *Metabolic Engineering, 9*, 68–86.

Bederman, I. R., Kasumov, T., Reszko, A. E., David, F., Brunengraber, H., & Kelleher, J. K. (2004). In vitro modeling of fatty acid synthesis under conditions simulating the zonation of lipogenic [13C]acetyl-CoA enrichment in the liver. *The Journal of Biological Chemistry, 279*, 43217–43226.

Bederman, I. R., Reszko, A. E., Kasumov, T., David, F., Wasserman, D. H., Kelleher, J. K., et al. (2004). Zonation of labeling of lipogenic acetyl-CoA across the liver: Implications for studies of lipogenesis by mass isotopomer analysis. *The Journal of Biological Chemistry, 279*, 43207–43216.

Board, M., & Newsholme, E. (1996). Hydroxycitrate causes altered pyruvate metabolism by tumorigenic cells. *Biochemistry and Molecular Biology International, 40*, 1047–1056.

Briscoe, D. A., Fiskum, G., Holleran, A. L., & Kelleher, J. K. (1994). Acetoacetate metabolism in AS-30D hepatoma cells. *Molecular and Cellular Biochemistry, 136*, 131–137.

Clarenbach, J. J., Lindenthal, B., Dotti, M. T., Federico, A., Kelleher, J. K., & Bergmann, K. (2005). Isotopomer spectral analysis of intermediates of cholesterol synthesis in patients with cerebrotendinous xanthomatosis. *Metabolism, 54*, 335–344.

Cobelli, C., Toffolo, G., & Foster, D. M. (1992). Tracer-to-tracee ratio for analysis of stable isotope tracer data: Link with radioactive kinetic formalism. *The American Journal of Physiology, 262*, E968–E975.

D'Adamo, A. F., & Haft, D. E. (1965). An alternative pathway of alpha-ketoglutarate catabolism in the isolated, perfused rat liver. *The Journal of Biological Chemistry, 240*, 613–617.

Dietschy, J. M., & Brown, M. S. (1974). Effect of alterations of the specific activity of the intracellular acetyl CoA pool on apparent rates of hepatic cholesterogenesis. *Journal of Lipid Research, 15*, 508–516.

DiStefano, J., III (2014). *Dynamic systems biology modeling and simulation* (1st ed.). London: Elsevier.

Draper, N. R., & Smith, H. (1981). *Applied regression analysis* (2nd ed.). New York: John Wiley & Sons.

Fenselau, A., Wallis, K., & Morris, H. P. (1975). Acetoacetate coenzyme a transferase activity in rat hepatomas. *Cancer Research, 35*, 2315–2320.

Hellerstein, M. K., & Neese, R. A. (1992). Mass isotopomer distribution analysis: A technique for measuring biosynthesis and turnover of polymers. *The American Journal of Physiology, 263*, E988–E1001.

Hellerstein, M. K., Wu, K., Kaempfer, S., Kletke, C., & Shackleton, C. H. (1991). Sampling the lipogenic hepatic acetyl-CoA pool *in vivo* in the rat. Comparison of xenobiotic probe to values predicted from isotopomeric distribution in circulating lipids and measurement of lipogenesis and acetyl-CoA dilution. *The Journal of Biological Chemistry, 266*, 10912–10919.

Hetenyi, G. J. (1979). Correction factor for the estimation of plasma glucose synthesis from the transfer of 14C-atoms from labelled substrate *in vivo*: A preliminary report. *Canadian Journal of Physiology and Pharmacology, 57*, 767–770.

Hildebrandt, L. A., Spennetta, T., Elson, C., & Shrago, E. (1995). Utilization and preferred metabolic pathway of ketone bodies for lipid synthesis by isolated rat hepatoma cells. *The American Journal of Physiology, 269*, C22–C27.

Holleran, A. L., Briscoe, D. A., Fiskum, G., & Kelleher, J. K. (1995). Glutamine metabolism in AS-30D hepatoma cells. Evidence for its conversion into lipids via reductive carboxylation. *Molecular and Cellular Biochemistry, 152*, 95–101.

Holleran, A. L., Fiskum, G., & Kelleher, J. K. (1997). Quantitative analysis of acetoacetate metabolism in AS-30D hepatoma cells with 13C and 14C isotopic techniques. *The American Journal of Physiology, 272*, E945–E951.

Jazmin, L. J., O'Grady, J. P., Ma, F., Allen, D. K., Morgan, J. A., & Young, J. D. (2014). Isotopically nonstationary MFA (INST-MFA) of autotrophic metabolism. *Methods in Molecular Biology, 1090*, 181–210.

Kelleher, J. K., & Bryan, B. M. (1985). A 14CO2 ratios method for detecting pyruvate carboxylation. *Analytical Biochemistry, 151*, 55–62.

Kelleher, J. K., Kelly, P. J., & Wright, B. E. (1979). Amino acid catabolism and malic enzyme in differentiating dictyostelium discoideum. *Journal of Bacteriology, 138*, 467–474.

Kelleher, J. K., Kharroubi, A. T., Aldaghlas, T. A., Shambat, I. B., Kennedy, K. A., Holleran, A. L., et al. (1994). Isotopomer spectral analysis of cholesterol synthesis: Applications in human hepatoma cells. *The American Journal of Physiology, 266*, E384–E395.

Kelleher, J. K., & Masterson, T. M. (1992). Model equations for condensation biosynthesis using stable isotopes and radioisotopes. *The American Journal of Physiology, 262*, E118–E125.

Kharroubi, A. T., Masterson, T. M., Aldaghlas, T. A., Kennedy, K. A., & Kelleher, J. K. (1992). Isotopomer spectral analysis of triglyceride fatty acid synthesis in 3T3-L1 cells. *The American Journal of Physiology, 263*, E667–E675.

Krebs, H. A., Hems, R., Weidemann, M. J., & Speake, R. N. (1966). The fate of isotopic carbon in kidney cortex synthesizing glucose from lactate. *The Biochemical Journal, 101*, 242–249.

Lee, W. N., Bergner, E. A., & Guo, Z. K. (1992). Mass isotopomer pattern and precursor-product relationship. *Biological Mass Spectrometry, 21*, 114–122.

Lee, W. N., Byerley, L. O., Bergner, E. A., & Edmond, J. (1991). Mass isotopomer analysis: Theoretical and practical considerations. *Biological Mass Spectrometry, 20*, 451–458.

Lindenthal, B., Aldaghlas, T. A., Holleran, A. L., Sudhop, T., Berthold, H. K., von Bergmann, K., et al. (2002). Isotopomer spectral analysis of intermediates of cholesterol synthesis in human subjects and hepatic cells. *American Journal of Physiology. Endocrinology and Metabolism, 282*, E1222–E1230.

Martin, R. B. (1997). Disadvantages of double reciprocal plots. *Journal of Chemical Education, 74*, 1238–1250.

Metallo, C. M., Gameiro, P. A., Bell, E. L., Mattaini, K. R., Yang, J., Hiller, K., et al. (2012). Reductive glutamine metabolism by IDH1 mediates lipogenesis under hypoxia. *Nature, 481*, 380–384.

Nelson, D. P., & Homer, L. D. (1983). *Nonlinear regression analysis: A general program for data modeling using personal computers*. Bethesda, MD: Naval Medical Research Command.

Newsholme, E. A., Crabtree, B., & Ardawi, M. S. (1985). Glutamine metabolism in lymphocytes: Its biochemical, physiological and clinical importance. *Quarterly Journal of Experimental Physiology: An International Journal of the Physiological Society, 70*, 473–489.

Sabine, J. R., Kopelovich, L., Abraham, S., & Morris, H. P. (1973). Control of lipid metabolism in hepatomas: Conversion of glutamate carbon to fatty acid carbon via citrate in several transplantable hepatomas. *Biochemica Biophysica Acta, 296*, 493–498.

Schadewaldt, P., Schommartz, B., Wienrich, G., Brosicke, H., Piolot, R., & Ziegler, D. (1997). Application of isotope-selective nondispersive infrared spectrometry (IRIS) for evaluation of [13C]octanoic acid gastric-emptying breath tests: Comparison with isotope ratio-mass spectrometry (IRMS). *Clinical Chemistry*, *43*, 518–522.

Shoenheimer, R., & Rittenberg, D. (1935). Deuterium as an indicator in the study of intermediary metabolism. I. *Journal of Biological Chemistry*, *111*, 163–168.

Wolfe, R. R., & Chinkes, D. L. (2005). *Isotopic tracers in metabolic research* (2nd ed.). John Wiley and Sons, Inc.

Yoo, H., Antoniewicz, M. R., Stephanopoulos, G., & Kelleher, J. K. (2008). Quantifying reductive carboxylation flux of glutamine to lipid in a brown adipocyte cell line. *The Journal of Biological Chemistry*, *283*, 20621–20627.

Zhang, Y., Agarwal, K. C., Beylot, M., Soloviev, M. V., David, F., Reider, M. W., et al. (1994). Nonhomogeneous labeling of liver extra-mitochondrial acetyl-CoA. Implications for the probing of lipogenic acetyl-CoA via drug acetylation and for the production of acetate by the liver. *The Journal of Biological Chemistry*, *269*, 11025–11029.

Effect of Error Propagation in Stable Isotope Tracer Studies: An Approach for Estimating Impact on Apparent Biochemical Flux

Stephen F. Previs[1], Kithsiri Herath, Jose Castro-Perez, Ablatt Mahsut, Haihong Zhou, David G. McLaren, Vinit Shah, Rory J. Rohm, Steven J. Stout, Wendy Zhong, Sheng-Ping Wang, Douglas G. Johns, Brian K. Hubbard, Michele A. Cleary, Thomas P. Roddy

Merck Research Laboratories, Kenilworth, New Jersey, USA

[1]Corresponding author: e-mail address: stephen_previs@merck.com

Contents

Methods in Enzymology, Volume 561
ISSN 0076-6879
http://dx.doi.org/10.1016/bs.mie.2015.06.021

Abstract

Stable isotope tracers are widely used to quantify metabolic rates, and yet a limited number of studies have considered the impact of analytical error on estimates of flux. For example, when estimating the contribution of *de novo* lipogenesis, one typically measures a minimum of four isotope ratios, i.e., the precursor and product labeling pre- and posttracer administration. This seemingly simple problem has 1 correct solution and 80 erroneous outcomes. In this report, we outline a methodology for evaluating the effect of error propagation on apparent physiological endpoints. We demonstrate examples of how to evaluate the influence of analytical error in case studies concerning lipid and protein synthesis; we have focused on ^2H_2O as a tracer and contrast different mass spectrometry platforms including GC-quadrupole-MS, GC-pyrolysis-IRMS, LC-quadrupole-MS, and high-resolution FT-ICR-MS. The method outlined herein can be used to determine how to minimize variations in the apparent biology by altering the dose and/or the type of tracer. Likewise, one can facilitate biological studies by estimating the reduction in the noise of an outcome that is expected for a given increase in the number of replicate injections.

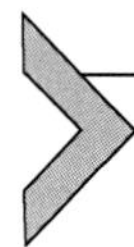

1. INTRODUCTION

1.1 Considering the Impact of Error Propagation

Stable isotopes have been used in biomedical research for nearly 80 years. Although the ability to develop high-throughput study designs can be realized via advances in instrumentation (Berthold, Wykes, Jahoor, Klein, & Reeds, 1994; Hachey, Wong, Boutton, & Klein, 1987), fundamental aspects of the tracer logic have not changed. Despite the fact that investigators have several choices regarding the use of specific tracers and analytical platforms, a common approach for estimating the synthesis of new material relies on establishing a precursor:product labeling ratio. For example, when ^2H_2O is used to quantify the contribution of *de novo* lipogenesis, one typically measures the labeling of water and palmitate, i.e., the precursor and product, respectively (Diraison, Pachiaudi, & Beylot, 1997; Leitch & Jones, 1993).

The straightforward scenario outlined here becomes complicated when one considers the potential error combinations. Namely, investigators will measure a minimum of four isotope ratios, i.e., the precursor and product labeling in a sample collected during a baseline period prior to tracer administration and again in a sample collected posttracer administration.

Recognizing that each measurement can (and likely does) have some component of error leads to a situation in which there are 81 possible values for the given analyses noted above. That is, since there are 3 possibilities for the error in a measured isotope ratio (i.e., positive, negative, and nil) and there are 4 terms being measured (i.e., the precursor and product labeling pre- and posttracer administration), there is 1 correct outcome and 80 possible errors (i.e., $3^4 - 1$).

Many reports regarding the pros and cons of the different tracer methods have centered on issues pertaining to the practicality of the approaches and/or the underlying assumptions regarding precursor:product labeling ratios (Casazza & Veech, 1984). Choosing a tracer method has immediate, and perhaps not so obvious, implications in regard to analytical options. For example, when ^2H_2O is used to study fatty acid synthesis, one can measure the precursor and product labeling using gas chromatography-quadrupole-mass spectrometry (GC-q-MS), liquid chromatography-quadrupole-mass spectrometry (LC-q-MS), GC-pyrolysis-isotope ratio mass spectrometry (GC-p-IRMS), or LC-Fourier transform-ion cyclotron resonance mass spectrometry-MS (LC-FT-ICR-MS) (Diraison et al., 1997; Herath et al., 2014; Leitch & Jones, 1991). We recognize that the decision to use one instrument versus another may be affected by many factors; it is not our intention to rank order the "best and worst" instruments (we do not believe that one can realistically make such statements). However, we outline an approach that allows one to examine the impact of analytical error(s) on conclusions regarding biological endpoints; we expect that this can be used to help guide a best-practice strategy in tracer studies regardless of the choice in mass spectrometer. This is especially important when considering *in vivo* studies since one has a limited ability to enrich precursor substrates. Although common sense suggests that reducing the coefficient of variation surrounding the analytics will improve the outcome, the magnitude is not obvious without a simulation of the condition.

1.2 How Can We Examine the Impact of Error Propagation?

In this report, we have outlined a practical and flexible approach for estimating the effect of error propagation in certain types of tracer-based studies. Using a conventional spreadsheet (e.g., MS Excel), one can program a standard tracer kinetic equation and add varying degrees of noise using a random number generator. This allows one to readily simulate the impact of different variables on the apparent outcome of a given experiment. While this approach may appear to lack some degree of mathematical rigor (Bevington & Robinson, 1992) and one should exercise caution with the indiscriminate use of random number generating functions, our approach is expected to be useful for biologists in many general cases.

It is important to acknowledge that other approaches have been used to examine how the error of a measurement(s) will impact flux estimates via propagation in a given mathematical model (Antoniewicz, Kelleher, & Stephanopoulos, 2006; Matthews et al., 1980; Stellaard et al., 2010; Yu, Sinha, & Previs, 2005). Although each example has merit, it is unfortunate that one cannot find many more examples in the literature especially since this topic is of critical importance to investigators and resources are generally becoming more limited.

Specifically, the approach outlined herein has proven helpful in our current work as we routinely find ourselves tackling a wide array of biological problems. In our experience, the development of more formal equations (Bevington & Robinson, 1992) becomes problematic since the goals of the various studies may change and since experimentalists are faced with design constraints. In particular, studies in human subjects are less limited with respect to the timing of sample collection and/or sample volume size, whereas studies in animal models are typically more restricted. Likewise, increasing the tracer dose can have a substantial impact on the cost of running a study in humans, whereas tracer costs are generally less problematic in animal models. The approach outlined here allows one to readily examine the impact of shifting sampling intervals, altering tracer doses and/or adapting methods to different types of mass spectrometers.

1.3 Application to Biological Problems

The first example that we have considered centers on measuring the contribution of *de novo* fatty acid synthesis, and it was assumed that one would administer ^2H_2O and then use one of various different mass spectrometry-based analyses. For example, the expected background labeling ratio for palmitate varies from ~18% (M+1/M0) for GC-q-MS, ~17% (M+1/M0) for LC-q-MS, ~0.6% (^{2}H−M+1/M0) for high-resolution FT-ICR-MS, and ~200 δ (^{2}H/H) for GC-p-IRMS. Note that an advantage of using high-resolution mass spectrometry is that by resolving the ^{13}C and ^{2}H isotopes, one can substantially lower the background labeling; in the case of palmitate, this is nearly 30-fold; conceptually, this is similar to the removal of ^{13}C that one gains by using GC-p-IRMS and measuring the $^2H/H$ ratio (Herath et al., 2014). We experimentally determined the precision of each platform by analyzing known standards of palmitate and found that the coefficients of variation (standard deviation divided by the mean) are typically $\leq 1\%$ for GC-q-MS, $\sim 1-2\%$ for LC-q-MS, $\sim 2-5\%$ for FT-ICR-MS, and $\sim 2-5\%$ for GC-p-IRMS. We then determined the effect(s) of those analytical errors on estimates of the apparent contribution of *de novo* lipogenesis.

Our second example centers on contrasting ^2H_2O, $H_2{}^{18}O$, or $^2H_3$-leucine in studies of protein synthesis. Briefly, it is possible to quantify protein synthesis by administering ^2H_2O or $H_2{}^{18}O$ and then allowing the subject to generate labeled amino acids or by directly administering a labeled amino acid (e.g., $^2H_3$-leucine) (De Riva, Deery, McDonald, Lund, & Busch, 2010; Lichtenstein et al., 1990; Rachdaoui et al., 2009). Although it is likely that the same instrument (e.g., LC-q-MS) can be used to support the analyses of any of these tracers, we show how a series of simulations can be used to determine the implications of natural isotopic background labeling. Namely, when ^2H_2O versus $H_2{}^{18}O$ versus $^2H_3$-leucine are administered, one generally measures a shift in the $M+1$ versus $M+2$ versus $M+3$ signals relative to the M0, respectively. As expected, the dramatic change in background has implications on error propagation. To demonstrate the magnitude of such effect(s), we contrasted the ability to measure the turnover of VLDL versus LDL-apoB100. Since the analytical considerations are the same for the respective proteins (i.e., the same peptide sequence is used to quantify the labeling), the experimental variable in this example is the half-life (VLDL-apoB100 turns over $\sim$10-fold faster than LDL-apoB100) which translates into different levels of enrichment for the analyte at the same time interval.

1.4 Relevance of the Examples Discussed Herein

Different factors can influence the choice of using one tracer method over another. First, when considering the logistics of running a study of lipid and protein synthesis, ^2H_2O is advantageous since it can be given orally to free-living subjects for prolonged periods, whereas ^{13}C-acetate (lipid synthesis) or $^2H_3$-leucine (protein synthesis) typically require continuous intravenous infusion (Di Buono, Jones, Beaumier, & Wykes, 2000; Lichtenstein et al., 1990; Taylor, Mikkelson, Anderson, & Forman, 1966). Second, it is generally accepted that ^2H_2O distributes equally in all cells; in contrast, other tracers can be affected by labeling gradients and therein lead to uncertainty regarding the true precursor labeling (Casazza & Veech, 1984; Cuchel et al., 1997; Dietschy & McGarry, 1974; Dietschy & Spady, 1984; Jeske & Dietschy, 1980; Lowenstein, Brunengraber, & Wadke, 1975).

NMR-based methods are also useful for measuring isotopic labeling (Duarte et al., 2014); however, we have not considered those analyses herein since they are generally slower throughput. It is important to note that NMR methods have unique advantages when one aims to determine the positional specificity of labeling (Jin et al., 2004); this is especially true since NMR can sort out complex labeling patterns, whereas mass spectrometry-based analyses must hope for some good fortune regarding the generation of fragment ions.

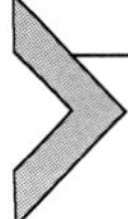

2. ANALYTICAL: INSTRUMENT SETTINGS AND DATA ACQUISITION

2.1 Water Labeling

The ^{2}H-labeling of plasma water is determined using GC-q-MS (Shah, Herath, Previs, Hubbard, & Roddy, 2010). Briefly, ^{2}H present in water is exchanged with hydrogen bound to acetone by incubating 10 μL of plasma or known standards in a 2-mL glass screw-top GC vial at room temperature for 4 h with 2 μL 10 N NaOH (Fisher Scientific) and 5 μL of acetone (Sigma–Aldrich). The instrument is programmed to inject 5 μL of headspace gas from the GC vial in a splitless mode using a 0.8-min isothermal run (Agilent 5973 MS coupled to a 6890 GC oven fitted with an Agilent DB-5MS column, 30 m × 250 μm × 0.15 μm, the oven is set at 170 °C, and helium carrier flow is set at 1.0 mL min^{-1}), acetone elutes at ~0.8 min, and the mass spectrometer is set to perform selected ion monitoring of *m/z* 58 and 59 (10 ms dwell time per ion) in the electron impact ionization mode. Note that this method will amplify the ^{2}H-labeling of water severalfold (Brunengraber, Kelleher, & DesRosiers, 1997). One expects a natural background M1/M0 ratio of ~3.5% which could increase to ~5.0% when body water is raised to ~0.5% via administration of ^{2}H-labeled water. We previously reported the analytical coefficient of variation of this assay (McCabe et al., 2006), and one expects this to be generally below 1%.

The ^{18}O-labeling of plasma water is determined using GC-q-MS (Brunengraber, McCabe, Katanik, & Previs, 2002). Briefly, 5 μL of plasma is reacted with PCl_5 to generate phosphoric acid, and 150 μL of TMS-diazomethane (Sigma) is then added to generate trimethyl phosphate. The solution is evaporated under a stream of nitrogen, and the residue is dissolved in 150 μL chloroform. Samples are analyzed using an Agilent 5973 MS coupled to a 6890 GC oven fitted with an Agilent DB5-MS column, 30 m × 250 μm × 0.15 μm, the oven is initially set at 100 °C and then programmed to increase at 35 °C per min to 250 °C, helium carrier flow is set at 1.0 mL min^{-1} (2 μL of sample is injected using a 40:1 split), trimethyl phosphate elutes at ~1.9 min, and the mass spectrometer is set to perform selected ion monitoring of *m/z* 140 and 142 (10 ms dwell time per ion) in the electron impact ionization mode. Note that this method will amplify the ^{18}O-labeling of water severalfold. One expects a natural background M2/M0 ratio of ~0.8% which could increase to ~2.8% when body water is raised to ~0.5% via administration of ^{18}O-labeled water. We previously reported the analytical coefficient of variation of this assay (Brunengraber et al., 2002), and one expects this to be generally below 1%.

2.2 GC-Quadrupole-MS Analyses of Palmitate

Palmitate is reacted with TMS-diazomethane to form its methyl ester. The ^{2}H-labeling of methyl palmitate is determined using an Agilent 5973N-MSD equipped with an Agilent 6890 GC system, a DB17-MS capillary column (30 m × 0.25 mm × 0.25 μm). The mass spectrometer is typically operated in the electron impact mode using selective ion monitoring of *m/z* 270 and 271 at a dwell time of 10 ms per ion (Patterson & Wolfe, 1993). Note that one expects a natural background M1/M0 ratio of ~18%; in our experience, the analytical coefficient of variation of this assay is typically ≤1.0%.

2.3 GC-Pyrolysis-IRMS Analyses of Palmitate

The ^{2}H-labeling of methyl palmitate is determined using a Delta V Plus isotope ratio mass spectrometer (Thermo) coupled to an Agilent 7890 Trace GC (Scrimgeour, Begley, & Thomason, 1999). Samples are analyzed using a splitless injection (1.5 μL injection volume at an inlet temperature of 230 °C) onto an Agilent DB-5MS column (30 m × 250 μm × 0.25 μm). Compounds eluting off of the column are directed into the pyrolysis reactor (heated at 1420 °C) and converted to hydrogen gas.

Isotope ratios are expressed in delta (δ) values using the equation:

$$^2\text{H} - \text{labeling}(\delta) = \left(R_{\text{sample}} - R_{\text{ref}}\right) / R_{\text{ref}} \times 1000$$

where R_{ref} is the isotope ratio of reference hydrogen gas, and R_{sample} is the isotope ratio of an unknown analyte. Ion chromatograms were processed using the Individual Background integration method found in the ISODAT 3.0 software along with corrections for H3+. The H3+ factor was determined daily by measuring the isotope ratio of the hydrogen reference gas at varying peak height; although 12 pressures were tested, this value was typically stable at ~3.3–3.4 ppm mV^{-1}. Note that one expects a natural background $^{2}H/^{1}H$ ratio of ~200δ; obviously this will vary depending on the reference gas that is used; in our experience, the analytical coefficient of variation of this assay is ~2.5%.

2.4 LC-Quadrupole-MS Analyses of Palmitate

The ^{2}H-labeling of palmitate is determined using a Waters-Micromass Quattro Micro triple quadrupole mass spectrometer equipped with an electrospray ion source and Agilent 1100 LC system. A Phenomenex Synegi Polar-RP (4.6 mm × 50 mm × 2.5 μm) column is used with solvent gradient 85–95% methanol/water with 0.3% triethylamine and formic acid in 5 min. The mass spectrometer is operated in negative ion electrospray ionization mode with −40 V cone voltage using selective ion monitoring of *m/z* 255 and 256 at a dwell time of 250 ms per ion. The cone and desolvation nitrogen gas were used at flow rates of 96 and 395 L/h, and the source and desolvation temperatures were set at 130 and 300 °C, respectively. Note that one expects a natural background M1/M0 ratio of ~17%; in our experience, the analytical coefficient of variation of this assay is typically ~1–2%.

2.5 High-Resolution LC-FT-ICR-MS Analyses of Palmitate

The ^{2}H-labeling of free palmitate is determined using a Bruker solariX hybrid Qq-FT-ICR 9.4T (Herath et al., 2014). Samples are analyzed using direct infusion (3.3 μL per min), and spectra are acquired in the negative ion mode with a capillary voltage of 2400 V. The quadrupole is set to isolate a 2 Da window (*m/z* 255.5–257.2), with the collision cell turned off, and the ICR cell is set for narrow band acquisition (1 Da window, *m/z* 255.7–256.7). In a typical run, 50 spectra are averaged (4.2 s per scan) at a resolving power of ~600,000 at *m/z* 256, and this resolves the M+1 ion into its ^{13}C and ^{2}H components (Herath et al., 2014). Note that one expects a natural background ^{2}H−M1/M0 ratio of ~0.6%; in our experience, the analytical coefficient of variation of this assay is typically ~2–5.0%.

3. SIMULATIONS OF LIPID FLUX

3.1 General Biochemical Overview

Palmitate in circulating triglycerides is derived from acetyl-CoA (i.e., newly made lipid, also referred to as *de novo* lipogenesis) and from recycled lipid (e.g., $\text{triglyceride}_{\text{adipose tissue}} \rightarrow \text{palmitate}_{\text{plasma}} \rightarrow \text{triglyceride}_{\text{liver secretion}}$) (Leitch & Jones, 1993). Following the administration of ^2H_2O, the appearance of ^2H in circulating triglyceride-bound palmitate can be used to estimate the contribution of *de novo* lipogenesis (Diraison et al., 1997; Leitch & Jones, 1993) according to Eq. (1):

$$de\,novo \text{ lipogenesis}(\%) = \Delta\text{palmitate labeling}/(\Delta\text{water labeling} \times n) \times 100 \quad (1)$$

where Δ palmitate labeling and Δ water labeling are determined by subtracting the experimentally measured natural background labeling from the labeling in a sample obtained at a point after administering ^2H_2O (at which time the pool of palmitate has reached a new steady-state labeling) and n is the number of exchangeable hydrogens, which is assumed to equal 22 (Diraison, Pachiaudi, & Beylot, 1996; Lakshmanan, Berdanier, & Veech, 1977; Lee et al., 1994; Previs et al., 2011).

3.2 General Simulations to Examine the Maximum Error Range

Since there are three possible outcomes for the error in a measured isotope ratio (i.e., positive, negative, and nil) and there are four terms in Eq. (1), there is 1 correct outcome and 80 (i.e., $3^4 - 1$) combinations of error when estimating the contribution of *de novo* lipogenesis using Eq. (1) for a given set of samples.

Figure 1 demonstrates the 81 outcomes (i.e., the apparent contribution of *de novo* lipogenesis) when one simulates the effect(s) of all possible errors assuming (i) a 5% contribution of *de novo* lipogenesis, (ii) that analyses are conducted using GC-q-MS, and (iii) that the coefficient of variation is exactly ±0.5% of the true value. These data were obtained by manually programming each combination of errors in MS Excel, and it was assumed that the baseline M1/M0 ratio was ~18% and that the posttracer ratio would increase to ~19.65%; assuming that water labeling would be analyzed using the acetone exchange method described earlier, it was expected that water

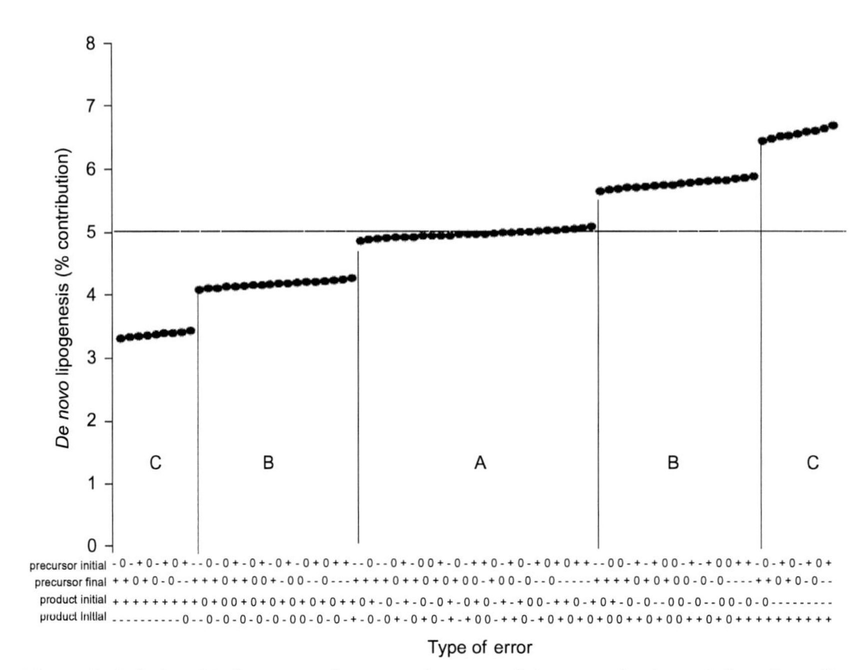

Figure 1 Relationship between the type of error and the magnitude and direction of its effect. Simulations determined the impact of background labeling and analytical error on estimates of palmitate synthesis. The expected value was 5% newly made palmitate (dotted line); according to Eq. (1), there are 81 possible outcomes, i.e., 1 correct value and 80 errors. Note that the nature of the analytical error is identical regardless of the type of mass spectrometer or dose of labeled water; the latter parameters impact the magnitude of the error on the apparent contribution of *de novo* lipogenesis. The data contained in the figure represent error that is introduced into palmitate synthesis (*y*-axis) when body water is enriched to ~0.5% and palmitate labeling is determined using GC-q-MS assuming a coefficient of variation of ~0.5% in M+1/M0. Vertical lines are drawn to facilitate visualization of the relationship between the type of error and its impact. The symbols shown on the *x*-axis demonstrate the nature of the error in the respective terms; i.e., "0," "−," and "+" refer to no error, a negative error, and a positive error at 0.5% of the expected isotope ratio, respectively. As shown, a substantial number of the errors tend to cancel each other (regions A and B), whereas a limited number of the errors (region C) have a sizeable impact on the data.

would be enriched ~0.5% in which case the baseline M1/M0 acetone would be ~3.5% and the posttracer value would be ~5.0%, i.e., $100 \times [(19.65 - 18.0)/22]/(5.0 - 3.5) = 5\%$. Note that the factor 22 represents "*n*" in Eq. (1).

The observations reported in Fig. 1 allow one to readily dissect how the type of error impacts the outcome. One of the first points learned is that when error is present only in measurements of the water labeling, there is virtually no impact on the apparent contribution of *de novo* lipogenesis; presumably this reflects the fact that even in cases when body water is labeled to ~0.5%, the analytical methods that one can use to measure the water labeling result in a large change over a relatively small background and thus reduce the impact of error associated with measurements of the precursor labeling (Shah et al., 2010; Yang et al., 1998).

Figure 1 also demonstrates that a substantial number of the errors tend to cancel each other. For example, approximately 1/3 of the errors yield a range of values for *de novo* lipogenesis that are virtually identical to the expected value (Region A) and approximately 4/5 of the errors yield a range of values for *de novo* lipogenesis between approximately ±20% of the expected values (Regions A and B combined). Nevertheless, although nearly 80% of the errors have minor to modest impact on the data, it is obviously not possible to predict the nature of a given error during the analysis of a sample set. Therefore, we then considered two different approaches for minimizing the propagation of error. First, one could simply administer more tracers. Second, one could analyze more replicates of a given sample. We outline a method that allows one to readily address these types of questions and we demonstrate the utility of this method in understanding how to minimize the propagation of error.

3.3 Flexible and Adaptable Simulations to Examine the Impact of Different Mass Spectrometers

The strategy outlined above presents a somewhat extreme scenario; in all cases, the error was ascribed a maximal value. For example, in cases where there was an error, it was assumed to equal ±0.5% of the measured ratio. It is certainly possible that one measurement could be in error of +0.5%, while another is in error of −0.1%. Therefore, we implemented a flexible system using randomly generated numbers to affect the error; the range in a given simulation was set to reach ±0.5%, ±1.0%, ±2.5%, or ±5.0% of a measured value. In addition, we aimed to readily adapt this to various mass spectrometry-based analyses, including the platforms noted earlier.

This can be easily programmed in MS Excel as shown in Table 1. One can then run numerous simulations and vary the random number for the noise, shown is an example in which we assumed a 1% coefficient of variation for each of six hypothetical sample sets. For example, to examine how a 5% coefficient of variation would impact the data, one can change the "10" to "50". Likewise, to examine how changing the natural background labeling would affect the outcome, one substitutes the "true values" in

Table 1 Outline of the Simulation Approach as Applied in Studies of Fatty Acid Biosynthesis

		M1/M0 Ratio						
		Error Value Assuming a Coefficient of Variation ±1.0% of a Measured Endpoint						
			Simulation Trial Output					
Measured Endpoint	**Expected Value**	**MS Excel Code**	**1**	**2**	**3**	**4**	**5**	**6**
Palmitate baseline	18.0	=18+RANDBETWEEN (−(18.0*10),(18.0*10))/1000	17.83	17.98	18.09	18.11	18.05	18.17
Palmitate posttracer	19.65	=19.65+RANDBETWEEN (−(19.65*10),(19.65*10))/1000	19.84	19.56	19.70	19.67	19.71	19.53
Water baseline	3.5	=3.5+RANDBETWEEN (−(3.5*10),(3.5*10))/1000	3.48	3.51	3.53	3.48	3.50	3.51
Water posttracer	5.0	=5+RANDBETWEEN (−(5.0*10),(5.0*10))/1000	5.01	5.02	4.98	4.96	4.97	4.97
Calculated % DNL	5.0		5.99	4.75	5.05	4.79	5.12	4.23

Using *a priori* knowledge of the expected natural background labeling and experimentally induced shifts in isotope ratios, one can program a set of equations in a conventional spreadsheet (e.g., MS Excel). The effect of analytical noise is added to determine the impact on a metabolic flux. In the example shown here, one assumes a constant error value (equal to ±1%) in a given measurement and determines the apparent contribution of *de novo* lipogenesis in six trials. This error value, and that of any other parameter, is easily varied to fit a particular experimental setting.

the respective column and in the MS Excel code, the example that is shown here assumes the analysis of methyl palmitate using GC-q-MS and water labeling determined via exchange with acetone.

Note that reports in the literature suggest that *de novo* lipogenesis may reach upward of ~20% in humans in certain situations so our simulations are somewhat conservative but within the range of reported data (Diraison et al., 1997; Leitch & Jones, 1993; Parks, Skokan, Timlin, & Dingfelder, 2008; Stanhope et al., 2009). Once the framework that is outlined in Table 1 is developed, it is possible to readily simulate other physiological conditions by simply adjusting the terms. For example, if one expects a 5% contribution of *de novo* lipogenesis and wants to know the impact of doubling the water tracer dose, one can change the "true value" for the palmitate posttracer term to 21.3 and the "water posttracer" term to 6.5.

Using this approach, we could immediately examine the impact of different mass spectrometer systems on estimates of *de novo* lipogenesis assuming a 5% contribution. We varied the assumption for body water labeling, to simulate a condition we might expect in humans; we considered that body water would be enriched to ~0.5%, whereas for simulations in rodents body water could be enriched to ~2.0%. Table 2 demonstrates the potential data sets that one would expect. While it is not surprising to see an increase in the noise of the estimate as the coefficient of variation increases or a reduction in the noise as the background labeling is reduced, the ability to easily run these types of simulations yields a quantitative estimate of the range and magnitude of the effect(s) under conditions where a single injection of given sample is made.

Several reports have shown that the ^{2}H-labeling of water can be increased to ~2% in rodents and other model systems (Bederman et al., 2006; Diraison et al., 1996; Gasier et al., 2009; Lee et al., 1994). Although the contribution of *de novo* lipogenesis is also often higher in rodents than humans, we ran additional simulations assuming a 5% contribution of *de novo* lipogenesis to enable an easy comparison with the data discussed earlier and therein allow readers to assess the impact of simply adding more tracers to a system. Table 2 demonstrates a substantial decrease in the spread of the observed errors as the dose of the tracer is increased fourfold. Specifically, increasing the precursor labeling allows for better estimates of *de novo* lipogenesis using "high-throughput" GC-q-MS and/or LC-q-MS analyses (Table 2).

From these general observations, it is clear that a seemingly good level of analytical reproducibility is associated with a sizeable propagation of error. Although one expects a 5% contribution of *de novo* lipogenesis, one can observe an apparent biological variation of nearly 10% even though the

Table 2 Effect(s) of Analytical Error on the Apparent Contribution of Palmitate Synthesis

Mass Spectrometry Platform	Analytical CV (%)	Ave.	SD	Biological CV (%)	Min	Max
Assuming ~0.5% enrichment of body water						
GC-q-MS	0.5	5.0	0.24	4.7	4.4	5.6
	1.0	5.0	0.47	9.4	3.8	6.01
	5.0	4.9	2.42	49.6	−0.1	11.9
LC-q-MS	0.5	5.0	0.21	4.1	4.5	5.6
	1.0	5.0	0.43	8.6	4.1	6.0
	5.0	4.8	2.14	44.3	−0.3	9.8
GC-p-IRMS	0.5	5.0	0.06	1.8	4.9	5.1
	1.0	5.0	0.13	2.5	4.8	5.8
	5.0	5.1	0.62	12.1	3.8	6.7
LC-FT-ICR-MS	0.5	5.0	0.06	1.1	4.8	5.1
	1.0	5.0	0.12	2.5	4.7	5.3
	5.0	5.1	0.69	13.6	3.9	6.8
Assuming ~2.0% enrichment of body water						
GC-q-MS	0.5	5.0	0.07	1.4	4.8	5.2
	1.0	5.0	0.14	2.9	4.6	5.4
	5.0	5.0	0.67	13.3	3.3	6.7
LC-q-MS	0.5	5.0	0.07	1.5	4.8	5.2
	1.0	5.0	0.14	2.8	4.7	5.3
	5.0	5.0	0.71	14.1	3.5	6.7
GC-p-IRMS	0.5	5.0	0.03	0.6	4.9	5.1
	1.0	5.0	0.06	1.2	4.9	5.1
	5.0	5.0	0.29	5.8	4.4	5.7
LC-FT-ICR-MS	0.5	5.0	0.03	0.6	4.9	5.1
	1.0	5.0	0.06	1.2	4.9	5.2
	5.0	5.0	0.32	6.4	4.4	5.8

All simulations were run under conditions where we expected that *de novo* lipogenesis would contribute ~5% to the pool of palmitate and assuming that body water was labeled to ~0.5% or ~2.0%, as is typically seen in studies involving humans or rodents, respectively. The data summarized here reflect the result of 250 simulations for a given condition.

analytical coefficient of variation may be ~1% (Table 2, e.g., GC-q-MS or LC-q-MS). Since GC-p-IRMS and high-resolution FT-ICR-MS remove the contribution of naturally occurring ^{13}C from the background, the absolute change in the palmitate (or product) labeling is greater using those platforms and therefore one can tolerate more analytical noise (Table 2).

3.4 Potential Usefulness in Guiding Study Designs

Additional simulations were run to determine the ability to quantify a 50% reduction in the contribution of *de novo* lipogenesis, e.g., suppose an inhibitor reduced the contribution of fatty acid synthesis from 5.0% to 2.5%, could this change be detected? This final round of simulations assumed that ~10 subjects would be studied in the different groups but that the analyst could run multiple injections of a given sample set. For example, there are cases in which it is not possible to increase the dose of the tracer and/or analysts have limited access to certain types of instruments. We simulated the outcome assuming the use of a GC-q-MS platform with ~1% coefficient of variation. Figure 2 demonstrates the ability to quantify a 5.0% versus a 2.5%

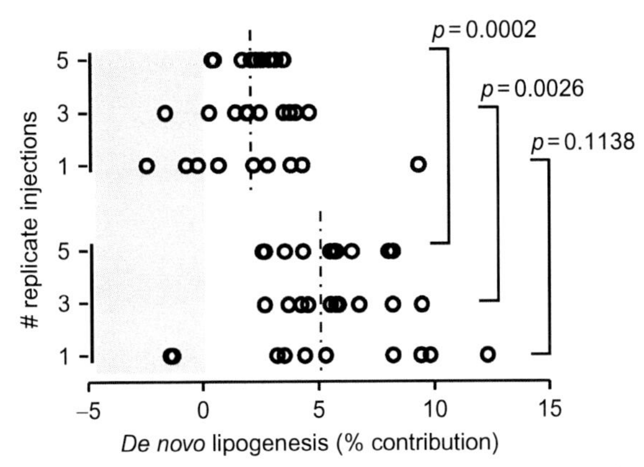

Figure 2 Effect of replicate injections on the propagation of analytical error when estimating *de novo* lipogenesis. Simulations were run to determine the impact of analytical error assuming that GC-q-MS (~1% coefficient of variation) would be used to analyze samples from subjects in which water labeling would be maintained at ~0.5%. The simulations assumed that 10 control and 10 "treated" subjects would be studied and that fatty acid synthesis would contribute ~5% and ~2.5%, bottom and top cluster, respectively. It was assumed that one would inject samples one, three, or five times, and the apparent contribution of fatty acid synthesis is shown for each condition. Comparisons were made across groups using a *t*-test to determine if increasing the number of injections would allow an investigator to find a difference in the contribution of *de novo* lipogenesis. The gray shaded box shows data points that are beyond the range of what is physiologically possible.

contribution of *de novo* lipogenesis when an analyst runs one, three, or five replicate injections of a given sample set. As noted, we assumed that 10 control subjects would be studied with a 5% contribution of fatty acid synthesis and 10 "treated" subjects would be studied where some novel compound would inhibit fatty acid synthesis by ~50%. As expected, taking the average of replicate injections reduces the variation in the apparent contribution of *de novo* lipogenesis and therein allows one to detect a difference between the groups. Again, the observation here is not surprising but the approach allows one to develop a quantitative assessment of the impact of analytical error before a study (or set of analyses) is run.

4. SIMULATIONS OF PROTEIN FLUX

4.1 General Biochemical Overview

The fractional synthetic rate (FSR) of a protein can be determined using Eq. (1) (in cases where samples are obtained during the initial, pseudo-linear, change in labeling) or from the exponential increase in protein labeling using Eq. (2):

$$\text{protein labeling}_{\text{time } t} = \text{final labeling} \times \left(1 - e^{-\text{FSR} \times \text{time } t}\right) \quad (2)$$

where "time *t*" is the time after exposure to a constant administration of a tracer and "final labeling" equals the asymptotic labeling of the protein (Foster, Barrett, Toffolo, Beltz, & Cobelli, 1993; Wolfe & Chinkes, 2005).

The primary goal here was to develop an approach that would allow us to readily determine the impact of analytical error on estimates of the FSR of VLDL-apoB100 and LDL-apoB100, where the FSR was assumed to equal 0.22 and 0.022 pools per hour, respectively. As noted below, we assumed that samples would be collected at specific time points, comparable to studies in the literature (Lichtenstein et al., 1990), and noise was then added to the perfect data points (using the random number generator in MS Excel). Simulations considered that the coefficient of variation might be ~1–3% for a given isotope ratio and the best-fit was determined using GraphPad Prism and applying Eq. (2).

4.2 Tracer Selection for Measuring Protein Synthesis

Figure 3A contains an example of the background labeling in the model peptide GFEPTLEALFGK which is generated following the tryptic digestion of either the fast or the slow turning over VLDL- and LDL-apoB,

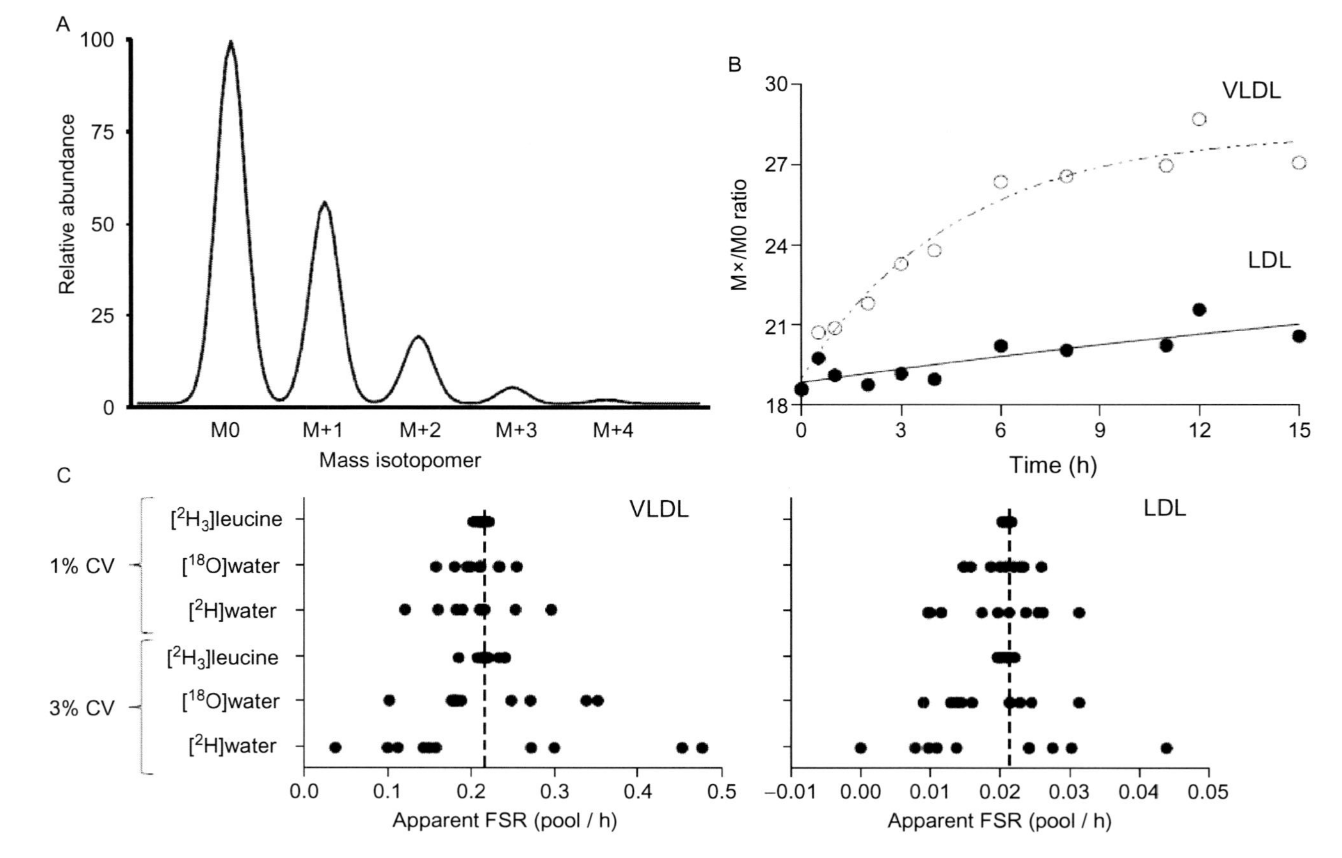

Figure 3 See legend on opposite page.

respectively (the transitions that are monitored when estimating isotope enrichment reflect the amino acid sequence PTLEALFGK, the $y9^+$ daughter ion) (Lassman et al., 2012; Zhou et al., 2012).

The first requirement for examining how error would propagate in cases where ^2H_2O, $H_2{}^{18}O$, or $^2H_3$-leucine is administered was to set some reasonable values for the simulations. Figure 3B demonstrates the change in labeling in the M+2/M0 ratio assuming an FSR of 0.22 or 0.022 per hour, comparable to VLDL-apoB100 or LDL-apoB100, respectively (Lichtenstein et al., 1990). Regardless of what tracer is administered the labeling of VLDL-apoB approaches a steady-state by ~15 h; however, the magnitude of the change in labeling will vary depending on the tracer (described below). In contrast, the labeling of the LDL-apoB yields a pseudo-linear change in the labeling over this time course.

We then set limits on the precursor:product labeling ratios; in cases where $^2H_3$-leucine is administered, investigators typically enrich the pool of free plasma leucine to ~10% and observe a steady-state labeling of protein-bound leucine of ~7% (note that there is some intracellular dilution) (Lichtenstein et al., 1990); therefore, we assumed that the steady-state labeling of the peptide GFEPTLEALFGK would be ~14% over the background, i.e., two leucine residues each at ~7% enrichment. In cases where $H_2{}^{18}O$ or ^2H_2O is used, we assumed a scenario in which the water was enriched to ~0.5% (comparable to the dose of tracer used in human studies); the expected change in peptide labeling reflects the water labeling times the number of copies of the precursor that are incorporated (n). For the simulations involving $H_2{}^{18}O$, we assumed $n \sim 9$, whereas in cases where ^2H_2O is used we assumed $n \sim 13$; as mentioned above, tryptic digestion of apoB

Figure 3 Isotope distribution analyses of apoB kinetics. LC–MS/MS analyses of the y9+ ion of GFEPTLEALFGK generate a background spectrum similar to that shown in (A); the natural isotopic abundance changes substantially when moving from the M+1 to the M+3 isotope peak. (B) The expected change in the M+2 isotope labeling assuming an FSR of 0.22 or 0.022, similar to what is expected for VLDL-apoB100 versus LDL-apoB100. The solid lines demonstrate the expected curve in cases where there is no analytical error, whereas the symbols demonstrate the effect of a randomly generated error in the isotope ratio with a coefficient of variation equal to ±3% of the true value. (C) The effects of analytical error (assuming a coefficient of variation equal to ±3% or 1% for any isotope ratio) on estimates of protein synthesis using ^2H_2O, $H_2{}^{18}O$, or $^2H_3$-leucine where one measures the change in labeling in the M+1, M+2, or M+3 isotopes, respectively. The data shown in (C) suggest what might be expected in studies of VLDL-apoB100 and LDL-apoB100, respectively (using one injection per sample in a respective time curve).

yields the peptide GFEPTLEALFGK, but our LC–MS/MS analysis measures the labeling of the $y9^+$ daughter ion (Kasumov et al., 2011; Zhou et al., 2012).

4.3 Rationale for Considering Analytical Error on Measurements of Peptide Labeling

To examine the impact of analytical error on estimates of protein synthesis, we ran simulations assuming that samples would be obtained over the time course of the experiment (i.e., Eq. 2). It was assumed that the labeling of a protein would be determined using proteome-based analyses. For example, previous studies of apolipoprotein flux required the isolation of a specific protein, followed by acid-catalyzed hydrolysis and subsequent analyses of a specific protein-bound amino acid (Lichtenstein et al., 1990). Although reliable kinetic data are obtained, the disadvantage of that approach is that it requires extensive sample preparation to ensure that a reasonably pure protein is hydrolyzed. Recent studies have demonstrated the ability to study protein flux following tryptic digestion of a mixture (Bateman, Munsell, Chen, Holtzman, & Yarasheski, 2007; Price et al., 2012; Zhou et al., 2011); the isotopic labeling of a protein-specific peptide(s) can be determined using LC–MS protocols, e.g., LC-q-MS. The advantage of this approach is that one can simultaneously study the kinetics of multiple proteins present in a complex mixture with a minimum degree of sample preparation/protein purification (Kasumov et al., 2011; Zhou et al., 2012).

4.4 Effect of Analytical Error on Respective Tracers

As noted, one can estimate the FSR by fitting the labeling curves (Eq. 2) or determining the initial rate of change in labeling (Eq. 1) (Foster et al., 1993). By contrasting VLDL-apoB against LDL-apoB, we were able to consider the impact of analytical error using both equations, and it was assumed that measurements would be made using an LC-q-MS since this instrument could handle the analyses of peptides. We first modeled a perfect data set assuming that the FSR was 0.22 or 0.022 per hour, and this was done starting at different background levels for the various mass isotopomers and tracer combinations (e.g., Fig. 3B and C). Since our simulations contrasted the use of ^2H_2O versus $H_2{}^{18}O$ versus $^2H_3$-leucine, we modeled the change in abundance of M + 1 versus M + 2 versus M + 3 labeled peptides, respectively.

A random number generator was used to add noise to the perfect data set, and the coefficient of variation was set to ±3% in any of the labeling ratios (Table 3). Obviously, the absolute error is greatest in the M + 1 isotope

Table 3 Outline of the Simulation Method as Adapted for Studies of Protein Synthesis **M1/M0 Ratio**

		Error Value Assuming a Coefficient of Variation ±3.0% of a Measured Endpoint						
			Simulation Trial Output					
Time (h)	**Expected Value**	**MS Excel Code**	**1**	**2**	**3**	**4**	**5**	**6**
0	56.8	=56.8+RANDBETWEEN (−(56.8*30),(56.8*30))/1000	56.4	56.8	56.7	58.7	55.8	58.0
1	58.1	=58.1+RANDBETWEEN (−(58.1*30),(58.1*30))/1000	60.0	56.9	58.5	58.3	58.9	58.3
2	59.1	=59.1+RANDBETWEEN (−(59.1*30),(59.1*30))/1000	58.2	60.1	57.7	61.1	57.9	59.0
3	59.9	=59.9+RANDBETWEEN (−(59.9*30),(59.9*30))/1000	59.2	58.7	57.9	61.1	58.6	59.1
4	60.6	...	62.2	59.0	59.9	61.0	59.4	59.4
5	61.1	...	63.0	61.4	59.8	60.3	60.4	62.2
6	61.6	...	59.5	61.9	62.9	60.6	59.6	60.6
7	61.9	...	62.6	62.0	63.5	60.3	60.3	61.0
8	62.2	...	63.5	60.8	61.3	62.2	61.7	60.8
9	62.4	...	63.7	61.5	60.7	62.4	64.6	62.6
10	62.6	...	60.6	64.4	62.3	60.7	64.5	64.7
11	62.7	...	64.6	61.2	62.8	63.0	64.0	60.6
12	62.8	...	63.2	64.6	63.6	62.1	63.5	62.7
13	62.9	...	64.7	62.7	62.1	61.8	62.6	61.9
14	63.0	...	63.6	64.2	61.8	63.0	63.5	62.0
15	63.1	...	63.2	64.3	63.5	61.9	61.4	62.4
Best-fit	0.22		0.21	0.11	0.18	0.17	0.15	0.17

The assumptions used here considered the problem of measuring the change in M1/M0 labeling of the VLDL-apoB peptide GFEPTLEALFGK (the transitions that are monitored when estimating isotope enrichment reflect the amino acid sequence PTLEALFGK, the $y9^+$ daughter ion). This example assumed that ^{2}H-labeling of water would approach 0.5% and that the expected turnover constant was 0.22 pools per hour. The best-fit of the simulation outputs can be determined using Eq. (2) and various software; we typically rely on GraphPad Prism (shown in the bottom row is an example of the best-fit estimate of the turnover for each of the six trials).

peak since it is ~56% of the M0 signal versus the M + 3 isotope peak which is ~4% of the M0 signal, i.e., the absolute error is 56×0.03 versus 4×0.03 (or 1.7 vs. 0.12), for the M + 1/M0 versus the M + 3/M0 isotope ratios making the baseline M + 1/M0 abundance vary from 57.7 to 54.3, while the M + 3/M0 varies from 4.1 to 3.9. Figure 3B and C demonstrates an example of how the M + 2/M0 ratio responds during these simulations in cases where $H_2{}^{18}O$ is used. By running multiple simulations of each condition, we could estimate variability in the FSR. As expected, studies which rely on ^2H_2O tend to have the worst fit of the data in spite of the fact that the analytical coefficient of variation might appear reasonable (i.e., ±3%, Fig. 3C). However, if one is able to improve the analytical reproducibility, there is a marked effect on the ability to fit the labeling curves (Fig. 3C). Note that the 10 simulations yield an apparent FSR with an 11% variation when the analytical coefficient of variation is ±1%. Comparable observations can be made when examining the impact on estimates of LDL-apoB100 kinetics.

A final set of simulations were run to examine the impact of using replicate injections and/or modulation of the tracer dose on the apparent FSR. For these tests, we only considered the use of ^2H_2O since this yielded the greatest variability in the model problem considered here. To build off of the experiments shown in Fig. 3, data shown in Fig. 4 considers two approaches for minimizing the impact of error propagation in the model of VLDL-apoB production. Clearly, the addition of more tracer improves the reproducibility in the apparent biological variability. This may be practical in some scenarios, e.g., studies in animal models. Likewise, depending on the protein of interest, it may be possible to identify target peptides and/or MS/MS fragment ions that lead to a greater change in labeling over the natural background, e.g., sequences which contain a higher abundance of heavily labeled amino acids. In addition, one can substantially improve the outcome by simply measuring replicate injections. As expected, analyzing samples in triplicate led to a sizeable reduction in the apparent biological variability (Fig. 4), which is most obvious in cases when lower amounts of tracer are administered.

4.5 Practical Implications for Modulating the Effect of Error

It is important to emphasize that the model problems used in the protein synthesis work reflect a number of variables that may not be immediately obvious; consequently, it is possible to further alter the outcome. First,

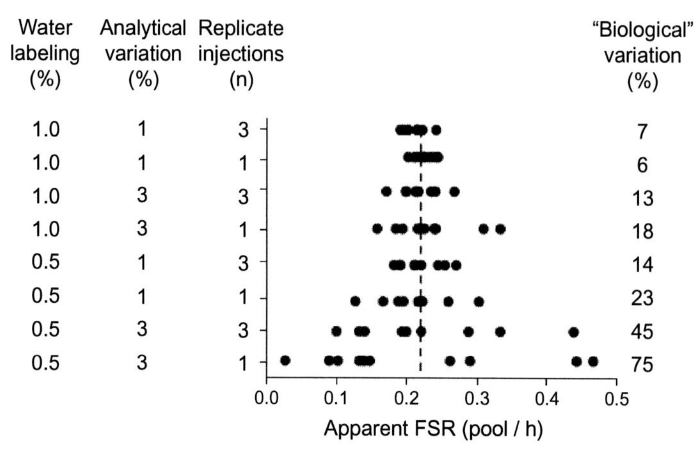

Figure 4 Minimizing the apparent biological variation by modulating tracer dose and analytical replicates. Simulations of VLDL-apoB were run assuminq that ^{2}H-water labeling was maintained at either 0.5% or 1.0% and that samples would be run as single or triplicate injections (the same frequency of sample collection as in Fig. 3). A total of 10 simulations were run per condition, the individual outcomes are shown in the plot, and the apparent biological variation was estimated. The expected FSR is equal to 0.22 pools per hour and is represented by the dashed line.

we have assumed that the error is consistent across the various isotope signals. It is possible that one could observe a decrease in the precision as the signal:noise approaches further extremes. Second, one could administer a more heavily labeled precursor (e.g., *ring*-[U-$^{13}C_6$]phenylalanine) and further push the shift in the spectrum so that measurements are made in a region where there is virtually no background labeling. Third, one can optimize the choice of peptides and focus attention on those that contain multiple copies of the precursor, namely, the (bio)polymerization reaction that constitutes protein synthesis will increase the expected change in labeling (Brunengraber et al., 1997; Papageorgopoulos, Caldwell, Shackleton, Schweingrubber, & Hellerstein, 1999). In cases where $^{2}H_2O$ is administered, this requires that one identify peptides which contain a substantial amount of nonessential amino acids since they are extensively labeled as compared to essential amino acids (Herath et al., 2011). Both of the caveats mentioned above somewhat limit the choice of peptides that one can target, and presumably studies that utilize $H_2{}^{18}O$ can target any peptide since the tracer should be incorporated more uniformly; however, investigators should probably consider identifying the largest peptides since they would contain the most ^{18}O (Zhou et al., 2012). Last, although it may appear that the use of labeled water will inevitably lead to a more variable estimate of protein flux, especially in cases where the FSR is low (e.g., LDL model in Fig. 3C), it is important to note that our simulations here only considered a 15-h protocol. Since body water has a relatively long half-life, it would be possible to collect samples for several days which therein further increases the signal:noise. Not only would this have the added benefit of reducing the error but it would allow investigators to examine flux in free-living conditions (Previs et al., 2004). The modeling method that is outlined here (Table 3) can be immediately modified in order to look for better solutions.

5. SUMMARY AND CONCLUSIONS

We have outlined a methodological approach that should allow investigators to readily examine the impact of analytical error in cases where one aims to estimate the rate of incorporation of a labeled precursor into a newly made product. Although commenting on the potential sources of analytical error is beyond the scope of this report (Antoniewicz, Kelleher, & Stephanopoulos, 2007; MacCoss, Toth, & Matthews, 2001; Matthews & Hayes, 1975), it is obvious that lowering the natural isotopic background and/or administering more tracers is advantageous when calculating rates of metabolic flux. Yang et al. clearly demonstrated this when developing the acetone-based exchange method for measuring the ^{2}H-labeling of water; they could minimize the natural background labeling and therein measure very low levels of ^{2}H in water by substituting [U-^{13}C]acetone (Yang et al., 1998). Hazey et al. applied a similar approach in developing a method to measure the ^{2}H-labeling of specific glucose-bound hydrogens (Hazey et al., 1997). In the absence of any simulations and/or calculations, one is not able to determine the magnitude of any modification on the analyses. For example, while it is obvious that lowering the analytical noise fivefold will improve the ability to estimate flux, investigators are at a loss in knowing whether it is sufficient to reliably interpret ^{2}H-based quantitation of *de novo* lipogenesis and/or protein synthesis.

We believe that there is no single best tracer-instrument combination because it depends on the signal:noise ratio of a particular analyte ion, *in vivo* protocols, and cost/benefit considerations, and in many cases investigators are limited to the instruments in their laboratory. For example, although GC-q-MS is highly reproducible as compared to LC-q-MS, peptide analyses can only be done using the latter system. Our studies demonstrate that high-resolution FT-ICR-MS is advantageous in certain situations (especially when ^{2}H-labeled tracers are administered), but the ability to administer more precursor makes a considerably less expensive GC-q-MS a realistic option for some lipid analyses.

While we did not discuss other possible instrument configurations (e.g., trap-based FT-ICR, Orbitraps, or Q-TOF analyzers), our experience is generally consistent with the literature (Kamphorst, Fan, Lu, White, & Rabinowitz, 2011). Namely, isotope ratios can be determined with a high degree of reproducibility in certain instances. Unfortunately, for trap-based instruments, we have encountered problems which are presumably related to limited trap volumes that have a dramatic impact on the ability to obtain reproducible data across a wide dynamic range; in some cases, it is possible to circumvent some of these problems (Ilchenko et al., 2013). We believe that Q-TOF instruments have great potential (Castro-Perez, Previs, et al., 2011; Castro-Perez, Roddy, et al., 2011; Wang et al., 2007), they offer similar reproducibility in terms of isotope ratio measurements as compared to a triple quadrupole mass spectrometer with the advantage of acquiring in full scan mode which permits the multiplexed assessment of product labeling (Wrona, Mauriala, Bateman, Mortishire-Smith, & O'Connor, 2005). We expect that newer trap-based mass spectrometers and Q-TOFs will yield similar performance as the Qq-FT-ICR mass spectrometers along with the advantage of newer developments in software to enable global proteomic and/or lipidomic flux measurements in a single run. Likewise, the concepts discussed herein may be of importance in the application of "parallel reaction monitoring" for quantifying tracer kinetics (Gallien, Kim, & Domon, 2015).

A final consideration has to do with spectral corrections to account for natural isotopic background labeling. We previously proposed a method that is suitable for handling complex mass isotopomer distribution patterns, including the dynamic changes in labeling that one observes when heavily substituted precursors are administered (Fernandez, DesRosiers, Previs, David, & Brunengraber, 1996). The matters discussed here are less impacted by an elaborate correction algorithm. For example, if we assume that a sample can be collected prior to isotope administration ($t=0$) and one only expects a shift in the M1/M0 ratio, then the $t=0$ sample can be simply subtracted from the later sample. In the case where one measures the temporal change in labeling, e.g., according to Eq. (2), one can simply fit the curve to the starting M1/M0 ratio. Again, it is not imperative that one subtract the $t=0$ sample from each subsequent data point. Although the matter of data correction is important, it is not critical in all cases.

In conclusion, the importance of the discussion here centers on the implementation of a simple and flexible method that can be adapted by many investigators with a limited background in mathematical modeling. This allows one to make a critical assessment of the qualitative and quantitative impact of error. Regardless of the instrument or analyte, one strategy that we apply in order to assess the magnitude of the intra-run variation is to measure the isotope labeling of an internal standard. For example, since heptadecanoic acid (C17:0) is often added as an internal standard for determining the concentration of endogenous fatty acids, it is possible to measure its natural background labeling in order to determine the intra-run variability of the isotope ratio. We hope that the discussion herein stimulates future studies of a similar nature.

REFERENCES

Antoniewicz, M. R., Kelleher, J. K., & Stephanopoulos, G. (2006). Determination of confidence intervals of metabolic fluxes estimated from stable isotope measurements. *Metabolic Engineering*, *8*, 324–337.

Antoniewicz, M. R., Kelleher, J. K., & Stephanopoulos, G. (2007). Accurate assessment of amino acid mass isotopomer distributions for metabolic flux analysis. *Analytical Chemistry*, *79*, 7554–7559.

Bateman, R. J., Munsell, L. Y., Chen, X., Holtzman, D. M., & Yarasheski, K. E. (2007). Stable isotope labeling tandem mass spectrometry (SILT) to quantify protein production and clearance rates. *Journal of the American Society for Mass Spectrometry*, *18*, 997–1006.

Bederman, I. R., Dufner, D. A., Alexander, J. C., & Previs, S. F. (2006). Novel application of the "doubly labeled" water method: Measuring CO_2 production and the tissue-specific dynamics of lipid and protein in vivo. *American Journal of Physiology. Endocrinology and Metabolism*, *290*, E1048–E1056.

Berthold, H. K., Wykes, L. J., Jahoor, F., Klein, P. D., & Reeds, P. J. (1994). The use of uniformly labelled substrates and mass isotopomer analysis to study intermediary metabolism. *The Proceedings of the Nutrition Society*, *53*, 345–354.

Bevington, P. R., & Robinson, D. K. (1992). *Data reduction and error analysis for the physical sciences*. New York: McGraw-Hill, Inc.

Brunengraber, H., Kelleher, J. K., & DesRosiers, C. (1997). Applications of mass isotopomer analysis to nutrition research. *Annual Review of Nutrition*, *17*, 559–596.

Brunengraber, D. Z., Mccabe, B. J., Katanik, J., & Previs, S. F. (2002). Gas chromatography-mass spectrometry assay of the O-18 enrichment of water as trimethyl phosphate. *Analytical Biochemistry*, *306*, 278–282.

Casazza, J. P., & Veech, R. L. (1984). Quantitation of the rate of fatty acid synthesis. *Laboratory and Research Methods in Biology and Medicine*, *10*, 231–240.

Castro-Perez, J. M., Previs, S. F., McLaren, D. G., Shah, V., Herath, K., Bhat, G., et al. (2011). In-vivo D2O labeling in C57Bl/6 mice to quantify static and dynamic changes in cholesterol and cholesterol esters by high resolution LC mass-spectrometry. *Journal of Lipid Research*, *52*, 159–169.

Castro-Perez, J. M., Roddy, T. P., Shah, V., McLaren, D. G., Wang, S. P., Jensen, K., et al. (2011). Identifying static and kinetic lipid phenotypes by high resolution UPLC/MS: Unraveling diet-induced changes in lipid homeostasis by coupling metabolomics and fluxomics. *Journal of Proteome Research*, *10*, 4281–4290.

Cuchel, M., Schaefer, E. J., Millar, J. S., Jones, P. J., Dolnikowski, G. G., Vergani, C., et al. (1997). Lovastatin decreases de novo cholesterol synthesis and LDL Apo B-100 production rates in combined-hyperlipidemic males. *Arteriosclerosis, Thrombosis, and Vascular Biology, 17*, 1910–1917.

De Riva, A., Deery, M. J., McDonald, S., Lund, T., & Busch, R. (2010). Measurement of protein synthesis using heavy water labeling and peptide mass spectrometry: Discrimination between major histocompatibility complex allotypes. *Analytical Biochemistry, 403*, 1–12.

Di Buono, M., Jones, P. J., Beaumier, L., & Wykes, L. J. (2000). Comparison of deuterium incorporation and mass isotopomer distribution analysis for measurement of human cholesterol biosynthesis. *Journal of Lipid Research, 41*, 1516–1523.

Dietschy, J. M., & McGarry, J. D. (1974). Limitations of acetate as a substrate for measuring cholesterol synthesis in liver. *The Journal of Biological Chemistry, 249*, 52–58.

Dietschy, J. M., & Spady, D. K. (1984). Measurement of rates of cholesterol synthesis using tritiated water. *Journal of Lipid Research, 25*, 1469–1476.

Diraison, F., Pachiaudi, C., & Beylot, M. (1996). In vivo measurement of plasma cholesterol and fatty acid synthesis with deuterated water: Determination of the average number of deuterium atoms incorporated. *Metabolism, 45*, 817–821.

Diraison, F., Pachiaudi, C., & Beylot, M. (1997). Measuring lipogenesis and cholesterol synthesis in humans with deuterated water: Use of simple gas chromatographic/mass spectrometric techniques. *Journal of Mass Spectrometry, 32*, 81–86.

Duarte, J. A., Carvalho, F., Pearson, M., Horton, J. D., Browning, J. D., Jones, J. G., et al. (2014). A high-fat diet suppresses de novo lipogenesis and desaturation but not elongation and triglyceride synthesis in mice. *Journal of Lipid Research, 55*, 2541–2553.

Fernandez, C. A., DesRosiers, C., Previs, S. F., David, F., & Brunengraber, H. (1996). Correction of C-13 mass isotopomer distributions for natural stable isotope abundance. *Journal of Mass Spectrometry, 31*, 255–262.

Foster, D. M., Barrett, P. H., Toffolo, G., Beltz, W. F., & Cobelli, C. (1993). Estimating the fractional synthetic rate of plasma apolipoproteins and lipids from stable isotope data. *Journal of Lipid Research, 34*, 2193–2205.

Gallien, S., Kim, S. Y., & Domon, B. (2015). Large-scale targeted proteomics using internal standard triggered-parallel reaction monitoring (IS-PRM). *Molecular & Cellular Proteomics, 14*, 1630–1644.

Gasier, H. G., Previs, S. F., Pohlenz, C., Fluckey, J. D., Gatlin, D. M., & Buentello, J. A. (2009). A novel approach for assessing protein synthesis in channel catfish, *Ictalurus punctatus*. *Comparative Biochemistry and Physiology Part B, Biochemistry & Molecular Biology, 154*, 235–238.

Hachey, D. L., Wong, W. W., Boutton, T. W., & Klein, P. D. (1987). Isotope ratio measurements in nutrition and biomedical research. *Mass Spectrometry Reviews, 6*, 289–328.

Hazey, J. W., Yang, D., Powers, L., Previs, S. F., David, F., Beaulieu, A. D., et al. (1997). Tracing gluconeogenesis with deuterated water: Measurement of low deuterium enrichments on carbons 6 and 2 of glucose. *Analytical Biochemistry, 248*, 158–167.

Herath, K., Bhat, G., Miller, P. L., Wang, S. P., Kulick, A., Andrews-Kelly, G., et al. (2011). Equilibration of (2)H labeling between body water and free amino acids: Enabling studies of proteome synthesis. *Analytical Biochemistry, 415*, 197–199.

Herath, K. B., Zhong, W., Yang, J., Mahsut, A., Rohm, R. J., Shah, V., et al. (2014). Determination of low levels of 2H-labeling using high-resolution mass spectrometry: Application in studies of lipid flux and beyond. *Rapid Communications in Mass Spectrometry, 28*, 239–244.

Ilchenko, S., Previs, S. F., Rachdaoui, N., Willard, B., McCullough, A. J., & Kasumov, T. (2013). An improved measurement of isotopic ratios by high resolution mass spectrometry. *Journal of the American Society for Mass Spectrometry, 24*, 309–312.

Jeske, D. J., & Dietschy, J. M. (1980). Regulation of rates of cholesterol synthesis in vivo in the liver and carcass of the rat measured using [3H]water. *Journal of Lipid Research*, *21*, 364–376.

Jin, E. S., Jones, J. G., Merritt, M., Burgess, S. C., Malloy, C. R., & Sherry, A. D. (2004). Glucose production, gluconeogenesis, and hepatic tricarboxylic acid cycle fluxes measured by nuclear magnetic resonance analysis of a single glucose derivative. *Analytical Biochemistry*, *327*, 149–155.

Kamphorst, J. J., Fan, J., Lu, W., White, E., & Rabinowitz, J. D. (2011). Liquid chromatography-high resolution mass spectrometry analysis of fatty acid metabolism. *Analytical Chemistry*, *83*, 9114–9122.

Kasumov, T., Ilchenko, S., Li, L., Rachdaoui, N., Sadygov, R. G., Willard, B., et al. (2011). Measuring protein synthesis using metabolic (2)H labeling, high-resolution mass spectrometry, and an algorithm. *Analytical Biochemistry*, *412*, 47–55.

Lakshmanan, M. R., Berdanier, C. D., & Veech, R. L. (1977). Comparative studies on lipogenesis and cholesterogenesis in lipemic BHE rats and normal wistar rats. *Archives of Biochemistry and Biophysics*, *183*, 355–360.

Lassman, M. E., McLaughlin, T. M., Somers, E. P., Stefanni, A. C., Chen, Z., Murphy, B. A., et al. (2012). A rapid method for cross species quantitation of apolipoproteins A1, B48 and B100 in plasma by UPLC-MS/MS. *Rapid Communications in Mass Spectrometry*, *26*, 101–108.

Lee, W. N., Bassilian, S., Ajie, H. O., Schoeller, D. A., Edmond, J., Bergner, E. A., et al. (1994). In vivo measurement of fatty acids and cholesterol synthesis using D2O and mass isotopomer analysis. *The American Journal of Physiology*, *266*, E699–E708.

Leitch, C. A., & Jones, P. J. (1991). Measurement of triglyceride synthesis in humans using deuterium oxide and isotope ratio mass spectrometry. *Biological Mass Spectrometry*, *20*, 392–396.

Leitch, C. A., & Jones, P. J. (1993). Measurement of human lipogenesis using deuterium incorporation. *Journal of Lipid Research*, *34*, 157–163.

Lichtenstein, A. H., Cohn, J. S., Hachey, D. L., Millar, J. S., Ordovas, J. M., & Schaefer, E. J. (1990). Comparison of deuterated leucine, valine, and lysine in the measurement of human apolipoprotein A-I and B-100 kinetics. *Journal of Lipid Research*, *31*, 1693–1701.

Lowenstein, J. M., Brunengraber, H., & Wadke, M. (1975). Measurement of rates of lipogenesis with deuterated and tritiated water. *Methods in Enzymology*, *35*, 279–287.

MacCoss, M. J., Toth, M. J., & Matthews, D. E. (2001). Evaluation and optimization of ion-current ratio measurements by selected-ion-monitoring mass spectrometry. *Analytical Chemistry*, *73*, 2976–2984.

Matthews, D. E., & Hayes, J. M. (1975). Systematic errors in gas chromatography-mass spectrometry isotope ratio measurements. *Analytical Chemistry*, *48*, 1375–1382.

Matthews, D. E., Motil, K. J., Rohrbaugh, D. K., Burke, J. F., Young, V. R., & Bier, D. M. (1980). Measurement of leucine metabolism in man from a primed, continuous infusion of L-[1-^{13}C]leucine. *The American Journal of Physiology*, *238*, E473–E479.

Mccabe, B. J., Bederman, I. R., Croniger, C. M., Millward, C. A., Norment, C. J., & Previs, S. F. (2006). Reproducibility of gas chromatography-mass spectrometry measurements of H-2 labeling of water: Application for measuring body composition in mice. *Analytical Biochemistry*, *350*, 171–176.

Papageorgopoulos, C., Caldwell, K., Shackleton, C., Schweingrubber, H., & Hellerstein, M. K. (1999). Measuring protein synthesis by mass isotopomer distribution analysis (MIDA). *Analytical Biochemistry*, *267*, 1–16.

Parks, E. J., Skokan, L. E., Timlin, M. T., & Dingfelder, C. S. (2008). Dietary sugars stimulate fatty acid synthesis in adults. *The Journal of Nutrition*, *138*, 1039–1046.

Patterson, B. W., & Wolfe, R. R. (1993). Concentration dependence of methyl palmitate isotope ratios by electron impact ionization gas chromatography/mass spectrometry. *Biological Mass Spectrometry*, *22*, 481–486.

Previs, S. F., Fatica, R., Chandramouli, V., Alexander, J. C., Brunengraber, H., & Landau, B. R. (2004). Quantifying rates of protein synthesis in humans by use of (H2O)-H-2: Application to patients with end-stage renal disease. *American Journal of Physiology. Endocrinology and Metabolism*, *286*, E665–E672.

Previs, S. F., Mahsut, A., Kulick, A., Dunn, K., Andrews-Kelly, G., Johnson, C., et al. (2011). Quantifying cholesterol synthesis in vivo using (2)H(2)O: Enabling back-to-back studies in the same subject. *Journal of Lipid Research*, *52*, 1420–1428.

Price, J. C., Holmes, W. E., Li, K. W., Floreani, N. A., Neese, R. A., Turner, S. M., et al. (2012). Measurement of human plasma proteome dynamics with (2)H(2)O and liquid chromatography tandem mass spectrometry. *Analytical Biochemistry*, *420*, 73–83.

Rachdaoui, N., Austin, L., Kramer, E., Previs, M. J., Anderson, V. E., Kasumov, T., et al. (2009). Measuring proteome dynamics in vivo. *Molecular & Cellular Proteomics*, *8*, 2653–2663.

Scrimgeour, C. M., Begley, I. S., & Thomason, M. L. (1999). Measurement of deuterium incorporation into fatty acids by gas chromatography isotope ratio mass spectrometry. *Rapid Communications in Mass Spectrometry*, *13*, 271–274.

Shah, V., Herath, K., Previs, S. F., Hubbard, B. K., & Roddy, T. P. (2010). Headspace analyses of acetone: A rapid method for measuring the 2H-labeling of water. *Analytical Biochemistry*, *404*, 235–237.

Stanhope, K. L., Schwarz, J. M., Keim, N. L., Griffen, S. C., Bremer, A. A., Graham, J. L., et al. (2009). Consuming fructose-sweetened, not glucose-sweetened, beverages increases visceral adiposity and lipids and decreases insulin sensitivity in overweight/ obese humans. *The Journal of Clinical Investigation*, *119*, 1322–1334.

Stellaard, F., Bloks, V. W., Burgerhof, H. G., Scheltema, R. A., Murphy, E. J., Romijn, H. A., et al. (2010). Two time-point assessment of bile acid kinetics in humans using stable isotopes. *Isotopes in Environmental and Health Studies*, *46*, 325–336.

Taylor, C. B., Mikkelson, B., Anderson, J. A., & Forman, D. T. (1966). Human serum cholesterol synthesis measured with the deuterium label. *Archives of Pathology*, *81*, 213–231.

Wang, B., Sun, G., Anderson, D. R., Jia, M., Previs, S., & Anderson, V. E. (2007). Isotopologue distributions of peptide product ions by tandem mass spectrometry: Quantitation of low levels of deuterium incorporation. *Analytical Biochemistry*, *367*, 40–48.

Wolfe, R. R., & Chinkes, D. L. (2005). *Isotope tracers in metabolic research: Principles and practice of kinetic analyses*. Hoboken, NJ: Wiley-Liss.

Wrona, M., Mauriala, T., Bateman, K. P., Mortishire-Smith, R. J., & O'Connor, D. (2005). 'All-in-one' analysis for metabolite identification using liquid chromatography/hybrid quadrupole time-of-flight mass spectrometry with collision energy switching. *Rapid Communications in Mass Spectrometry*, *19*, 2597–2602.

Yang, D., Diraison, F., Beylot, M., Brunengraber, D. Z., Samols, M. A., Anderson, V. E., et al. (1998). Assay of low deuterium enrichment of water by isotopic exchange with [U-$^{13}C_3$]acetone and gas chromatography-mass spectrometry. *Analytical Biochemistry*, *258*, 315–321.

Yu, L., Sinha, A. K., & Previs, S. F. (2005). Effect of sampling interval on the use of "doubly labeled" water for measuring CO_2 production. *Analytical Biochemistry*, *337*, 343–346.

Zhou, H., Li, W., Wang, S. P., Mendoza, V., Rosa, R., Hubert, J., et al. (2012). Quantifying apoprotein synthesis in rodents: Coupling LC-MS/MS analyses with the administration of labeled water. *Journal of Lipid Research*, *53*, 1223–1231.

Zhou, H., McLaughlin, T., Herath, K., Lassman, M. E., Rohm, R. J., Wang, S.-P., et al. (2011). Development of lipoprotein synthesis measurement using LC-MS and deuterated water labeling. *Journal of the American Society for Mass Spectrometry*, *22*, 182.

AUTHOR INDEX

Note: Page numbers followed by "*f*" indicate figures and "*t*" indicate tables.

A

B

C

D

E

H

I

J

K

L

M

P

Q

R

S

T

X

Y

Z

SUBJECT INDEX

Note: Page numbers followed by "*f*" indicate figures, "*t*" indicate tables, "*b*" indicate boxes and "*ge*" indicate glossary.

L

M

N

T

V

W

X

Z

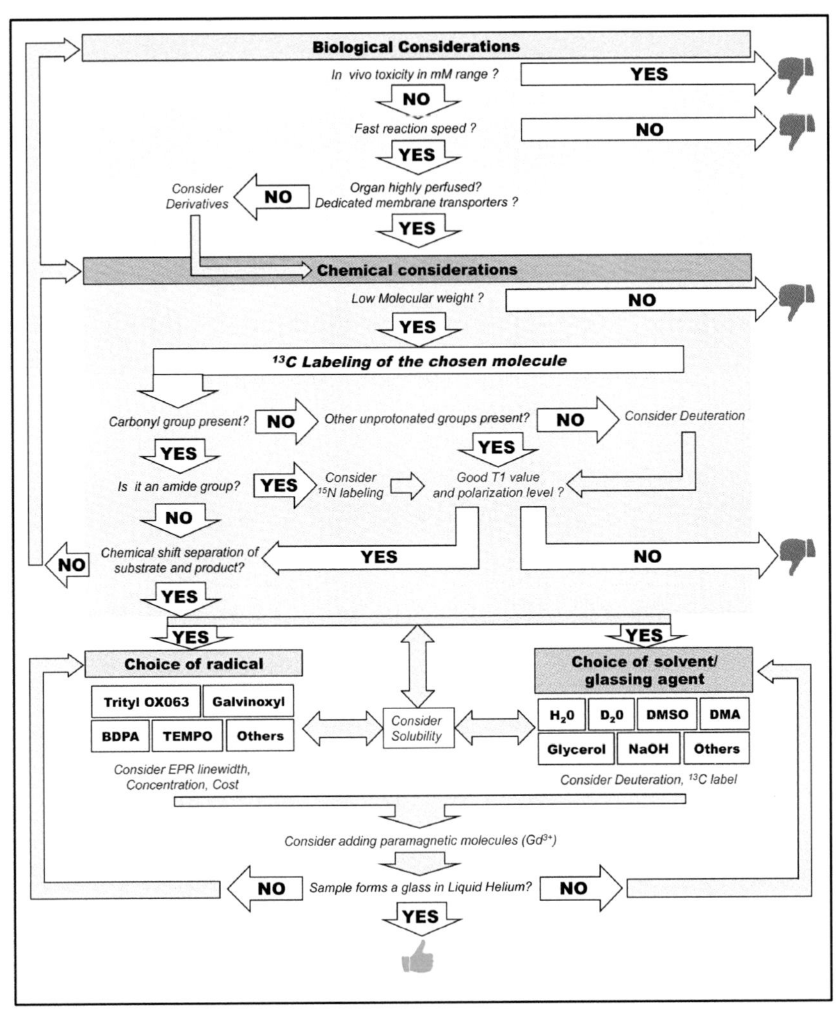

Myriam M. Chaumeil *et al.*, Figure 1 Designing a ^{13}C DNP probe for applications to the study of living systems: a brief guide. The figure summarizes Section 3 in a graphical manner. Briefly, when designing a ^{13}C DNP probe, one has to take into consideration biological parameters, such as reaction speed and delivery, as well as chemical parameters, such as ^{13}C label positioning, and deuteration and then come the choices of the radical and solvent/s for polarization, before final testing for low-temperature glassing.

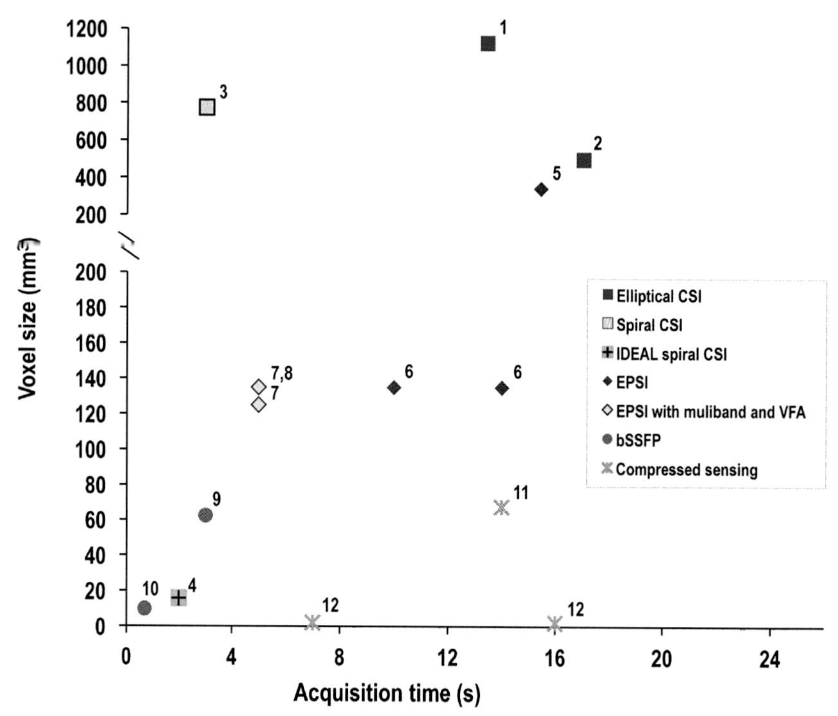

Myriam M. Chaumeil *et al.*, Figure 4 Voxel size and acquisition time for 2D slices. Summary of voxel sizes and acquisition times that can be achieved when acquiring a 2D slice using the different sequences found in the literature. References: 1, Golman et al. (2008); 2, Kohler et al. (2007); 3, Lau et al. (2010); 4, Wiesinger et al. (2012); 5, Cunningham et al. (2007); 6, Chen, Albers, et al. (2007); 7, Larson et al. (2010); 8, Larson et al. (2008); 9, von Morze et al. (2013); 10, Leupold et al. (2009); 11, Hu et al. (2008); 12, Hu et al. (2010).

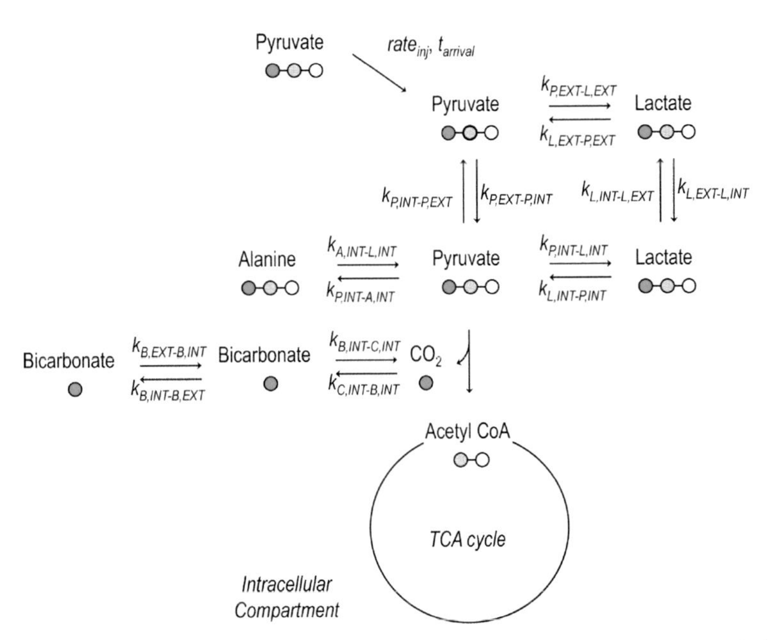

Myriam M. Chaumeil *et al.*, Figure 5 Schematic of pyruvate uptake and metabolism. Illustration of parameters that need to be included in modeling pyruvate delivery and conversion. These include the rate of injection ($rate_{inj}$), the delivery of the hyperpolarized substrate ($t_{arrival}$), its transport across membranes (e.g., $k_{P,EXP-P,INT}$ is the rate for pyruvate transport from the extra- to intracellular space and $k_{P,INT-P,EXT}$ the reverse) and its conversion through multiple enzymatic reactions. Here, we depict the conversion of pyruvate into lactate, alanine, bicarbonate, and acetyl-CoA (e.g., $k_{P,INT-L,INT}$ is the rate for the pyruvate (P)-to-lactate (L) conversion in the intracellular space). Note: C1 position: green/dark; C2 position: blue/light.

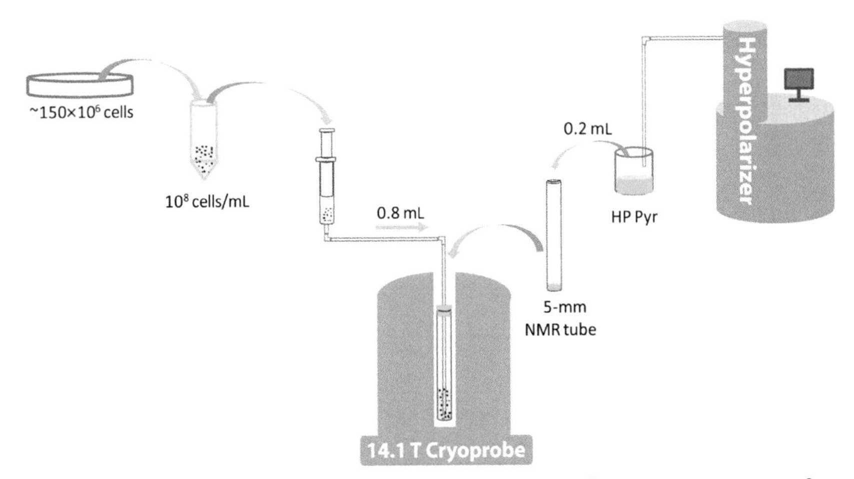

Lloyd Lumata *et al.*, Figure 2 Schematic of a highly efficient system to perform hyperpolarized ^{13}C NMR experiments in cultured cells. The cells must be rapidly harvested for the experiment without changing their metabolism. Mixing of the cells with the hyperpolarized agent is obviously essential. The preferred method is placing a small volume of the HP solution into the bottom of the tube and subsequently injecting a large volume of cells into it to cause turbulent mixing. Injecting a small volume of HP solution into a large volume of cells does not accomplish this goal. To record the initial kinetics, mixing of the cells and the imaging agent should take place inside the magnet with the experiment already queued.

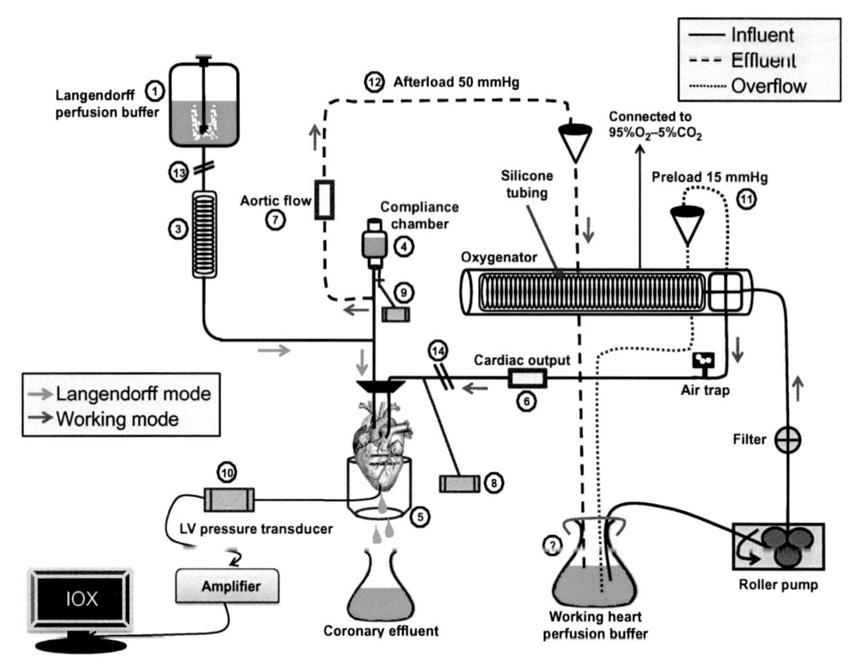

Matthieu Ruiz *et al.*, Figure 1 Schematic overview of the semi-recirculating working heart. See Section 2.1 for details. The number refers to the following items: **1**: Langendorff buffer reservoir, **2**: working mode perfusion buffer reservoir, **3**: helical glass coil, **4**: compliance chamber, **5**: opened heart chamber, **6**: electromagnetic flow probes (atrial inflow), **7**: electromagnetic flow probes (aortic outflow), **8** and **9**: pressure transducers (preload and afterload pressures), **10**: pressure transducer (intraventricular pressures), **11**: preload line (15 mmHg), **12**: afterload line (50 mmHg), **13**: three-way valve (aortic line), and **14**: three-way valve (atrial line). Abbreviation: LV = Left Ventricle.

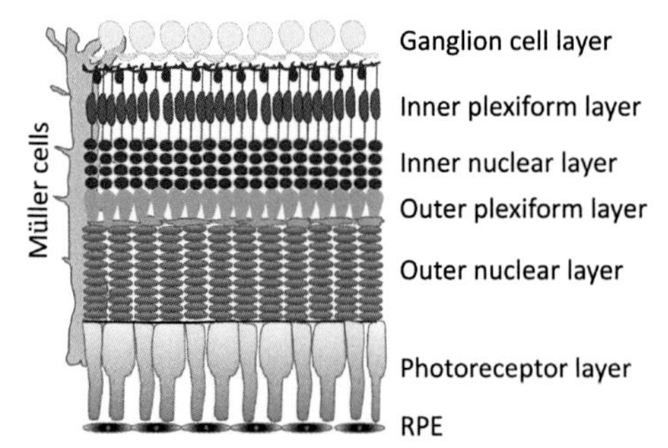

Jianhai Du *et al.*, Figure 1 Laminar structure of vertebrate retina. The photoreceptor layer contains the outer segments and cell bodies of rods and cone cell bodies and the outer nuclear layer contains their nuclei; the inner nuclear layer contains nuclei of bipolar, amacrine, and horizontal and Müller cells; the inner plexiform layer contains dendrites and synapses of the inner retinal neurons and ganglion cells; the ganglion cell layer consists of ganglion cells; and Müller glia radiate across all layers of the retina. RPE is the retinal pigment epithelium.

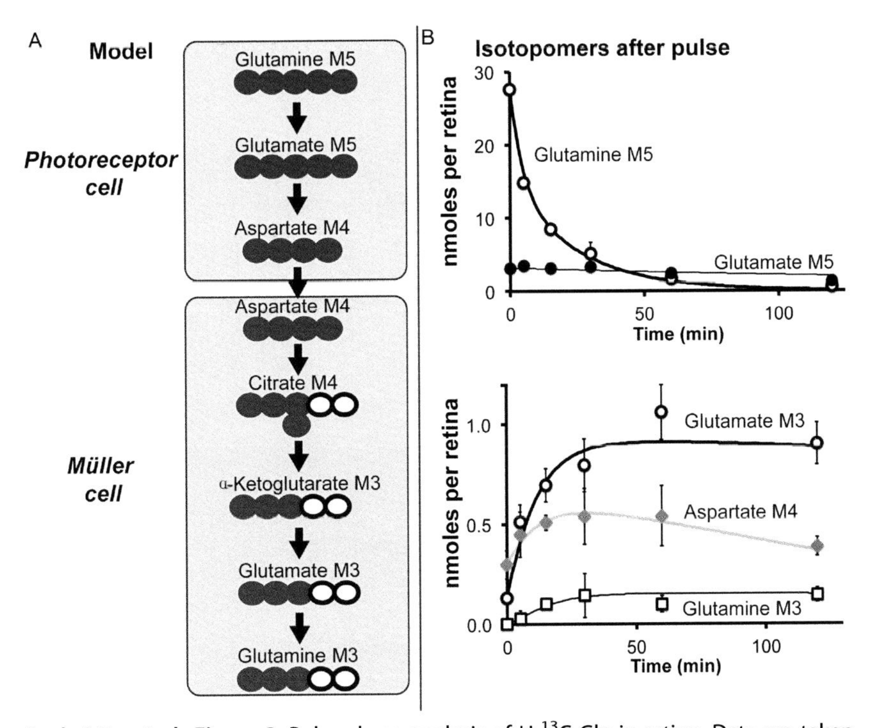

Jianhai Du *et al.*, Figure 3 Pulse-chase analysis of U-^{13}C Gln in retina. Data are taken from the authors' previous study (Lindsay et al., 2014). (A) Schematic model for the role of aspartate as a carrier of oxidizing power between retinal neurons and glia. Red circles represent the ^{13}C carbons, and black circles represent the ^{12}C carbons. (B) ^{13}C labeling of Aspartate, Glutamate, and Glutamine from the pulse of U-^{13}C Glutamine. (Upper) The M5 Glutamine and Glutamate are derived directly from the pulse of 5 m*M* U-^{13}C Glutamine. After 5 min, the medium was changed to 5 m*M* unlabeled Lac with no added Gln. The retinas were subsequently harvested at the indicated times after the pulse. (Lower) The M4 Aspartate derived from oxidation of Glutamate via the TCA. The M3 Glutamate is made by further oxidation via citrate, and M3 Glutamine is made only in MCs by Glutamine synthetase ($n=6$).

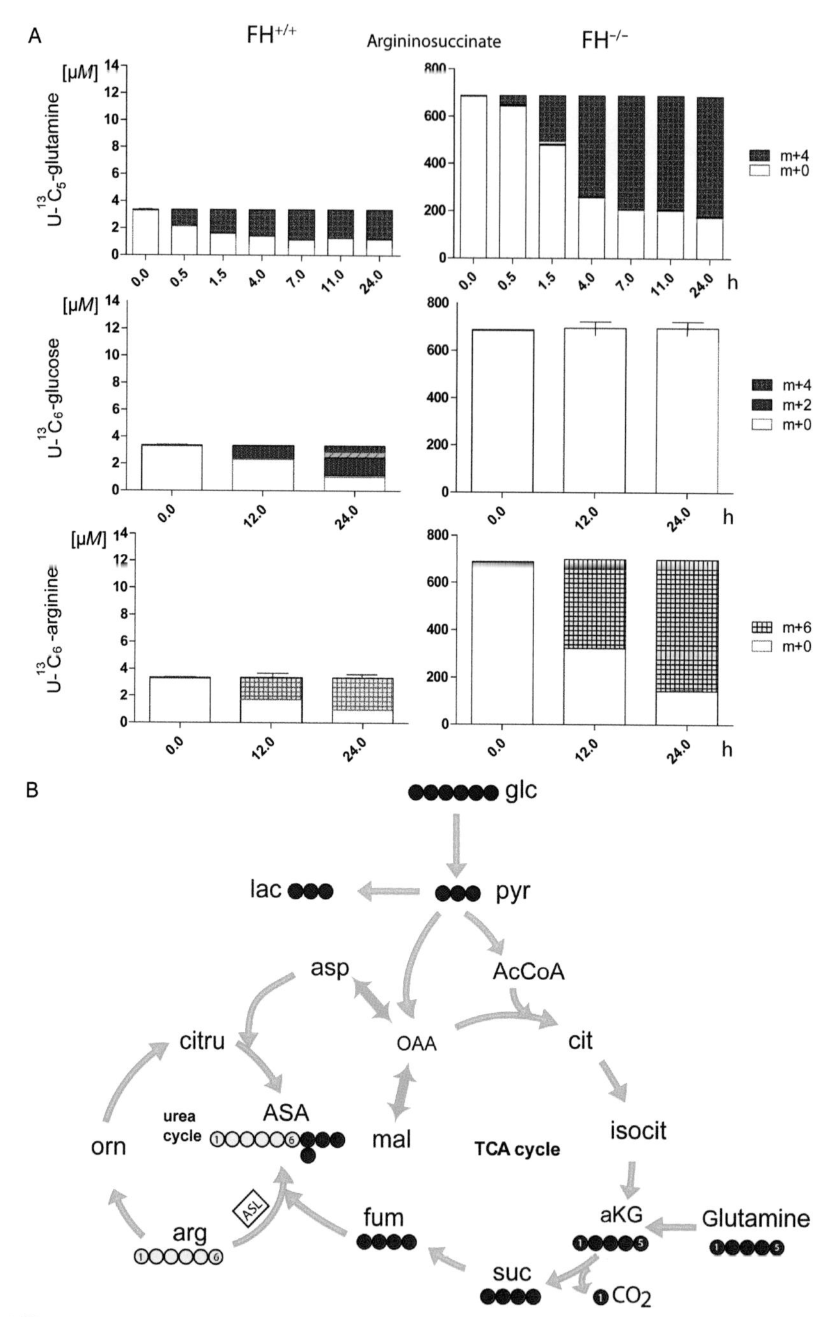

Gillian M. Mackay *et al.*, Figure 2 (A) The labeling patterns derived from U-$^{13}C_5$-glutamine, U-$^{13}C_6$-glucose, and U-$^{13}C_6$-arginine over time and the concentrations of the urea cycle metabolite, argininosuccinate (ASA) in FH-proficient or deficient cells. (B) The labeling patterns derived from U-$^{13}C_5$-glutamine, U-$^{13}C_6$-glucose, and U-$^{13}C_6$-arginine demonstrated an unexpected synthesis of ASA from arginine and fumarate by the reverse activity of ASL.

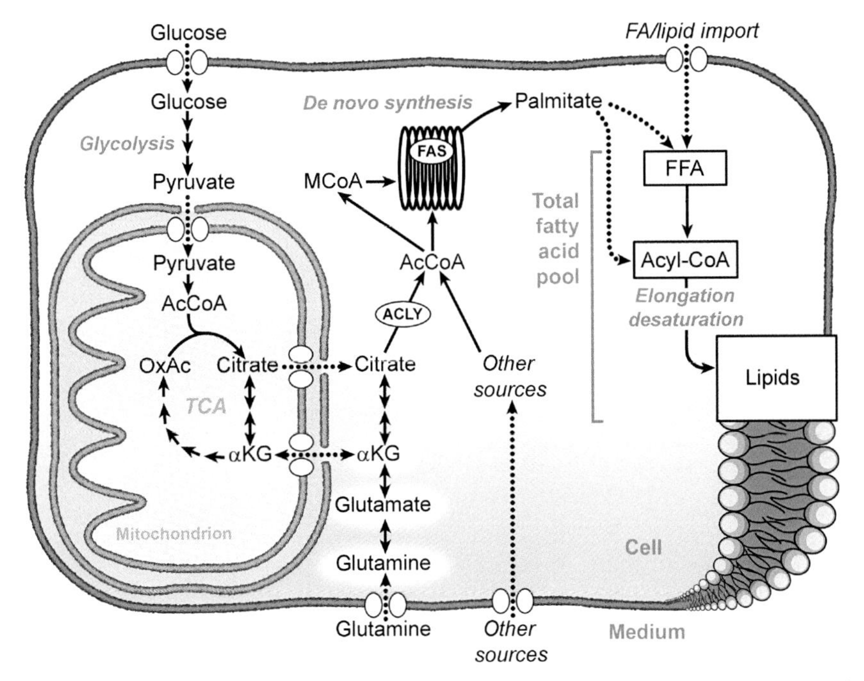

Sergey Tumanov *et al.*, Figure 1 Fatty acid synthesis by cells. Glucose, glutamine, and other substrates are precursors for the production of cytosolic acetyl-CoA. Acetyl-CoA in turn is the two-carbon donor for fatty acid synthesis. The resulting palmitate is subjected to elongation and desaturation reactions to produce a variety of fatty acids that are required for proper cellular functioning. Abbreviations: AcCoA, acetyl-CoA; OxAc, oxaloacetate; αKG, α-ketoglutarate; ACLY, ATP citrate lyase; MCoA, malonyl-CoA; FAS, fatty acid synthase; FFA, free (nonesterified) fatty acid.

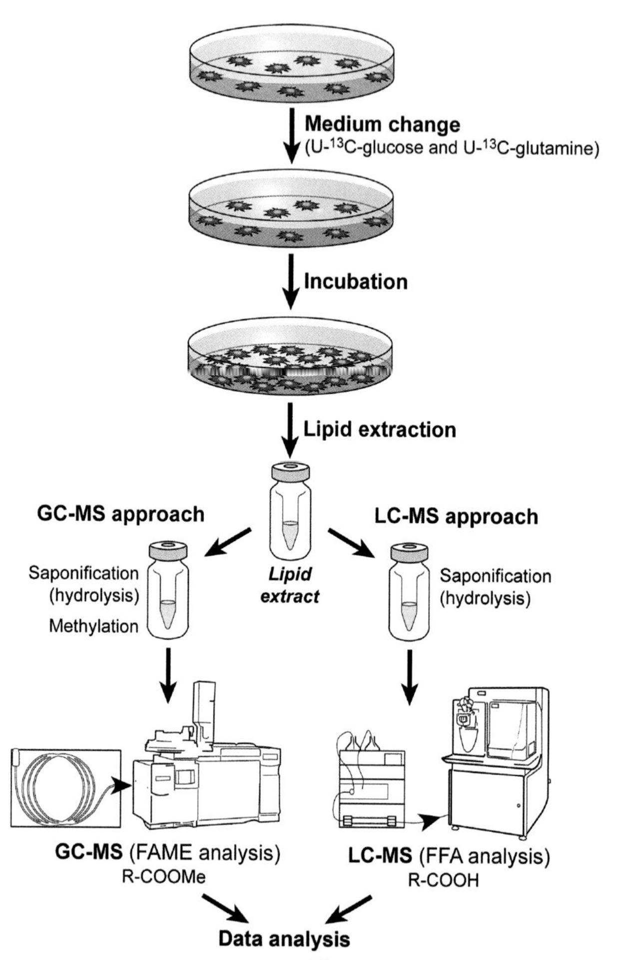

Sergey Tumanov *et al.*, Figure 2 Fatty acid ^{13}C-isotope incorporation in cultured cells. Cells are incubated in medium containing U-^{13}C-glucose and U-^{13}C-glutamine. Then a whole-cell lipid extraction is performed followed by hydrolysis and methylation (GC-MS) or hydrolysis only (LC-MS) of fatty acids, followed by mass spectrometry analysis. Abbreviations: GC-MS, gas chromatography-mass spectrometry; LC-MS, liquid chromatography-mass spectrometry.

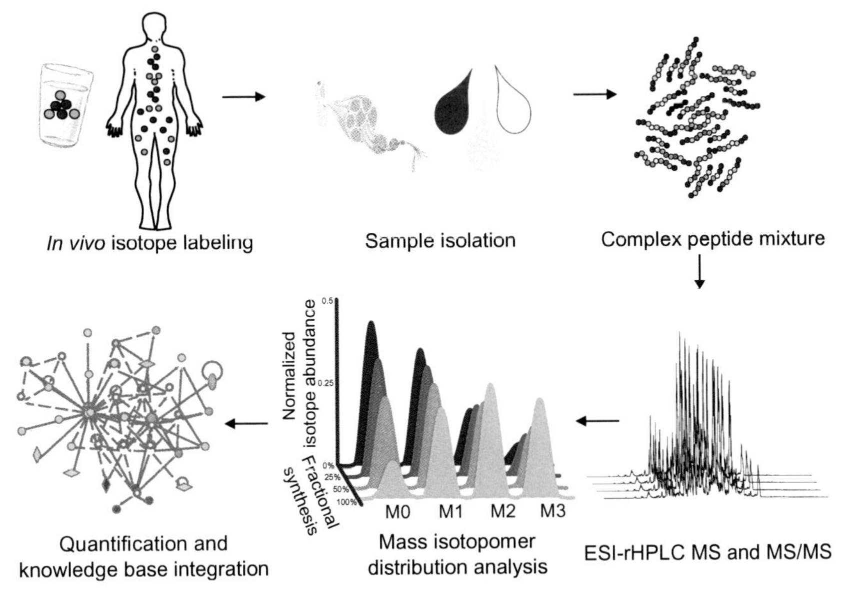

W.E. Holmes *et al.*, Figure 1 Experimental approach for measurement of *in vivo* protein dynamics.

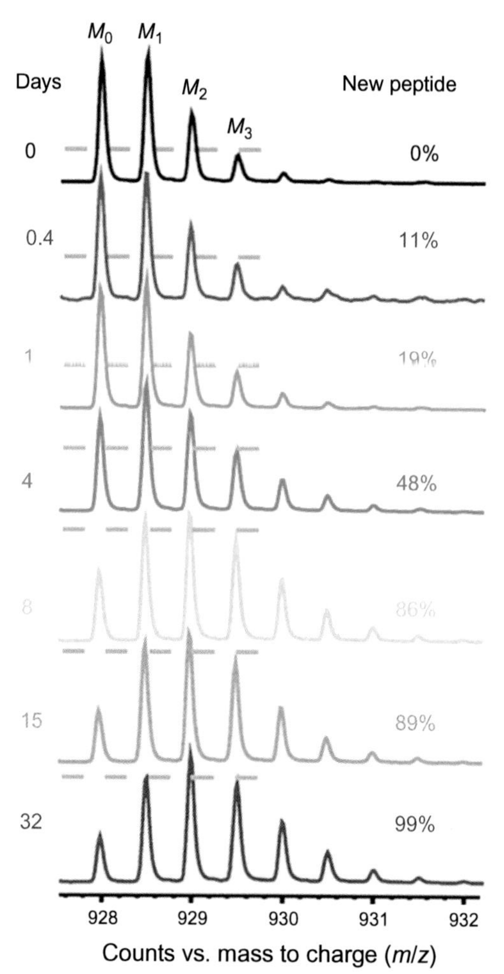

W.E. Holmes *et al.*, Figure 4 The relative abundances of peptide isotopomers (M_0 to M_4) change during metabolic labeling with ^2H_2O. The plot illustrates the change in isotope pattern of a tryptic peptide (FEDGDLTLYQSNAILR, $n = 29$) as it approaches 100% new during 32 days of continuous label. Isotope abundance relative to baseline (excess M_x or EM_x) changes as label is incorporated into this peptide; EM_0 and EM_1 become progressively more negative, while EM_2, EM_3, and EM_4 become more positive. *Adapted from Price, Khambatta, et al. (2012). © The American Society for Biochemistry and Molecular Biology.*

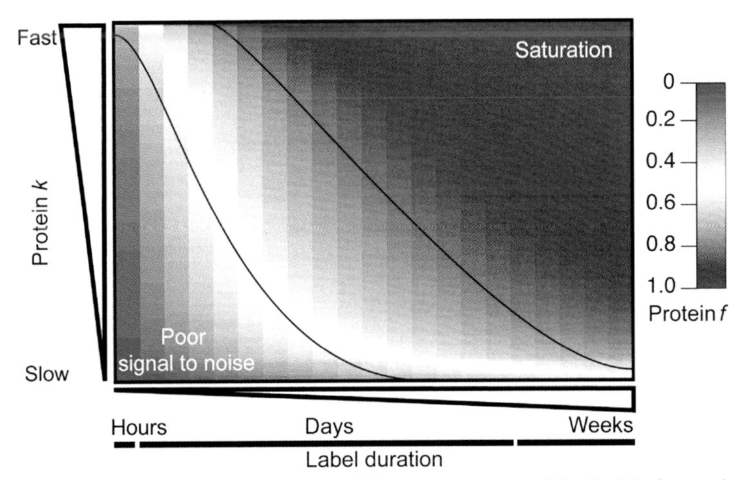

W.E. Holmes *et al.*, Figure 7 Depiction of the target range of f suitable for analysis by mass spectrometry. Slower turnover proteins (lower k) require more time to incorporate label to be measurable while faster turnover proteins (higher k) may become fully labeled after a certain duration of label exposure. The "sweet spot" is shown visually as the area between the two curves.

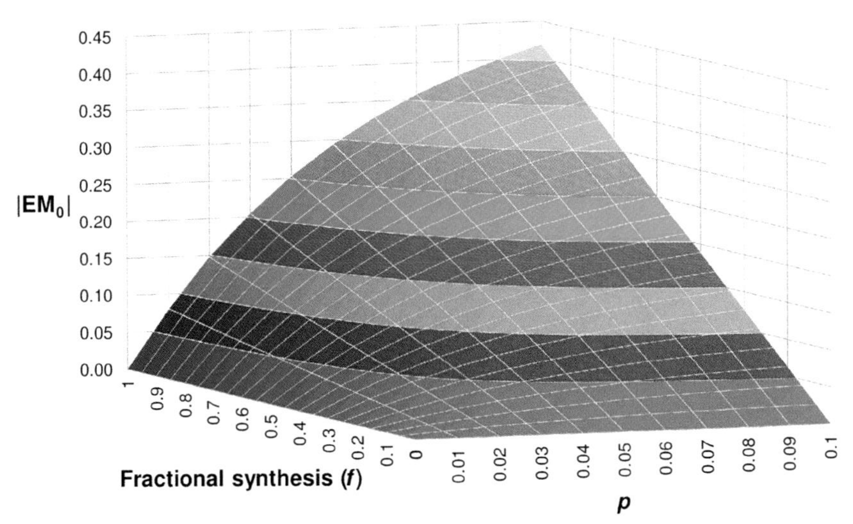

W.E. Holmes *et al.*, Figure 10 The relationship between precursor pool enrichment (p), fractional synthesis (f), and EM_0^* for peptide VLEDLRSGLF having mass = 1147, $n = 17$.

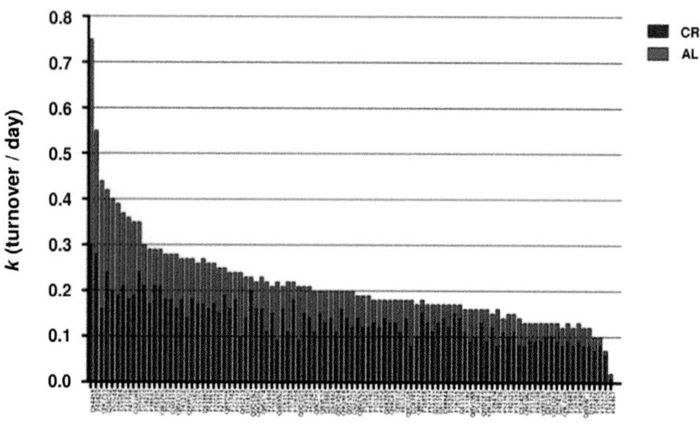

W.E. Holmes *et al.*, Figure 15 Global changes in protein turnover across the proteome can be detected by dynamic proteomics techniques. Long-term calorie restriction in 18-month-old rats reduces protein turnover across the proteome in liver, compared to *ad libitum*-fed rats. More than 300 proteins are shown. *Adapted from Price, Khambatta, et al. (2012).*

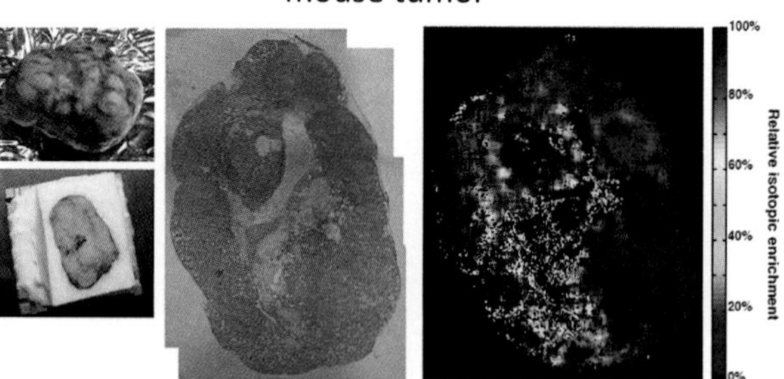

W.E. Holmes *et al.*, Figure 16 Kinetic mass spectrometry imaging or *in situ* kinetic histochemistry. (A). Lipid fluxes are visualized across tissue histologic specimens after *in vivo* heavy water labeling of a tumor-bearing mouse (triple-negative mammary cancer). Flux rates of the lipid molecule shown here reveal hot spots and cold spots within the anatomic bounds of the tumor. *Adapted from Louie et al. (2013).*

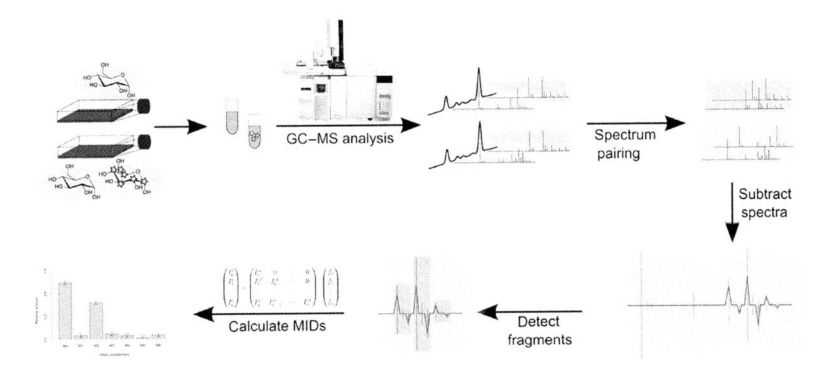

Daniel Weindl *et al.*, Figure 1 The NTFD algorithm operates on GC–EI-MS measurements of an isotopically enriched and nonenriched sample. For each compound, mass spectra from both measurements are paired. Isotopically enriched fragments are determined from characteristic patterns in the difference spectrum. For each of these enriched fragments the MID is determined.

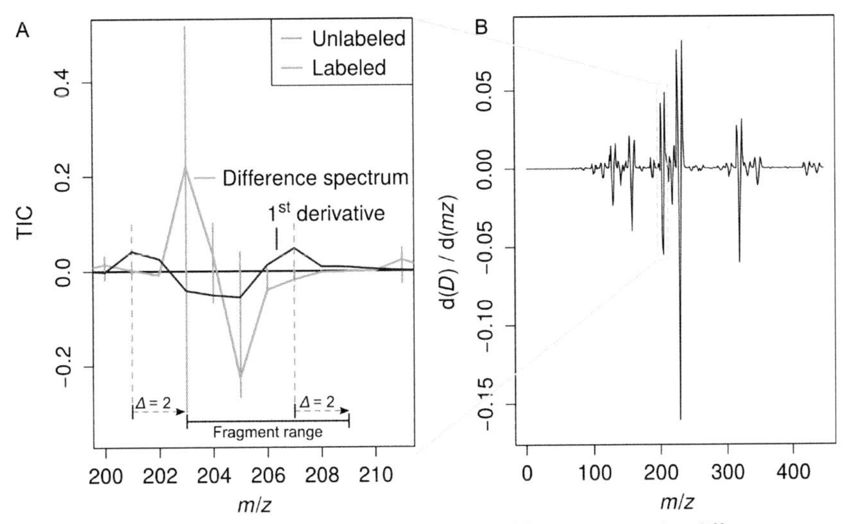

Daniel Weindl *et al.*, Figure 2 NTFD detects labeled fragments in the difference spectrum of mass spectra derived from a potentially isotopically enriched and nonenriched compound. (A) Mass spectrum of a fragment of the unlabeled compound and its labeled analogue, normalized to total ion current (TIC). Isotopic enrichment leads to characteristic peak patterns in the difference spectrum and its first derivative. Fragment boundaries are detected in the first derivative of the smoothed difference spectrum. The boundaries are shifted by 2 units due to the applied smoothing algorithm. (B) First derivative $\left(\frac{dD}{d(m/z)}\right)$ of the full difference spectrum of glutamine 4 TMS. Two positive peaks, separated by a negative peak, mark an isotopically enriched fragment.

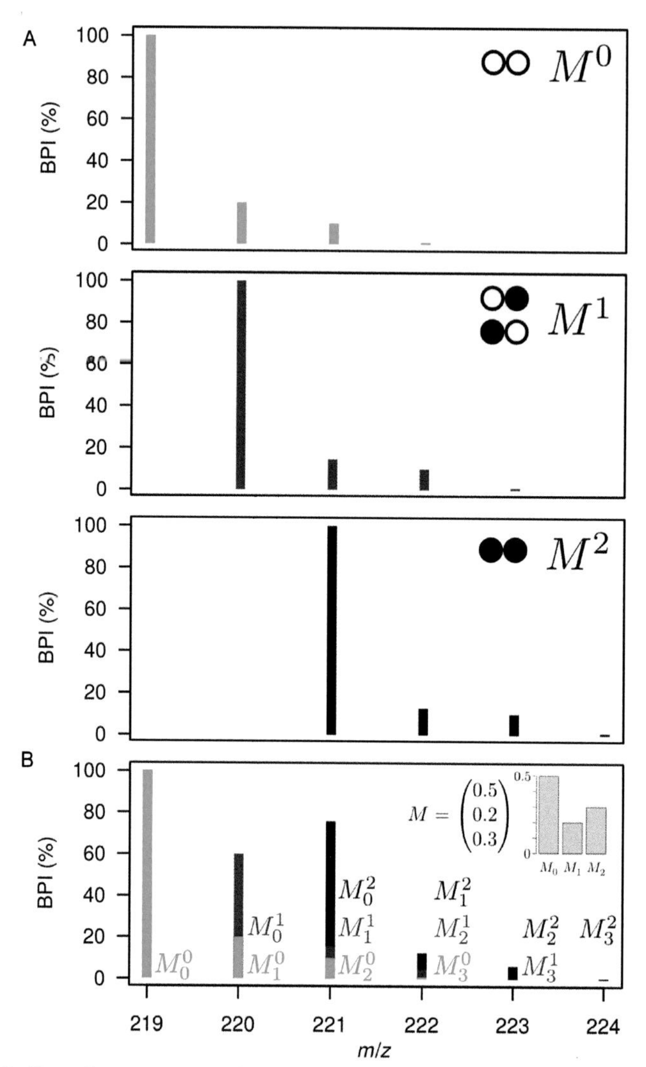

Daniel Weindl *et al.*, Figure 3 Artificial isotopic enrichment and natural isotope abundance determine isotopic peak patterns in a mass spectrum after stable isotope labeling experiments. Stable isotope labeling of a compound with two enrichable positions. (A) The mass spectra of all three mass isotopomers that can arise from stable isotope labeling show isotopic peaks due to the natural isotope abundance in the nonenriched positions. The natural isotope contribution, i.e., the relative intensity of the M+1 peak, decreases with increasing artificial enrichment. The relative M+0 intensity increases by the same value. Filled circle, heavy isotope; empty circle, light isotope. Signal intensity is scaled to base peak intensity (BPI). (B) After a stable isotope labeling experiment the mass spectrum of the given compound is a mixture of the natural MIDs of the three artificial mass isotopomers. This spectrum is the average of M^0, M^1, and M^2, weighted by the corrected MID M.

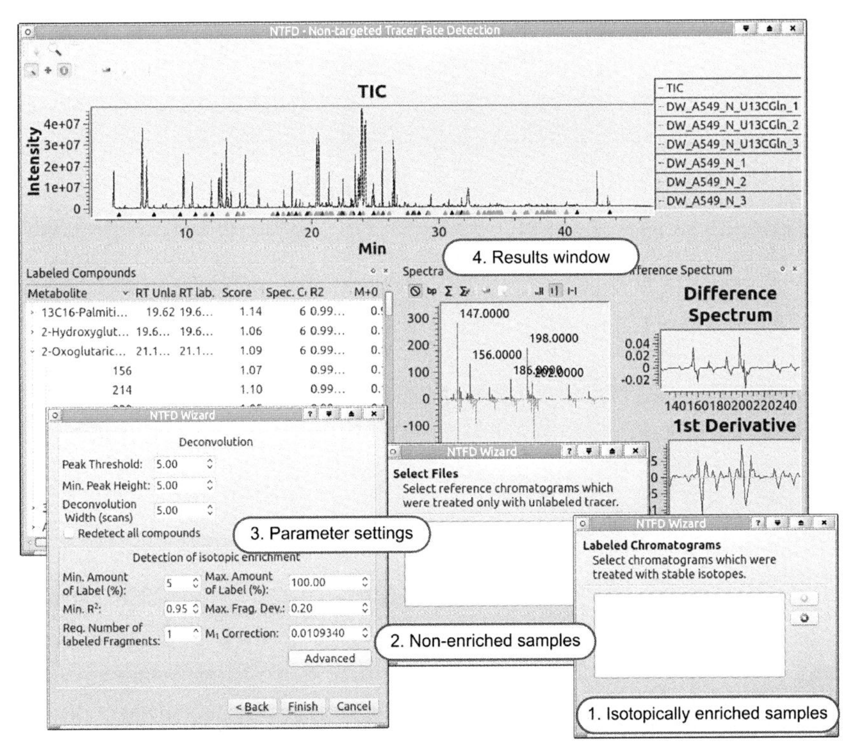

Daniel Weindl *et al.*, Figure 4 Graphical user interface of the NTFD implementation (Hiller et al., 2010). The program is freely available for Linux and Windows operating systems. After GC–MS measurement, (1) the files of the isotopically enriched and (2) the nonenriched samples need to be selected, (3) parameters for the detection and quantification of isotopic enrichment need to be adjusted, then (4) NTFD will determine all isotopically enriched compounds and fragments along with their MIDs.

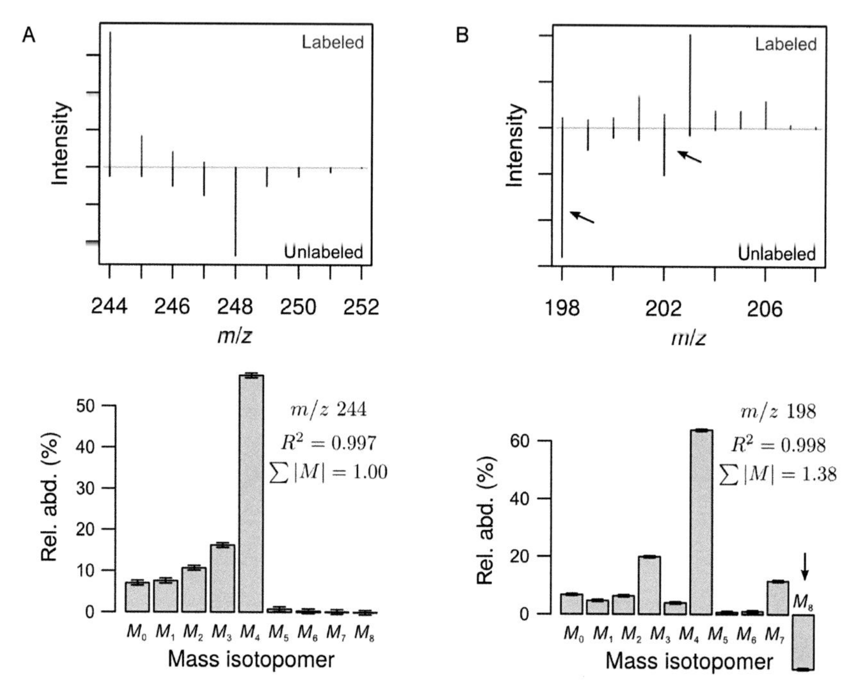

Daniel Weindl *et al.*, Figure 6 Quality measures of MID determination. MIDs of 2-oxoglutaric acid 2TMS 1MeOX determined after incubating mammalian cells with [U-^{13}C]glutamine. (A) Fragment *m/z* 244: High quality mass spectra without fragment overlap yield well-determined MIDs (small confidence intervals, $R^2 \approx 1.00$, $\sum|M| \approx 1.00$). (B) Overlapping fragments impair calculated MIDs. Fragments *m/z* 198 *m/z* 202 are overlapping causing a high negative M_9 abundance in the determined MID. Such MIDs are unusable, despite low confidence intervals and high R^2.

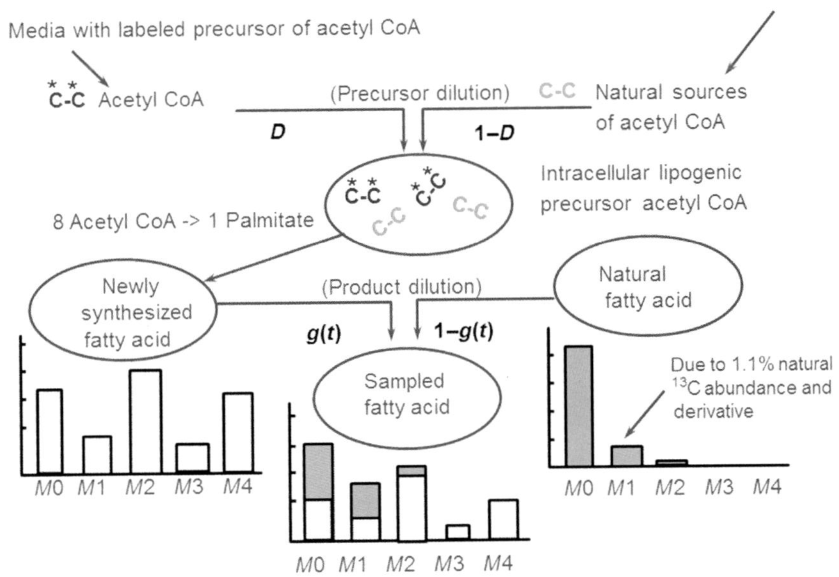

Joanne K. Kelleher and Gary B. Nickol, Figure 3 Classic ISA model for biosynthesis of a polymer. Label enters at acetyl CoA and mixes with endogenous sources of cytosolic acetyl CoA to form the lipogenic precursor pool. D and $g(t)$ are the two unknown parampeters. Following biosynthesis, the newly synthesized palmitate is diluted with preexisting pools of palmitate of natural abundance. The goal of ISA is to estimate the values of D and $g(t)$ from the sampled palmitate pool.

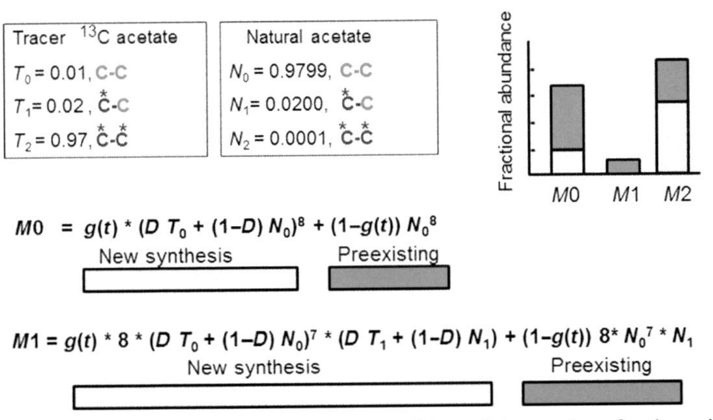

Joanne K. Kelleher and Gary B. Nickol, Figure 4 ISA model equations for the palmitate biosynthesis. The fractional abundance of the three isotopomers of C–C subunits from the tracer and from the natural abundance acetate present in the cell is inputs to the model. The first 2 of the 17 model equations are shown with the sections of the equations representing new synthesis and preexisting palmitate marked in yellow and green, respectively.

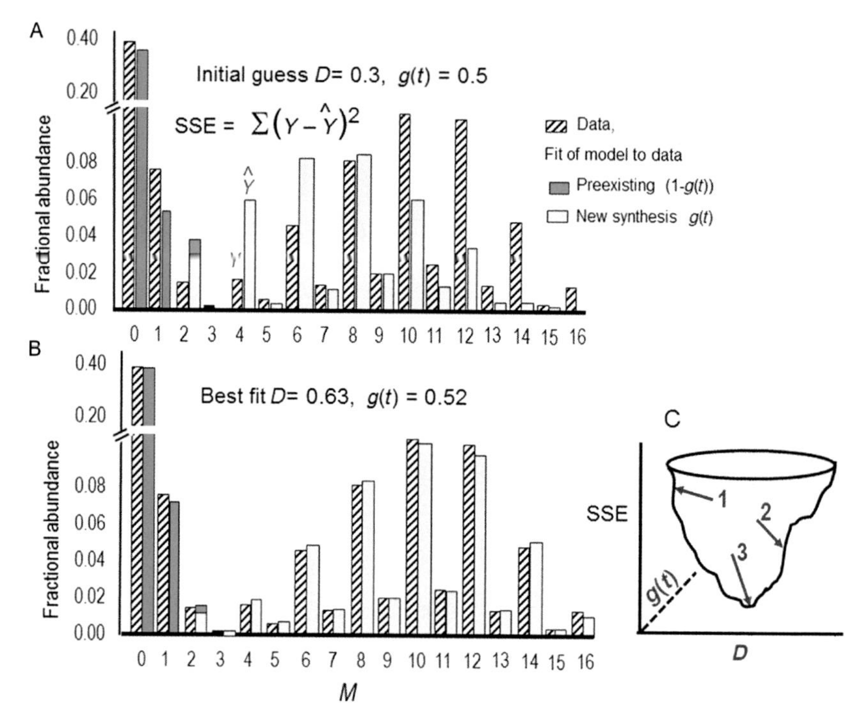

Joanne K. Kelleher and Gary B. Nickol, Figure 5 Solving for ISA parameters using nonlinear regression. Experimental data and model fits for hepatoma cells in medium 2 m*M* containing [1,2-^{13}C]acetate (2 m*M*). GC/MS data converted to fractional abundance values are plotted as stripped bars. ISA output simulations for each mass isotopomer plotted to the right of data. (A) Initial simulation with best guess starting values. (B) Best fit results found with ISA. (C) Plot of error *Z*-axis versus the two parameters, *D* and *g*(*t*).

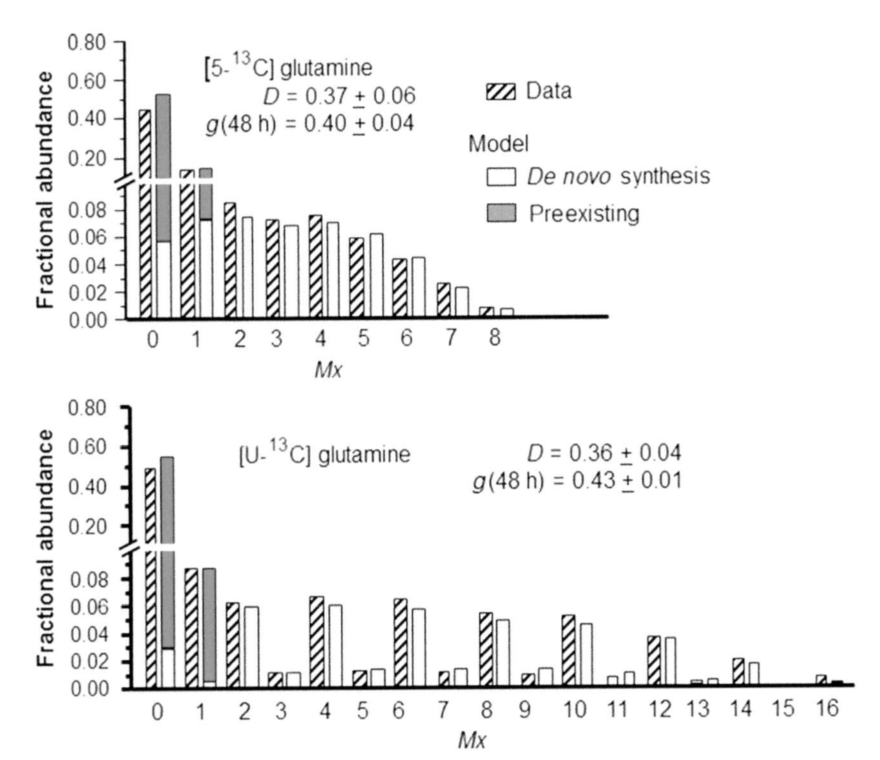

Joanne K. Kelleher and Gary B. Nickol, Figure 7 HepG2 human hepatoma cells-labeling patterns and ISA parameter estimates. Cells incubated with either [5-^{13}C]glutamine or [U-^{13}C]glutamine. Striped bars represent actual data plotted alongside model results indicating relative amounts of newly synthesized palmitate and preexisting at 48 h for best fit parameter values.

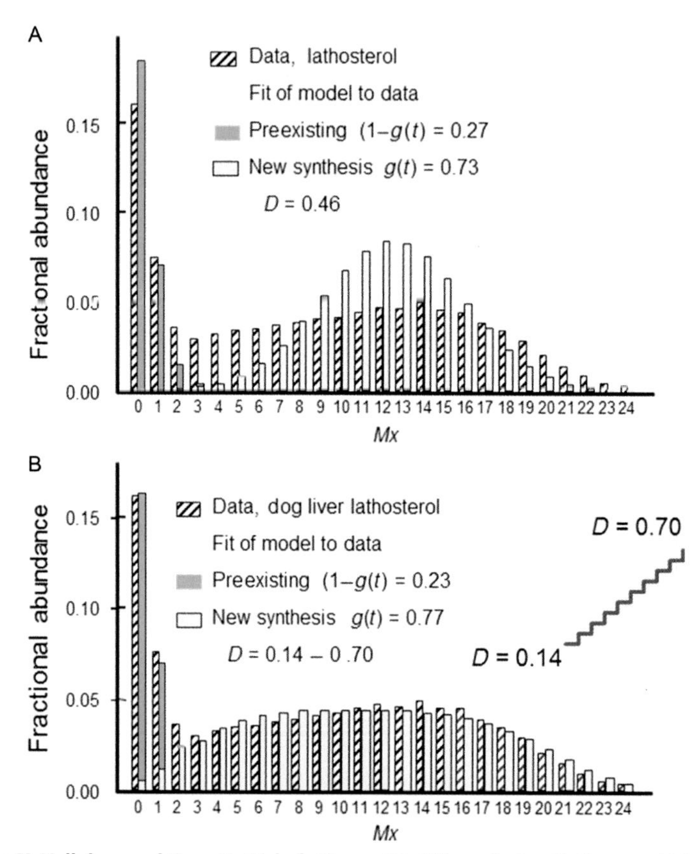

Joanne K. Kelleher and Gary B. Nickol, Figure 10 ISA analysis of lathosterol biosynthesis *in vivo*. (A) Lathosterol isotopomer profile from *in vivo* labeling using [1,2-^{13}C]acetate, where stripped bars representing GC/MS data are plotted alongside the best fit from the standard model using a constant value for *D*. (B) Data and best fit results for modified ISA model varying *D* between two values selected by the model in linear steps.

Edwards Brothers Malloy
Thorofare, NJ USA
September 22, 2015